Electrochemistry

Volume 8

$40.00

WILLIAM F. MAAG LIBRARY
YOUNGSTOWN STATE UNIVERSITY

A Specialist Periodical Report

Electrochemistry
Volume 8

A Review of Recent Literature

Senior Reporter
D. Pletcher *Department of Chemistry, University of Southampton*

Reporters
J. Grimshaw *The Queen's University of Belfast*
N. A. Hampson *Loughborough University of Technology*
A. J. S. McNeil *Loughborough University of Technology*
C. J. Pickett *ARC Unit of Nitrogen Fixation, Brighton*
J. Robinson *University of Southampton*
D. J. Schiffrin *Wolfson Centre for Electrochemical Science, Southampton*

The Royal Society of Chemistry
Burlington House, London W1V 0BN

ISBN 0-85186-067-2
ISSN 0305-9979

Copyright©1983
The Royal Society of Chemistry

All Rights Reserved
No part of this book may be reproduced or transmitted in any form
or by any means – graphic, electronic, including photocopying, recording,
taping, or information storage and retrieval systems – without
written permission from The Royal Society of Chemistry

Filmset in Monophoto Times by Mid-County Press, London, SW15
and printed by Adlard and Son Ltd., Bartholomew Press, Dorking

Made in Great Britain

Foreword

Electrochemistry is now too broad a subject for the complete range of the fundamental and applied branches to be comprehensively reviewed in a single publication. It is therefore inevitable that many interesting topics will be covered only intermittently. I believe, however, that the present volume contains five reviews which are topical and relate to major thrusts of current research.

The principal hope of the new Senior Reporter is to return the frequency of this series of Specialist Periodical Reports to annual publication, with reviews which discuss the literature of the immediate past. I am therefore particularly grateful to all the authors who have kept close to schedule and who have striven to include the literature of 1981. The exception is the chapter on organic electrochemistry, a regular in earlier volumes which had fallen well behind in the period reviewed. The present chapter covers the five years 1976–80, and this extended period has necessitated the greater selectivity in the choice of references and a briefer treatment than otherwise would be desirable.

D. Pletcher

Contents

Chapter 1 The Electrochemistry of Porous Electrodes:
Flooded, Static (Natural) Electrodes   1
By N. A. Hampson and A. J. S. McNeil

  1 Introduction   1

  2 The Development of Theories   3
    The Single-pore Model   3
    The Macrohomogeneous Model   6
    Theoretical Modelling of Porous Electrode Systems   7

  3 Work on Individual Electrode Systems   18
    Aluminium   18
    Cadmium   19
    Carbon   23
    Copper   23
    Iron   24
    Lead   25
    Lead Dioxide   28
    Lithium   34
    Mercuric Oxide   35
    Molybdenum   35
    Nickel   35
    Platinum   39
    Ruthenium   39
    Silicon Carbide   40
    Silver   40
    Titanium Carbide   43
    Zinc   44

  4 Discussion and Conclusions   52

Chapter 2 Electrode Processes in Molten Salts   54
By J. Robinson

  1 Introduction   54

  2 Metal Halide Systems   55
    Metal-deposition Processes   55
      Chloride Melts   55
      Fluoride Melts   58
      Iodide Melts   59

Inorganic Electrochemistry in Chloride Melts	59
3 Aluminium Halide Based Melts	61
Alkali-metal Chloroaluminates	62
Deposition of Metals	62
The Chlorine-evolution Reaction	62
Inorganic Electrochemistry	64
Spectroelectrochemistry	64
Cryolite Melts	66
Room-temperature Chloroaluminate Melts	66
Inorganic Electrochemistry	69
Organic Electrochemistry	70
Organometallic Electrochemistry	71
4 Nitrate Melts	72
5 Carbonate Melts	75
6 Other Molten-salt Systems	76

Chapter 3 The Electrochemistry of Transition-metal Complexes 81
By C. J. Pickett

1 Introduction	81
2 The Early Transition Metals	81
Titanium, Zirconium, and Hafnium	81
Vanadium, Niobium, and Tantalum	83
3 Chromium, Molybdenum, and Tungsten	84
Complexes with M=O and Related Groups	84
Molybdenum–Sulphur–Iron Complexes	86
Complexes of Dinitrogen and its Derivatives	89
Organometallic Compounds	93
Other Studies of Metals of Group 6A	98
4 Manganese, Technetium, and Rhenium	99
Manganese	99
Technetium and Rhenium	100
5 Iron, Ruthenium, and Osmium	102
Iron	102
Iron Porphyrins and Related Compounds	102
Iron–Sulphur Compounds	106
Organometallic Compounds	107
Ruthenium and Osmium	112
6 Cobalt, Rhodium, and Iridium	116
Cobalt	116
Compounds that are Related to the Cobalamins	116
Cobalt Complexes and Dioxygen	119

　　　　　Other Studies of Cobalt Complexes　　　　121
　　　　　Rhodium and Iridium　　　　　　　　　　121

7　**Nickel, Palladium, and Platinum**　　　　　　122
　　　　Nickel　　　　　　　　　　　　　　　　　122
　　　　Palladium and Platinum　　　　　　　　　124

8　**Conclusions**　　　　　　　　　　　　　　　124
　　　　Organometallic Complexes　　　　　　　　124
　　　　Complexes with Co-ordinated Nitrogen and Oxygen
　　　　　Groups　　　　　　　　　　　　　　　　125
　　　　The Influence of Ligands upon the Redox Properties
　　　　　of Transition-metal Complexes　　　　　　125
　　　　Photoelectrochemistry and the Electrochemistry of
　　　　　Modified Electrodes　　　　　　　　　　125

Chapter 4　The Electrochemistry of Oxygen　　　126
By D. J. Schiffrin

1　**Introduction**　　　　　　　　　　　　　　　126

2　**A General Mechanism for the Reduction of Oxygen**　127

3　**Studies on Different Electrode Materials**　　　　127
　　　　Carbon　　　　　　　　　　　　　　　　　127
　　　　Cobalt　　　　　　　　　　　　　　　　　131
　　　　Copper and its Alloys　　　　　　　　　　132
　　　　Gold　　　　　　　　　　　　　　　　　　133
　　　　Iron and Steel　　　　　　　　　　　　　135
　　　　Lead　　　　　　　　　　　　　　　　　　135
　　　　Mercury　　　　　　　　　　　　　　　　136
　　　　Nickel　　　　　　　　　　　　　　　　　137
　　　　Palladium　　　　　　　　　　　　　　　137
　　　　Platinum　　　　　　　　　　　　　　　　138
　　　　Silver　　　　　　　　　　　　　　　　　143

4　**Co-ordination Compounds of Transition Metals**　144

5　**Oxides and Mixed Oxides of the Transition Metals**　157

6　**The Electrochemistry of Oxygen in Molten Salts and in
　　Fuel Cells at High Temperatures**　　　　　　160

7　**Non-aqueous Solvents and Reactions of Superoxide Ion**　161

8　**The Biochemistry of Oxygen**　　　　　　　　165
　　　　Cytochrome P-450　　　　　　　　　　　　166
　　　　Cytochrome Oxidase　　　　　　　　　　　167

9　**Applications**　　　　　　　　　　　　　　　168

10　**Final Remarks**　　　　　　　　　　　　　170

Chapter 5 Organic Electrochemistry – Synthetic Aspects 171
By J. Grimshaw and D. Pletcher

Part I General Topics and Reductions 171
By J. Grimshaw

1 General Topics 171

2 Reduction of Aliphatic and Aromatic Compounds 173
General 173
Hydrocarbons 176
Halogen-containing Compounds 178
Carbonyl Compounds 185
Activated Olefins 190
Oxygen- (as Further Functional Groups), Sulphur-,
 and Selenium-containing Compounds 193
Nitro- and Nitroso-compounds 197
Other Nitrogen-containing Compounds 199
Phosphorus- and Arsenic-containing Compounds 203

Part II Oxidations 204
By D. Pletcher

1 Aliphatic Compounds 204
Carboxylic Acids 204
Alcohols and Aldehydes 210
Nitrogen-containing Compounds 214
Hydrocarbons and Alkyl Chains 216
Miscellaneous 222

2 Aromatic Compounds 227
Hydrocarbons: Substitution in the Ring and Side-Chain 227
Phenols 236
Amines 239
Miscellaneous 241

3 Heterocyclic Compounds 244

Abbreviations

THF	tetrahydrofuran
DMF	NN'-dimethylformamide
dmpe	1,2-bis(dimethylphosphino)ethane
dppe	1,2-bis(diphenylphosphino)ethane
cat	catecholato(2−) anion
OEP	2,3,7,8,12,13,17,18-octaethylporphinato(2−) dianion
TPP	5,10,15,20-tetraphenylporphinato(2−) dianion
salen	NN'-ethylenebis(salicylideneiminate)
en	1,2-diaminoethane
trien	triethylenetetra-amine
bipy	2,2′-bipyridyl
py	pyridine
edta	ethylenediaminetetra-acetato(4−) anion
phen	o-phenanthroline
EXAFS	extended X-ray absorption fine structure

1
The Electrochemistry of Porous Electrodes: Flooded, Static (Natural) Electrodes

BY N. A. HAMPSON AND A. J. S. McNEIL

1 Introduction

The theory of electrochemistry that is presented in the standard textbooks has been obtained by considerations of ideal electrodes, and generally confirmed by experiments with mercury and amalgam electrodes, which present the nearest approach to the ideal situation. The ideal solid electrode is smooth, of accurately known surface area and crystal orientation, structurally perfect, and strain-free. The practical electrochemistry that is encountered in industry is concerned with electrodes that are rough and which present a large number of differently oriented crystal faces to the electrolyte solution. Often, these surfaces are fissured, and they may even contain phase demarcations. The need to present the maximum surface area to the reacting electrode/electrolyte interface has inevitably resulted in the development of quite porous electrodes, such as are commonly found in the electrical storage-battery industry. The lead–acid cell and the Leclanché cell, the two best known to commerce, both contain porous electrodes. Indeed, the authors do not recall a single example of a storage cell consisting of two non-porous solid electrodes.

Porous electrodes can be subdivided into five distinct classes:

(i) a prefatory class of rough electrodes, in which the surface area is somewhat increased over the projected, geometric area of the electrode; all 'plane' electrodes are rough to some extent;
(ii) porous or granular electrodes, produced by a specialized process of electrodeposition;
(iii) hydrophobic gas electrodes, whose operation depends critically upon the establishment of a three-component (solid–liquid–gas) interphase;
(iv) flow-through electrodes, with forced input of reactants; and
(v) 'natural', flooded, static porous electrodes.

Rangarajan[1] has presented a brief review of the theory and operations of porous electrodes, with a classification of the operating models. The authors are not aware of any review of gas electrodes [class (iii)] that has been made since the survey of fuel cells by Bockris and Srinivasan[2] in 1969. Newman and Tiedemann have on two occasions[3,4] reviewed the subject of flow-through electrodes [class (iv)]. Work on

[1] S. K. Rangarajan, *Curr. Sci.*, 1971, **40**, 175.
[2] J. O'M. Bockris and S. Srinivasan, 'Fuel Cells: their Electrochemistry', McGraw-Hill, New York, 1969, Ch. 5.
[3] J. S. Newman and W. Tiedemann, *AIChEJ.*, 1978, **21**, 25.
[4] J. S. Newman and W. Tiedemann, 'Advances in Electrochemistry and Electrochemical Engineering', ed. H. Gerischer and C. W. Tobias, Wiley–Interscience, New York, 1977, Vol. 11.

both of these classes of porous electrodes over the past decade will be reviewed in the next volume in this series.

Although the basic theories have remained unchanged since the review in 1966 by de Levie,[5] a substantial amount of theoretical development and confirmatory experimental work has been carried out in connection with the major porous electrodes of electrotechnology. It is timely, then, to review the progress made in understanding the fundamental electrochemistry of flooded porous electrodes [class (v)]. Although the majority of the papers discussed in this Report are from the main scientific sources, any of the most important contributions that appear uniquely in the published proceedings of symposia have also been included.

It is worthwhile, at this point, to indicate briefly the major avenues of approach that have been followed in studies of porous electrodes. The more important electrochemical relationships are noted, since these are not found in standard works on electrode kinetics.

There are, fundamentally, two approaches which can be taken in order to deal with the porous electrode. First of all, porous electrodes can be considered as extensions of planar electrodes of known electrode-kinetic behaviour. This is the discrete-pore-model approach; historically, it provided the first explanation of the behaviour of porous electrodes. It was developed to a high degree, notably by Frumkin,[6] Winsel,[7] and de Levie[5] (the review by de Levie[5] contains a thorough account of the early work). Differences from the relationships for the plane electrode arise because, in the ideal porous electrode (*i.e.* with circular pores), the current, instead of arriving normally to the plane of the electrode, arrives parallel to it. This consideration engenders the concept of the penetration depth, the interlinking of ohmic, concentration, and activation polarizations, and the 'halving' of the time-dependent (transient) responses.

The other approach (which, if anything, has been more successful than the pore model) is the macrohomogeneous model. This was first effectively used by Newman and Tobias;[8] the porous electrode was considered to be an 'average' of the solid electrode and the electrolyte. Thus, the effective conductance of the porous electrode was the weighted volume average of the respective conductances; diffusion coefficients were similarly averaged, and so on. In this case, however, the electrochemistry cannot be taken from that of the plane electrode; the potential–current relationship must be obtained from the porous electrode by measurement. This is the least satisfying aspect of the macrohomogeneous approach. A possible solution might be to use the pore model to establish the electrochemistry; in general, however, it is clearly better to use experimental methods. The macrohomogeneous model will clearly be the more useful for the electrochemical engineer; however, the single-pore approach is still proving useful in developing our understanding of porous electrodes.

---

[5] R. de Levie, *Adv. Electrochem. Electrochem. Eng.*, 1967, **6**, 329.
[6] A. N. Frumkin, *Zh. Fiz. Khim.*, 1949, **23**, 1477.
[7] A. Winsel, *Z. Elektrochem.*, 1962, **66**, 287.
[8] J. S. Newman and C. W. Tobias, *J. Electrochem. Soc.*, 1962, **109**, 1183.

2 The Development of Theories

The Single-pore Model.—We summarize the state of the theory of porous electrodes at the time of the review by de Levie[5] in 1966.

In spite of the clearly evident need to consider porosity of the electrode in relation to electrode-kinetic investigations of many types, it is not customary to do so. This arises from two causes. One is that, to a first approximation, porous and rough electrodes behave as smooth electrodes of enhanced surface area. The other is that porosity is difficult to incorporate into electrode kinetics, because of lack of definition of the porous electrode. Simplification of the porous electrode to give, for example, a parallel array of pores of uniform diameter is an obvious first extension from the planar electrode. In considering this model, the pore is essentially one-dimensional and the resistance of the electrolyte is uniformly distributed along its length. The simplest approach[6,9,10] is to assume that the pore is of uniform cross-section and completely filled with solution. Frumkin[6] considered the curvature of the equipotential surfaces within the pore; however, by replacing the equipotential surfaces by the mean values which lies in planes perpendicular to the axis of the pore, the problem is avoided, as the model becomes essentially one-dimensional.

de Levie[5] has shown that the transmission-line representation leads to the expression:

$$E_x = E_0 \cosh(\rho x - \rho l)/\cosh \rho l \quad (1)$$

for the potential drop $(E_0 - E_x)$ over the distance x from the mouth of the pore $(x=0)$ within a pore of length l. The important constant ρ has the form:

$$\rho = (R_\Omega/R_D)^{\frac{1}{2}} \quad (2)$$

where R_Ω is the ohmic resistance of the solution within the pore for unit pore length (as distinct from that of the bulk solution) and R_D is the charge-transfer resistance for unit length. The quantity ρ^{-1} has the dimensions of length and is called the 'penetration depth'.[6] The current at the mouth of the pore (total current) is given by:

$$i_{x=0} = -(R_\Omega)^{-1}(dE/dx)_{x=0} = (\rho E_0 \tanh \rho l)/R_\Omega \quad (3)$$

The pore behaves as a resistance, according to:

$$E_0/i_{x=0} = R_{\text{porous}} = (R_\Omega R_D)^{\frac{1}{2}} \coth \rho l \quad (4)$$

If we assume a cylindrical pore, the value of R_Ω in the pore can be calculated as:

$$(R_\Omega)^{-1} = \pi a^2 \kappa \quad (5)$$

for a radius a cm and solution conductance $\kappa \, \Omega^{-1} \, \text{cm}^{-1}$, and with R_D expressed in terms of the exchange-current density i_0, using:

$$R_D = RT/2zF\pi a i_0 \quad (6)$$

The penetration depth becomes:

$$1/\rho = (a\kappa RT/2zFi_0)^{\frac{1}{2}} \quad (7)$$

[9] V. S. Daniel-Bek, *Zh. Fiz. Khim.*, 1948, **22**, 697.
[10] O. S. Ksenzhek and V. V. Stender, *Dokl. Akad. Nauk SSSR*, 1959, **106**, 487.

Hence, the penetration depth decreases with decreasing κ and pore radius and with increasing i_0. A number of workers have attempted to get exact solutions for the current–potential characteristics of a circular pore, assuming linearity of the characteristic,[6,7,11–13] but, without a detailed knowledge of the structure of a pore, there seems little point in this, particularly as the more complex situations yield solutions which agree with the simple ones to within 5%, provided that the pores are of significant depth.

The simple use of the specific charge-transfer resistance for the impedance of the electrode surface only holds for small polarizations, generally of the order of a few millivolts. For more significant polarizations, when the planar electrode (under charge-transfer control) is expected to follow Erdey-Gruz kinetics,[14] the Tafel equation can be applied in the region for which the overpotential (η) exceeds some tens of millivolts. For a cylindrical pore, the Tafel slope $dE/d(\ln i)$ is double that which would characterize the corresponding planar electrode.

The fact that the form of the kinetic equations for an electrode involves the product of both charge-transfer and mass-transport characteristics clearly emphasizes these two as equally important modes of limitation of the current for a porous electrode. Changes of concentration within the pores of the electrode can obviously be just as important as the Tafel behaviour in limiting the current. de Levie catalogues the early work[5] and shows that a limitation due to diffusion results in a current–potential relationship that is similar to the doubled (planar) Tafel behaviour of the charge-transfer-limited system.

The penetration depth is a function of both i_0 and the concentration at the opening of the pore; even at the mouth of the pore, the concentration ($c_{x=0}$) is different from that of the bulk (c_b), and the current into the porous electrode results from Fick's first law, as:

$$i_{x=0} = (c_b - c_{x=0}) \pi a^2 z F D / \delta \tag{8}$$

where δ is the thickness of the diffusion layer, provided that the front of the electrode can be treated as flat. In terms of porous parameters:

$$i_{x=0} = \rho c_b D' \tanh \rho l / (1 + \rho \delta \tanh \rho l) \tag{9}$$

where D' is an 'effective' coefficient for diffusion, expressed as $D' = \pi a^2 z F D$ in terms of the diffusion coefficient, D. When the potential becomes very large, the penetration depth is small and the diffusion-limited current becomes that of a flat electrode of the same dimensions as the projected area of the porous one. This illustrates why the use of flooded porous electrodes is more effective for slow electrode reactions, where transport of mass is less significant.

The case $S^r \rightleftharpoons S^o + ze^-$ (where S^r and S^o are reduced and oxidized species, respectively) was considered by Austin and Lerner,[15] who established the expression:

$$i_{x=0} = (2\pi a i_0/\rho)\{(c_{x=0}^r/c_b^r) \exp(\alpha z F E/RT) - (c_{x=0}^o/c_b^o) \exp[(\alpha-1)zFE/RT]\} \tanh \rho l \tag{10}$$

[11] J. Euler and W. Nonnenmacher, *Electrochim. Acta*, 1960, **2**, 268.
[12] J. Euler, *Naturwissenschaften*, 1958, **45**, 537; *Electrochim. Acta*, 1962, **7**, 205.
[13] P. Drossbach, *Z. Electrochem.*, 1952, **56**, 599.
[14] T. Erdey-Gruz and M. Volmer, *Z. Phys. Chem., Abt. A*, 1930, **150**, 203.
[15] L. G. Austin and H. Lerner, *Electrochim. Acta*, 1964, **9**, 1469.

where α is the cathodic charge-transfer coefficient and c^r and c^o refer to concentrations of reduced and oxidized species, respectively, either in the bulk solution (c_b) or at the mouth of the pore $(c_{x=0})$.

The penetration depth, ρ^{-1}, accordingly has a maximum value at:

$$E = (RT/zF) \ln \left[(1-\alpha) c_b^r D^r / \alpha c_b^o D^o \right] \qquad (11)$$

which is at or near the standard potential of the redox system: D^r and D^o are diffusion coefficients of the reduced and oxidized species, respectively.

Generally, impedance measurements are made with small perturbations of amplitude, so that the rate equation may be considered linear. Thus, for a frequency ω, the pore exhibits an impedance:

$$Z_0 = E_0/i_{x=0} = (R_\Omega Z_P)^{\frac{1}{2}} \coth \rho l \qquad (12)$$

where Z_P is the impedance per unit pore length.

For a semi-infinite pore, the phase angle of Z_0 (the impedance of the pore) is half the planar impedance (since R_Ω is a real quantity), and $|Z_0|$ is proportional to $Z_P^{\frac{1}{2}}$. Thus a 'squaring' operation is a simple way of correlating the impedance of a porous electrode with that of the corresponding flat electrode. The phase angle of the impedance of the pore is a function of the depth of the pore, and deep pores clearly contribute more significantly to the impedance of the electrode than do shallow ones.

The part of the impedance of a porous electrode that is due to double-layer charging is given by:

$$Z_0 = (1-j)(R_\Omega/2\omega C)^{\frac{1}{2}} \coth (1+j)/(\omega R_\Omega C/2)^{\frac{1}{2}} \qquad (13)$$

where $j = \sqrt{-1}$ and C is the capacitance per unit length of pore.[5,7,10–17] When the pore is infinite in length it exhibits a phase shift of 45° between current and potential, and it appears as a simple Warburg impedance. The reciprocal penetration depth:

$$\rho = (1+j)(\omega R_\Omega C/2)^{\frac{1}{2}} \qquad (14)$$

increases with ω; at sufficiently high values of ω, shallow pores will behave as inifnitely long ones. Using the equation for the impedance (Z) of an electrode reaction that is controlled by double-layer charging and the electrode reaction:

$$1/Z = R_D^{-1} + j\omega c \qquad (15)$$

the charge-transfer resistance is given by:

$$1/R_D = (2\pi a i_0 z F/RT) \{ \alpha \exp \left[\alpha z F E/RT \right] - (\alpha - 1) z F E/RT] \} \qquad (16)$$

and the familiar semicircle in the diagram in the complex plane for the planar electrode becomes lemniscate.

If the diffusion processes that are linked with the electrode reaction can be described in terms of semi-infinite linear diffusion, as was done by Winsel,[7] then a Warburg dependance with a phase angle of 45° for the surface impedance of the pore becomes a phase angle of $22\frac{1}{2}°$ for a semi-infinite pore.

[16] O. S. Ksenzhek, *Zh. Fiz. Khim.*, 1963, **37**, 2007.
[17] E. A. Grens and C. W. Tobias, *Z. Electrochem.*, 1964, **68**, 236.

The transient responses of porous electrodes are extended in the time domain, and steady states may take a considerable time to achieve. Equations show that the double-layer charging process is time-dependent, with a time constant of $l^2 R_\Omega C$. The charging process travels down the pore and reaches different depths l at different times; deviations from simple semi-infinite behaviour must be considered where the charging process reaches the bottom end of the pore. The problem of coupling the double-layer charging process with the transfer of charge at an electrode and the attendant diffusion is formidable. Double-layer charging is a much faster process than diffusion, and it has been suggested that the processes are separable.[18] Ksenzhek[19] solved the Fick equation:

$$D^{-1}(\partial c/\partial t) = (\partial^2 c/\partial x^2) - (c/D'Z') \qquad (17)$$

for a galvanostatic pulse to show that the potential increased linearly with time, at a rate that is proportional to the concentration (c) of electro-active species, providing an effective concentration impedance Z', as was observed for the oxidation of hydrogen on porous nickel.[20]

The Macrohomogeneous Model.—de Levie[5] primarily considered the single-pore model, thinking of the porous electrode as a combination of single pores, essentially without cross-links. The macroscopic current is then the sum of the contributions of all the single pores. Prediction of the overall performance of the electrode, however, would require detailed knowledge of the distribution of particle sizes, which would be difficult to determine as well as to relate to practical (*e.g.* sintered) electrodes. In contrast, the macroscopic or continuum approach describes the whole electrode–electrolyte system as two continua (one of the electrode matrix and the other of the solution, filling in all the void fraction of the electrode). Both phases are assumed to be homogeneous, isotropic, and spatially complementary. The avoidance of detailed structural description (*e.g.* for tortuosity factor) is reflected in mathematical simplicity, although the mathematical expressions are analogous to those of the single-pore model.

Newman and Tobias[8] were the first to consider explicitly the continuum approach, their initiative being followed by Grens and Tobias[17] and by Micka.[21] de Levie[5] gives a brief summary of this new development as it stood in 1966.

The macrohomogeneous model, developed by Newman and Tobias,[8] disregards the actual structural detail of the pores and regards the porous electrode as the superposition of two continua, *i.e.* the solution and the matrix. The current density in the electrolyte solution that is due to a flux N_i of species i is:

$$i_2 = F \sum_i z_i N_i \qquad (18)$$

For the material balance,

$$\partial \varepsilon c_i / \partial t = aj_{in} - \nabla \cdot N_i \qquad (19)$$

---

[18] O. S. Ksenzhek, *Zh. Fiz. Khim.*, 1962, **36**, 243.
[19] O. S. Ksenzhek, *Zh. Fiz. Khim.*, 1964, **38**, 1846.
[20] O. S. Ksenzhek and E. A. Kalinowskii, *Zh. Prikl. Khim.*, 1963, **37**, 541.
[21] K. Micka, *Collect. Czech. Chem. Commun.*, 1964, **30**, 223.

where ε is porosity, c_i is the concentration of species i, and j_{in} is the pore-wall flux of species i, averaged over interfacial area a. For electroneutrality,

$$\sum_i z_i c_i = 0 \tag{20}$$

and

$$\nabla \cdot i_1 + \nabla \cdot i_2 = 0 \tag{21}$$

where i_1 and i_2 are current densities in the matrix and in the electrolyte in the pores, respectively.

The electrochemistry is expressed by the general relationship:

$$\nabla \cdot i_2 = ai_0[\exp(\alpha_a \eta_s F/RT) - \exp(-\alpha_c \eta_s F/RT)] \tag{22}$$

where α_a and α_c are transfer coefficients in the anodic and cathodic directions, respectively, and η_s is the surface overpotential.

Transport processes for the matrix phase are introduced, using Ohm's law:

$$i_1 = -\sigma \nabla \Phi_1 \tag{23}$$

where σ and Φ_1 are the conductivity and electric potential of the matrix, respectively.

For a dilute solution, within the pores, the solutes move by processes of diffusion, dispersion, migration, and convection:

$$N_i/\varepsilon = -(D_i + D_a)\nabla c_i - z_i u_i F c_i \nabla \Phi_2 + (vc_i/\varepsilon) \tag{24}$$

where D_a is a dispersion coefficient, u_i is the mobility of species i, Φ_2 is electric potential in solution, and v is the fluid velocity.

These seven equations (18)—(24) are solved together, after various simplifications and adjustments, as required by the special features of the systems considered.

Theoretical Modelling of Porous Electrode Systems.—Theoretical developments cannot be separated into two distinct streams, springing from either the single-pore or the continuum theory. These approaches are complementary rather than exclusive, and investigators have employed both, in theoretical as well as experimental work. By the early 1970s, as Newman and Tiedemann[3] remarked, the basic theoretical groundwork had been developed to the point where it could begin to be applied to almost any electrode system. What appears to be a distinctive feature of the work of the past decade is that investigators have begun to study specific electrode systems, adapting the general theoretical principles to their particular electrochemical character.

In general, it is true to say that theoretical development since the time of the review by de Levie[5] has been due as much to investigations of specific electrode systems as to pure theoretical elaboration. However, all the theoretical development hitherto was of porous electrodes that were electrochemically functioning, but structurally invariant. No treatment had been made of the practical situation in which an electrode is structurally transformed during operation, with inevitable consequences for its electrochemistry.

The work of Alkire, Grens, and Tobias[22] was an early theoretical analysis which

---

[22] R. C. Alkire, E. A. Grens, and C. W. Tobias, *J. Electrochem. Soc.*, 1969, **116**, 1328.

took into account the variation of structure of the electrode with the extent of reaction, and considered a simple class of structural change, that of porous metal anodes undergoing electrochemical dissolution. This is a situation which is not often encountered in practice; nevertheless, the work is interesting since two processes now affect the current, these being concentration polarization and structural changes to the matrix. In the first process, transient change is relatively fast, but the second usually occurs over a very much longer period.

The authors took a macrohomogeneous approach, and predicted the external polarization of the electrode and the distribution of the porosity of the electrode as a function of the extent of reaction. Equations were developed to represent the pseudo-steady-state that prevails during dissolution, after appropriate gradients of concentration were fully established within the pores. To do this, the model took account of two phenomena that occur during dissolution of the electrode, namely a change in the specific surface area and convective fluid flow that arises from the consequent volumetric changes.

The change in structure was included in the model by assuming long circular pores, which simply increased in diameter with dissolution; otherwise the established equations[23] were applied to the macroporous model of Grens and Tobias.[24,25] Results were calculated for porous copper anodes in sulphuric acid. The authors show that, as dissolution proceeds, the distribution of current becomes more uniform, since increased porosity aids penetration of the current within the pores. In addition, induced convection due to removal of metal also tends to make the distribution of current more uniform.

However, a much more realistic situation, which embraces most of the situations in batteries, is that of porous electrodes of the second kind. A series of papers has been presented by Newman, Bennion, and co-workers,[26–30] considering an electrode that contains a sparingly soluble active species that is dispersed as small crystallites. It is assumed that this active material is stored within the electrode in the form of small dispersed crystallites of low solubility. The model corresponds to lead/lead sulphate or cadmium/cadmium hydroxide negative electrodes in practical batteries. During operation, the active material dissolves and is transported by diffusion to an electrochemically active site. After reaction, the products diffuse from supersaturated to saturated regions, where they precipitate as crystallites.

To a certain point, this model is similar to that of Newman and Tobias,[8] treating the electrode as a one-dimensional continuum. Dunning and Bennion[27] introduced the concept of a mass-transfer coefficient to account for the diffusion in solution of active species between active metal sites and reactant salt crystallites. Average values were given to the parameters mean diffusion length, crystallite size, and pore diameter.

---

[23] J. S. Newman, 'Advances in Electrochemistry and Electrochemical Engineering', ed. C. W. Tobias and P. Delahay, Interscience, New York, 1967, Vol. 5.
[24] E. A. Grens and C. W. Tobias, *Ber. Bunsenges. Phys. Chem.*, 1964, **68**, 236.
[25] E. A. Grens and C. W. Tobias, *Electrochim. Acta*, 1965, **10**, 761.
[26] J. S. Dunning, D. N. Bennion, and J. Newman, *J. Electrochem. Soc.*, 1971, **118**, 1251.
[27] J. S. Dunning and D. N. Bennion, *Proc. Adv. Battery Tech. Symp.*, 1969, **5**, 135.
[28] H. Gu, Ph.D. dissertation, University of California (*Diss. Abstr.*, 1977, **B38**, 801).
[29] J. S. Dunning, D. N. Bennion, and J. Newman, *J. Electrochem. Soc.*, 1973, **120**, 906.
[30] H. Gu, D. N. Bennion, and J. Newman, *J. Electrochem. Soc.*, 1976, **123**, 1364.

Electrochemistry of Flooded, Static Porous Electrodes 9

The model predicted diffusion-limited currents to be possible within the porous electrode under certain conditions. In this way, high polarizations may be obtained during discharge at limiting currents before all the theoretically available active material is consumed. Even with idealized assumptions, including constant solution properties, the model predicted significant redistributions of reactants and products during cycling, so that the oxidized species tends to be concentrated away from the mouth of the pore and towards the backing plate. These changes produced variations in the total polarizations of the electrode from cycle to cycle. The model points out a possible mode of failure, caused by an internal mass-transfer limitation.

In the paper by Dunning *et al.*,[29] an electrode model was developed for a single circular pore configuration which considered more realistic, complex situations, such as varying solution properties, convection arising from differing densities of reactant and product species, the complexing of the sparingly soluble salt with the bulk electrolyte anion, and finally the effects of local transfer of mass between salt crystallites and metal surfaces. In the earlier model,[26,27] the transport of the sparingly soluble active salt was covered by a single expression, incorporating constant transport coefficient and redox overpotential. Dunning *et al.*[29] have now derived a more detailed description of the transport process and of the effects of crystallization resistance and of complexing.

Two possible situations were considered. The first, solid-film model assumes that the salt covers the metal and that the ions move by solid-state transport through this film. The solution-diffusion model considers the alternative case where transport of the salt species between its own solid phase and the metal is through the solution. Whereas a solid-film model engenders a linear overpotential–time relationship, with an almost uniform distribution of current, the solution-diffusion model can generate very non-uniform distributions of current, often with maxima within the pores. In these calculations, a pseudo-steady-state approximation was used to calculate the time-dependent results. Dunning and Bennion noted the novel prediction by the solution-diffusion model that, in certain cases, the interior of a porous electrode might become exhausted before the front face.

Generally, anodic failure is caused either by blocking of pores or cover passivation, and cathode failure is caused by limited transport of mass, which leads to current limitation, these effects being critically dependent on the mass-transfer coefficient. The operation of the model was illustrated in terms of the $Cd/Cd(OH)_2$ couple in potassium hydroxide and of the $Ag/AgCl$ couple in potassium chloride. The general behaviour of the cadmium anode is considered to be best expressed by a solution-diffusion version of the model at low current densities and by a solid-film version at high current densities.

In this model,[29] a pseudo-steady-state approximation was used to calculate time-dependent results, so that the time-dependent concentration term was eliminated from the conservation equation. A new steady-state solution was derived for each time required. However, an accurate account of the temporal variation of electrolyte concentration is important in modelling the behaviour of an electrode. In extending this work,[29] Gu *et al.*[30] followed the macrohomogeneous approach, so as to model practical electrodes more realistically; for example, with regard to tortuosity factor.

The application of this model was restricted to that class of porous electrodes in which transport of reactants involves a diffusion step in solution. The operation of the model was illustrated in terms of Ag/AgCl porous electrodes, where AgCl is the sparingly soluble reactant. The model predicts variations in the concentration of KCl electrolyte within the electrode during cycling, increasing during discharging, and *vice versa*. In the case of discharging, the increasing strength of the electrolyte increases the electrical conductivity of the solution as well as increasing the solubility of AgCl. In the model, the effect of concentration of electrolyte upon the penetration depth of the reaction is an important determinant of the performance of the electrode. High reaction rates within the electrode are possible, even under conditions of rapid discharge, and over 90% utilization of the electrode charge is predicted if that charge has been stored uniformly at a slow charging rate.

Katan *et al.*[31] tested the theory developed by Bennion and co-workers[26,27,29,30] for the case of the particle-bed Ag/AgCl electrode. The model's predictions of temporal variations of electrode potential and of distribution of silver chloride accorded well with observations. The time to onset of pore blockage, with a consequent failure to accumulate charge, and the depth of penetration of the reaction were predicted with reasonable accuracy. The model could also be used to show how the governing rate processes could be shifted by the occurrence of blockages of the first and second kinds. The observations are covered in the section on silver.

An early attempt to model the porous electrode of the second kind was made in 1973 by Gidaspow and Baker.[32] The model of Dunning *et al.*, in its early development,[26] could predict electrode failure due to internal mass-transfer limitation of a sparingly soluble reactant, but did not consider failure by blockage of the pores, which is commonly observed in lead–acid batteries[33] and in cadmium electrodes.[34] Gidaspow and Baker took account of transformation of one solid phase into another, and the structure of their model electrode thus changes with its state of charge or discharge, with pores opening or closing depending on the molar volumes of solid reactants and products.

Overpotential was calculated, using the Ksenzhek[35] model and the assumption that a solid-phase reaction takes place in the electrode pores, producing a highly resistive phase. Differential mass balances for the solid reactant and product yielded dimensionless groups that characterized structural changes occurring within the electrode. Numerical methods were used to study the effects on overpotential (at constant current) of exchange current, molar density ratios, and fraction of active material. The predicted trends agreed fairly well with Shepherd's[36] empirical correlations for the galvanostatic discharging of lead–acid cells. Grens[37] has shown that the assumption of a uniform concentration of electrolyte has a very restricted validity in the modelling of porous electrodes, but concluded that the assumption of a uniform conductivity within the pores was nevertheless relevant to a broader set of conditions.

[31] T. Katan, H. Gu, and D. N. Bennion, *J. Electrochem. Soc.*, 1976, **123**, 1370.
[32] D. Gidaspow and B. S. Baker, *J. Electrochem. Soc.*, 1973, **120**, 1005.
[33] G. W. Vinal, 'Storage Batteries' (4th edn.), Wiley, New York, 1951.
[34] P. Bro and H. Y. Kang, *J. Electrochem. Soc.*, 1971, **118**, 519.
[35] O. S. Ksenzhek, in 'Fuel Cells: Their Electrochemical Kinetics', Consultants Bureau, New York, 1966, p. 1.
[36] C. M. Shepherd, *J. Electrochem. Soc.*, 1965, **112**, 657.
[37] E. A. Grens, *Electrochim. Acta*, 1970, **15**, 1047.

Kunimatsu[38] has presented a theoretical treatment for a change of overpotential, $\eta(t)$, with time on a porous electrode, in response to a galvanostatic transient. The electrode was assumed to comprise a flat part that is penetrated by N cylindrical pores, of different sizes. The impedance at the electrode/solution interface was considered as a parallel combination of impedances of the flat surfaces and those of the pores. The impedance of the pores was estimated on the basis of a one-dimensional transmission line. It was shown that the feature of the $\eta(t)$ versus t curve was affected not only by the time-constant, τ, of the electron-transfer reaction, but also by the time-constants of the pores, τ_k, for $k=1, 2, \ldots N$. The differential capacitance, C, of the double layer (as well as τ) was determined from the linear relation between $\ln(I/\dot\eta)$ and time (where $\dot\eta$ is the derivative of η with respect to time) in the region of time later than τ_N (the largest of the individual pore time-constants), on the basis of the expression:

$$\ln(I/\dot\eta) = \ln C + t/\tau \qquad (25)$$

when τ is larger than τ_N.

Kunimatsu[38] treated the distribution of current and potential in the pores when the electrode was polarized by a current step function. These distributions were found not only to depend on the time-constant of the pore (as defined by the product of resistance of the solution inside the pore and the double-layer capacitance of the wall of the pore) but were also sensitive to the time-constant of the electron-transfer reaction. The current density was distributed evenly throughout the whole internal pore surface in concentrated solutions, whilst in dilute solutions the current density was at first high near the mouth of the pore. This distribution of current density became more heterogeneous as the time-constant of the electron-transfer reaction was reduced. This heterogeneity was the cause of the deviation from linearity of the curve of $\ln(I/\dot\eta)$ versus time at short times.

Rangarajan[39] has given an analytical solution for the macrohomogeneous model of a flooded porous electrode under galvanostatic and sinusoidal perturbations. Earlier works by Ksenzhek,[16,18,19] Grens and Tobias,[24,25] Gurevich and Bagotskii,[40,41] and Winsel[7] were discussed and extended to obtain a satisfactory solution, using diffusion as the only mode of transport of mass and a simple potential distribution, in conjunction with Poisson's equation. A number of special cases are discussed and impedance elements are illustrated as functions of frequency. The results shown in the paper, where comparable, agree well with those of others, obtained by numerical methods. The impedance calculations cited in the paper are now very relevant, when the methods for rapidly obtaining the impedances of cells that incorporate porous electrodes are becoming available. The correspondence of the 'steady-state' polarization curves that arise from numerical and analytical solutions of the macrohomogeneous model has been convincingly demonstrated by Austin.[42]

Bennion and Newman et al.[26,29,30] considered electrode systems in which the

---

[38] K. Kunimatsu, J. Res. Inst. Catal., Hokkaido Univ., 1972, **20**, 1, 20 (Electroanal. Abstr., 1973, **11**, 415, 416).
[39] S. K. Rangarajan, J. Electroanal. Chem. Interfacial Electrochem., 1969, **22**, 89.
[40] I. G. Gurevich and V. S. Bagotskii, Elektrokhimiya, 1965, **1**, 1102.
[41] I. G. Gurevich and V. S. Bagotskii, Electrochim. Acta, 1964, **9**, 1151.
[42] L. G. Austin, Electrochim. Acta, 1969, **14**, 639.

reactants were sparingly soluble. The contrasting situation (in which the electrode reactants are completely insoluble and the exchange reaction at the electrode is fast, in the absence of significant concentration polarization) has been considered theoretically by Tiedemann and Newman.[43] In this case, the constraint of the flow of current is due solely to the ohmic resistance of the electrode. This is a situation which might well occur in electrodes of the second kind that are undergoing reaction at sufficiently high rates. The model used was the macrohomogeneous one,[8] and it was developed in terms of a reacting zone at the front of the electrode, separating the completely discharged electrode from the partially discharged inner regions. A further zone in the deep inner regions separated the unreacted matrix from the discharging region. The positions of both of these zones were time-dependent. Optimization was investigated as a function of a number of interesting parameters for a limiting value of the overpotential. It is shown that the thickness and the porosity of the electrode must be considered in relation to the developing conductivities of the component phases, porous and otherwise. The results of the calculations are applied to the LiAl electrode.

The completely ohmic-limited electrode is an interesting special case, albeit an extreme one. A general mechanism has been discussed by Simonsson,[44,45] applied to the special case of the lead dioxide electrode discharging in sulphuric acid. The distribution of current within the porous matrix was first established. The macrohomogeneous method of Newman and Tobias[8] was used, together with the polarization equation of Erdey-Gruz and Volmer,[14] which was found by experiment[44] to apply to the (rather crude) relationship between the initial overpotential and the applied current. Using a matrix equation for the relationship between conductivity and potential, a differential equation was obtained and solved to give the current – distance relationship at a series of values of the total current. The current profiles were checked by direct analysis of partially discharged electrodes and a discussion of the theoretical and actual profiles was quite convincingly presented to show that the electrode reaction occurs at the outer surface of the porous electrode at high discharge rates. At low discharge rates, the initial distribution of current is more uniform; however, the outer layers still tend to have a better utilization. These conclusions are not really unexpected, and might have been extracted from the early paper by Frumkin.[46] What is interesting, however, is the correspondence between the calculation and experimental profiles, which were surprisingly accurate for the conditions corresponding to intermediate degrees of discharge.

A few papers[47–50] have been published that concern the changes in concentration which occur within the porous matrix, and thereby influence the electrochemical behaviour. Alkire and Place[47] have investigated the transient behaviour during depletion of a reactant. The result was obtained by using a one-dimensional model,

---

[43] W. Tiedemann and J. Newman, *J. Electrochem. Soc.*, 1975, **122**, 1482.
[44] D. Simonsson, *J. Electrochem. Soc.*, 1973, **120**, 151.
[45] D. Simonsson, *J. Appl. Electrochem.*, 1973, **3**, 261.
[46] A. N. Frumkin, *Zh. Fiz. Khim.*, 1949, **23**, 1477.
[47] R. Alkire and B. Place, *J. Electrochem. Soc.*, 1972, **119**, 1687.
[48] R. Alkire and R. Plichta, *J. Electrochem. Soc.*, 1973, **120**, 1060.
[49] I. F. Danilenko, L. S. Kalashnikova, and I. N. Taganov, *Elektrokhimiya*, 1979, **15**, 57.
[50] O. S. Ksenzhek, E. M. Shimbel, and V. Z. Moskovskii, *Elektrokhimiya*, 1978, **14**, 510.

for which a pseudo-steady-state calculation for transport was performed numerically for both potentiostatic and galvanostatic polarizations. It was shown that, for the former, the reactivity of the leading edge of the reactive zone passed through a minimum, due to a 'trade-off' interaction between the diffusion polarization and the charge-transfer polarization. The results showed similar features to a system modelled by Dunning, Bennion, and Newman[26] in which transport of mass is limited between crystals of sparingly soluble reactants. Alkire and Plichta[48] considered the nature of convection within the matrix of working porous electrodes. A driving force for convection within the porous structure results, for which the authors set up a one-dimensional steady-state theoretical model. The contribution of this convective component to the total current density was calculated for three cases. The results are used to argue that porous electrodes, of enhanced behaviour, might result if they were to contain a fine substructure for the electrochemistry, whilst attendant convection occurs within a larger pore structure (a system which occurs not infrequently with fuel-gas electrodes).[51]

A recent elegant mathematical attack on the diffusion-controlled porous electrode has been presented by Russian workers;[49] the concentration profiles are described and expressions for the current that is available from such electrodes are given. Though the work does not contain any experimental verification, the development of the equations is of interest. On a connected topic, Ksenzhek et al.[50] have established a mathematical model for the operation of porous electrodes with poorly soluble reactants. It was concluded that, during the operation of such electrodes, changes occur in the distributions of potential, current, and materials across the electrode. The results are used to predict the possibility of using organic compounds whose solubilities are of the order of (poorly soluble) quinones as fuels for a power source.

Winsel[52] has considered the basic theoretical methods of treating transport problems in porous electrodes, and their practical importance for fuel cells and accumulators. The important features of the $PbO_2/PbSO_4$ electrode are the diffusion of H_2SO_4 and of lead ions, the latter process being hindered at high current densities by the formation of $PbSO_4$ crystals at the surface of the lead.

A paper which appears to be the most complete attempt, to date, to describe the behaviour of porous electrodes was presented in 1973 by Simonsson.[53] This deals with the special case of the porous lead dioxide that is used (in dilute sulphuric acid) as the positive electrode of a lead–acid cell. Simonsson[53] used the method of Newman and Tobias[8] with an improved model based on that due to Stein.[54] The electrochemistry that was found and reported in an earlier contribution was used, with a simple equation, to describe the development of $PbSO_4$ in the electrode. Convection and diffusion were also included, the former, however, only in the sense of the electrolyte being 'squeezed' from the electrode by phases that develop within the pores. The change in the reversible electrode potential with concentration was incorporated, from experimental values.[55] The mathematical solutions to the

---

[51] K. Micka, Collect. Czech. Chem. Commun., 1969, **34**, 3205.
[52] A. Winsel, Ber. Bunsenges. Phys. Chem., 1975, **79**, 827.
[53] D. Simonsson, J. Appl. Electrochem., 1974, **4**, 109.
[54] W. Stein, dissertation, Technische Hochschüle Aachen, 1959.
[55] W. H. Beck, K. P. Singh, and W. F. K. Wynne-Jones, Trans. Faraday Soc., 1959, **55**, 331.

equations were obtained by using an iteration procedure. The results were interesting, for they modelled the overpotential–time curves for the discharge of the lead dioxide positive electrode to a degree never hitherto approached. The equations were also solved to demonstrate the rate of reaction and the concentration profiles within the electrodes during the discharge reaction at various rates. Simonsson recognizes certain shortcomings in his treatment; nevertheless, this paper appears to constitute a major contribution. We refer to this paper and to some subsequent work in the section on PbO_2.

A connection of the single-pore concept with a practical porous structure has been attempted by Szpak and Katan[56] for the case of the reaction-rate profile. A segmented-pore (0.07 cm) structured silver electrode (1.1 cm wide) was used, with a 'pore' thickness of 58 μm and a value of the spacing (0.03 cm) between segments to give a total electrode length of 0.6 cm. In its disposition, the cell resembled a thin-layer cell in which the current was put in on a plane that is parallel to the segments of the electrode; the flat segments were topped with a glass cover-slide. The AgCl/Ag electrode was studied, monitoring the current flowing out of each segment by measuring the potential drop across fixed resistors, in series. The observed profiles of the reaction rate qualitatively confirmed the expected distributions of current,[45,46] arising from a gradually increasing thickness of the reaction layer (defined analogously with the diffusion layer) which terminated when the pores became blocked. The results of this study are subjective, owing to the method that was used to measure the currents in the segments, and the method gives only the relative sequence of currents. Nevertheless, the work qualitatively confirms the ideas developed previously.

The kinetics of multiple-electron-transfer reactions at porous electrodes have been examined by Ateya.[57] Using the Sn^{IV}/Sn^{II} electrode as a model, the porous matrix was considered to be highly interlinked and to be of high electronic conductivity and uniformly small pore diameter. In this way, radial diffusion could be neglected (as being very fast). A polarization equation for a process, comprising the two one-electron steps in the exchange reaction at a planar electrode, was used.[58] The expected doubling of the normal Tafel slope for the controlling single-electron step was predicted. Unfortunately, experimental work was confined to forced convective conditions, which makes this of limited interest for the present Report.

Pollard and Newman[59] have investigated the transient behaviour of porous electrodes with high exchange-current densities. As is usual for Newman, a macrohomogeneous model was used to study the distribution of reactions with non-steady rates. The transfer current for short times was given as an expansion in time, using a dimensionless form. At somewhat longer times, the current response can be decomposed into a time-dependent and a time-independent part. For a completely reversible electrode reaction, the distribution of current throughout the electrode could be calculated. One result which arises from this paper is that, at high exchange-current densities (*i.e.* for fast reactions), the resulting non-uniform distributions of reactions can have a significant effect on the behaviour of a porous

[56] S. Szpak and T. Katan, *J. Electrochem. Soc.*, 1975, **122**, 1063.
[57] B. G. Ateya, *J. Electroanal. Chem. Interfacial Electrochem.*, 1977, **76**, 315.
[58] K. J. Vetter, 'Electrochemical Kinetics', Academic Press, New York, 1967, pp. 152, 481.
[59] R. Pollard and J. Newman, *Electrochim. Acta*, 1980, **25**, 315.

electrode. Thus it is shown that the depth of penetration for the relatively slow PbO_2/H_2SO_4 electrode is considerably greater than for the fast $LiAl/Li^+$ electrode at high temperatures.

In the special case of an electrical current passing through a porous metal diaphragm, the potential gradient in the electrolyte can be sufficient to drive an electrochemical reaction. A variety of important electrochemical systems (*e.g.* corrosion and electroless plating) involve these bipolar electrodes, which are distinguished by localized anodic and cathodic regions coexisting within the same conducting metallic phase. Alkire[60] has derived a one-dimensional model for a simplified bipolar electrode system to determine the parameters that affect the distribution of reaction rates. In the absence of concentration and electrode-resistance effects, the distribution of electrochemical reactions depended solely on ξ, the dimensionless reaction-velocity parameter. For $\xi > 1$, electronic co-conduction *via* electrode reactions became appreciable. A high value of χ, a dimensionless electrode-resistance parameter, resulted in a spatial variation of potential in the electrode phase. In the same way, the value of the dimensionless diffusion parameter determined the extent of concentration overpotential. If ξ was small, electrochemical reaction did not occur, and the roles of electrode resistance and concentration variation became insignificant. Although this study was of a simplified biporous electrode system, Alkire[60] points out that changes in the limiting assumptions make the model applicable to a range of realistic systems.

The case of electrodes which become covered with a layer of insoluble products is one that is often encountered in practical battery technology. For a lead dioxide electrode which is reduced in sulphuric acid, the front matrix of the electrode is completely covered with $PbSO_4$, whereas the inside regions remain partially free of $PbSO_4$. On re-oxidizing the electrode in H_2SO_4, the reaction initially occurs in the inner regions, and the front of the electrode simply behaves as a porous medium. Although this situation has not been considered since 1969, that work is rigorous, and its conclusions are pertinent to the present Report. Oldham and Topol[61-63] have considered both potentiostatic and galvanostatic experiments. A theory is derived for porous media, using four models. These are the 'gel model', where the only effect of the matrix is to cause a retardation of the diffusing ions; the 'diluent model', which considers dilution of the solution; the 'uniform tortuosity' model, which is analogous to the uniform-pore model; and finally the 'random-tortuosity' model, in which the tubules are all of the same cross-sectional area but are aligned in a perfectly random fashion. Solutions for the current at any time, corresponding to each of the four models, between two parallel electrodes were given; it was suggested that the differences between the currents predicted by the models would provide a means of distinguishing between the various structures. The theoretical predictions[61] were tested in a further paper.[62] Using glass-fibre and paper-fibre media, between mercury electrodes, the currents corresponding to potentiostatic steps were measured. The response with the paper-fibre medium fitted the random-tortuosity model better than either of the others. It was interesting that none of the four models

[60] R. Alkire, *J. Electrochem. Soc.*, 1973, **120**, 900.
[61] K. B. Oldham and L. E. Topol, *J. Phys. Chem.*, 1967, **71**, 3007.
[62] K. B. Oldham and L. E. Topol, *J. Phys. Chem.*, 1969, **73**, 1455.
[63] K. B. Oldham and L. E. Topol, *J. Phys. Chem.*, 1969, **73**, 1462.

fitted the results for glass fibre. The galvanostatic measurements were considered in a final paper.[63] Only three of the four pore models are considered; the random-tortuosity model is incompatible with their galvanostatic treatment. These effects of the medium are becoming technologically important, not only in connection with battery electrodes themselves, but also, in their own right, with the introduction of advanced batteries in which the electrolyte solution is absorbed onto a static phase. Further work is urgently needed.

The Russian school have not been inactive in the study of porous electrodes, but have concentrated their (generally theoretical) studies on hydrophobic three-component electrode systems. They have presented comparatively few studies of flooded electrodes.

Shortly after the review by de Levie,[5] Gurevich and Bagotskii[64-67] presented a theoretical treatment of a porous electrode that was polarized from one side only, but with diffusional transport of reactants and products at both sides. Ohmic losses, and activation and concentration polarization, were accounted for. Explicit solutions for the through-thickness distribution of electrochemical reaction were obtained for particular cases of low and high polarization. Gurevich and Bagotskii also considered a variant on this situation where the diffusional supply of reagent is from either the front[64] or rear[67] of the electrode.

The same authors[68] analysed the dependence of energy losses within a porous electrode, comprising activation polarization, and internal transport and ohmic losses. For systems with diffusional supply of reactants, polarization passes through a minimum value, where total energy losses are minimal, as the thickness of the electrode is increased. No such minimum was found if there was a forced supply of reactants.

Chirkov[69] derived a theoretical analysis of simple models of waterproof electrodes that are completely filled with electrolyte. The treatment was extended to the three-component situation, with the introduction of gas. Formulae for maximum and minimum conductance were described. Chirkov[70] developed this approximate analysis to provide more precise conductance data, using a method of consecutive approximations.

Percolation theory is an idealized treatment of a fluid phase penetrating (wetting) a solid lattice, and has been applied to many other systems besides porous electrodes. Chirkov[71,72] has provided a graphical method for estimating disruption points in soaking lattices of links and nodes that can be applied in the theory of electrochemical current sources.

Vol'fkovich and Sosenkin[73] have made an experimental study of the distribution of liquid in porous electrodes; under conditions of capillary equilibrium, the liquid phase was uniformly distributed within a structurally uniform porous body. For

---

[64] I. G. Gurevich and V. S. Bagotskii, *Electrochim. Acta*, 1967, **12**, 593.
[65] I. G. Gurevich and V. S. Bagotskii, *Elektrokhimiya*, 1967, **3**, 795.
[66] I. G. Gurevich and V. S. Bagotskii, *Elektrokhimiya*, 1967, **3**, 915.
[67] I. G. Gurevich and V. S. Bagotskii, *Elektrokhimiya*, 1967, **3**, 1405.
[68] I. G. Gurevich and V. S. Bagotskii, *Elektrokhimiya*, 1969, **5**, 1297.
[69] Yu. G. Chirkov, *Elektrokhimiya*, 1972, **8**, 1074.
[70] Yu. G. Chirkov, *Elektrokhimiya*, 1972, **8**, 1187.
[71] Yu. G. Chirkov, *Elektrokhimiya*, 1976, **12**, 889.
[72] Yu. G. Chirkov, *Elektrokhimiya*, 1976, **12**, 895.
[73] Yu. M. Vol'fkovich and V. E. Sosenkin, *Elektrokhimiya*, 1978, **14**, 5.

non-uniform porous structures, the distribution of the liquid was determined by the distribution of pore volumes, in terms of the equilibrium capillary pressures. Chirkov and Chernenko[74] have shown that, in serial models of porous media, the surface layer where filling of the pores with liquid is non-uniform can spread over large distances, comparable in scale with the size of the porous body.

A number of workers have attempted to relate electrochemical measurements to the physical structure of the electrode. A common purpose has been to derive a relationship so that simple and rapid electrochemical measurements, such as of double-layer capacitance or impedance, can provide structural information, possibly *in situ*, and during the course of reaction.

Lyklema[75] has considered the structure of the electrical double-layer on porous non-conducting oxide electrodes. The treatment equally applies to other systems in which the surface charge is not confined to the surface proper, but may extend towards the interior. Penetration of the superficial layer by counter-ions is considered, its extent depending primarily on radius, valency, and (to a lesser extent) on concentration and specific adsorption potential. The distribution of potential inside the surface layer is described according to the Poisson–Langmuir equation, for which computer solutions are given.

Armstrong et al.[76] developed an impedance technique for the rapid and reliable assessment of the surface areas of porous graphite specimens. The method was found to be more suitable for comparative measurements than for absolute determinations of total surface area.

Lago and Plachco[77] have presented a model of the diffusional process in a two-component electrochemical system that they were able to verify experimentally. This model provided the basis for a rapid and simultaneous determination of porosity and tortuosity of porous media, as well as estimating molecular diffusivities. The treatment is, however, restricted to non-conducting porous media, with pores of such a size that superficial diffusivity becomes negligible.

Ohms and Wiesener[78] have determined physical parameters of porous electrodes, using a porosimeter which recorded the depth of penetration of mercury into the test sample at pressures up to 1000 atmospheres. Measurements on lead storage electrodes were in agreement with figures obtained by other methods. McHardy et al.[79] used a simple a.c. technique to measure double-layer capacitances of porous electrodes. A small alternating potential (± 15 mV at 1 Hz) was applied potentiostatically to the test electrode, and the phase and amplitude of the alternating current were used to develop Lissajous' figures on an oscilloscope screen. Values of double-layer capacitance for porous carbon and gold agreed with data for smooth electrodes. The technique allows changes in the interfacial area between porous electrode and electrolyte to be monitored.

Keiser et al.[80] have derived a theoretical treatment for the frequency dependence of the impedance of a single pore, and its variation for different geometries. This is an

---

[74] Yu. G. Chirkov and A. A. Chernenko, *Elektrokhimiya*, 1978, **14**, 529.
[75] J. Lyklema, *J. Electroanal. Chem. Interfacial Electrochem.*, 1968, **18**, 341.
[76] R. D. Armstrong, D. Eyre, W. P. Race, and A. Ince, *J. Appl. Electrochem.*, 1971, **1**, 179.
[77] M. E. Lago and F. P. Plachco, *Electrochim. Acta*, 1972, **17**, 1707.
[78] D. Ohms and K. Wiesener, *Chem. Tech. (Leipzig)*, 1975, **27**, 229.
[79] J. McHardy, J. M. Baris, and P. Stonehart, *J. Appl. Electrochem.*, 1976, **6**, 371.
[80] H. Keiser, K. D. Beccu, and M. A. Gutjahr, *Electrochim. Acta*, 1976, **21**, 539.

area of considerable importance because of the increasing interest in the impedance of porous systems. The authors commenced with a right circular cylindrical pore, and treated it as a transmission-line ladder network, where each step of the ladder represented as the now familiar Sluyters plots of imaginary component against real variations in radius of the cylinder along the length of the pore. The overall input impedance of the pore could thus be determined for a variety of pore geometries, using a computer to perform the iterative calculations involved. The results were represented as the now familiar Sluyters plots of imaginary component against real component. The theoretical impedance curves for the reversible reaction form a family, with an intensifying minimum developing from a normal Warburg slope of 45°. The intensity of the minimum depends on the extent of the inner surface of the pore compared with the surface diameter of the pore. These calculations indicated that pores with larger interior volumes and smaller orifices have larger capacitance and consequently smaller capacitative reactance. These results for a single pore were used to predict the behaviour of a macroscopic porous electrode, and the theoretical model was tested against experimental measurements and microscopic observations on porous nickel electrodes.

A contribution to the methods for characterizing porous electrodes has been presented by Tilak, Rader, and Rangarajan[81,82] in connection with the double-layer capacitance. It is proposed that the double-layer resistance and the double-layer capacitance of porous electrodes can best be found by a relaxation technique, using perturbations of the exponential form $i(t) = \Delta i \exp(-t/\tau)$ and $i(t) = i \exp(1 - t/\tau)$. The method provides a means of estimating the effective surface area via the double-layer capacitances, in the presence of a faradaic reaction.

The consideration of electrokinetic flow through narrow capillaries is a pertinent theoretical problem whose results are applicable to diverse systems. Olivares et al.[83] have extended the classic work of Rice and Whitehead[84] to higher surface potentials, while still preserving the convenient analytic properties of their results.

Candy et al.[85] have measured the impedances of gold powder and Raney-gold porous electrodes over a wide frequency range, and demonstrated the value of impedance data for determining the characteristics of porous electrodes. When the penetration depth of the signal approached the pore depth, the shape of the pore had little influence on impedance. Thus, techniques for the measurement of impedance could be applied to structurally complex pore electrodes, and they give an evaluation in terms of an equivalent electrode with cylindrical pores. The values determined in this way for a Raney-gold electrode were in good agreement with values from other methods. This work is, however, limited to systems where no electrochemical reaction is occurring.

3 Work on Individual Electrode Systems

Aluminium.—Aluminium is known to be a fuel of exceptionally high energy density.

[81] B. V. Tilak, C. G. Rader, and S. K. Rangarajan, *J. Electrochem. Soc.*, 1977, **124**, 1879.
[82] B. V. Tilak and S. K. Rangarajan, *Trans. Soc. Adv. Electrochem. Sci. Technol.*, 1978, **13**, 261.
[83] W. Olivares, T. L. Croxton, and D. A. McQuarrie, *J. Phys. Chem.*, 1980, **84**, 867.
[84] C. L. Rice and R. Whitehead, *J. Phys. Chem.*, 1965, **69**, 4017.
[85] J-P. Candy, P. Fouilloux, M. Keddam, and H. Takenouti, *Electrochim. Acta*, 1981, **26**, 1029.

The free energy of both combustion and conversion into the hydrated oxide amounts to 42 614 kJ kg^{-1}.[86] Its promise as an electrochemical fuel is supported by a high current capacity and a very negative electrode potential. However, attempts to utilize aluminium in chemical energy sources have all failed, primarily due to the tendency of the metal to undergo violent chemical reaction with water. One response to this has been to eliminate water altogether, by using dried non-aqueous solvents. Another approach has been to modify the properties of aluminium by alloying, so as to retain an electrochemically active metal that shows little chemical reactivity with water.

Despić et al.[87] have shown that alloying aluminium with small ($<0.2\%$) additions of indium, gallium, or thallium, singly or in combination, results in three important benefits. First, the open-circuit electrode potential is made considerably more negative than that for pure oxidized aluminium. Secondly, the metal shows very little polarization up to quite high current densities. Thirdly, the rate of corrosion is reduced, at both open-circuit and working potentials.

Despić et al.[86] have examined the behaviour of these aluminium alloys in air cells. Because aluminium could not be reclaimed from its aqueous salt solutions, the batteries had to be of a replaceable-plate type. The air electrodes were of the active-carbon–metal-grid–plastic-backing type; 2M-NaCl was used as the electrolyte. The characteristics of the aluminium–indium alloy in such a cell resembled those of the Leclanché cell in open-circuit voltage and polarization, but with much higher current densities attainable. Qualitative estimates of the rate of evolution of hydrogen indicated relatively low corrosion rates, and so high faradaic yields. The aluminium electrode ultimately failed by drying out, which could be prevented to some extent by using a porous plastic diaphragm. The energy density for the experimental cell was calculated to be 80 W h kg^{-1}, for 30 mA cm^{-2} and 0.75 V output. On the basis of these figures, a 30 kW h traction battery of 10 kW nominal power was conceived, with an energy density of 200 W h kg^{-1}.

In a separate communication, Drazić et al.[88] report on the operation of two types of battery, using the alloys Al–0.1%In–0.05%Ga and Al–0.1%In–0.05%Ga–0.1%Tl. Both types exhibited the same energy density of 192 W h kg^{-1}, and power densities of 16 and 10 W kg^{-1}, respectively. However, the achievement of higher power density in the former case entailed a lower faradaic efficiency and corrosion of the aluminium. In both types of battery, the electrolyte (2M aqueous NaCl) had to be replaced after every 4—5 hours.

Cadmium.—Casey and Vergette[89] have measured the charge that is delivered by porous cadmium electrodes (formed by suitable impregnation of porous nickel electrodes) and have shown that the capacity is quantitatively dependent upon the free (pore) volume that is accessible to the solution. A 'choking index' was defined (arbitrarily) in terms of the normalized ratio of the deliverable charge to the current. This index enabled a nickel plaque with different loadings of cadmium to be

---

[86] A. R. Despić, D. M. Dražić, S. K. Zečević, and T. D. Grozdić, in 'Power Sources', ed. D. H. Collins, Academic Press, London, 1977, Vol. 6, p. 361.
[87] A. R. Despić, D. M. Dražić, M. M. Purenović, and N. Ciković, *J. Appl. Electrochem.*, 1976, **6**, 527.
[88] D. M. Dražić, A. R. Despić, M. Atanacković, S. Zečević, and I. Ilijev, *Meeting Int. Soc. Electrochem.*, *28th*, 1977, p. 370.
[89] E. J. Casey and J. B. Vergette, *Electrochim. Acta*, 1969, **14**, 897.

characterized. The characterization was linked with the distribution of pore sizes. Another aspect of the porous cadmium electrode that was tackled by these authors was the decarbonation of cadmium electrodes by a thermal cracking process at 350 °C. The parameters of the process need to be carefully evaluated and adhered to for successful operation.

The discharge profiles in a porous cadmium electrode have been investigated by Bro and Kang.[34] Porous cadmium electrodes were made by sintering loose powder compacts in 3M-HCl. The resulting porous cadmium electrodes were galvanostatically oxidized (discharged) in 30% KOH at 25 °C and the discharge profiles determined by chemical analysis of thin slices from the electrodes. Histograms were drawn of the discharge profiles, which were described by a hyperbolic cosine function:

$$q/q_0 = A[\cosh B(1-X)]/\cosh B \qquad (26)$$

while the mean efficiency of the electrode was described by a hyperbolic tangent function, where q/q_0 is the discharge density (A h cm^{-3}) normalized with that of the total (stoicheiometric) charge available: X is the fractional depth inwards from the front surface, A is the state of discharge at the front of the electrode, and B is a kinetic term. The given correlations provided a means of calculating the efficiency and the discharge profiles at any value of the current density. As in the work of Casey and Vergette,[89] the choking of the porous matrix was considered to be the discharge-terminating process, together with the depletion of the interior of the matrix of OH$^-$ ion; that is, the discharge efficiency was limited by both faradaic processes and transfer of mass. Dunning et al.[29] used the Cd/Cd(OH)$_2$ couple to illustrate their analysis of porous electrodes with sparingly soluble reactants.

An interesting paper by Will and Hess[90] describes morphological studies of the structure along the oxidized surface of a micropore formed from a cadmium chip (of millimetre size) that is covered with an electrolyte film (of micron thickness). Optical microscopy in situ was used to study structural changes during both the charging and discharging processes; additional observations were also made by scanning electron microscopy (SEM). It was found that, after the first discharge (oxidation), 60% of the theoretical capacity is recoverable; however, this figure falls to 12% after 10 cycles of charge and discharge. Discharging leads to the formation of large crystals of β-Cd(OH)$_2$ plus some γ-phase. The large crystals are difficult to reduce, and are responsible for the loss in capacity. The effective capacity resides in the smaller crystals in the pore, which react by a solid-state mechanism. Passive films, formed by the continued cycling of the single-pore electrode, are composed of CdO. Prolonged reduction is effective in transforming the large crystals of Cd(OH)$_2$ into cadmium.

In a series of four papers, Selånger[91-94] has developed a macrohomogeneous model for the porous cadmium electrode. Using this, a one-dimensional model, which included the effects of variation in composition of the electrolyte

---

[90] F. G. Will and H. J. Hess, *J. Electrochem. Soc.*, 1973, **120**, 1.
[91] P. Selånger, *J. Appl. Electrochem.*, 1974, **4**, 249.
[92] P. Selånger, *J. Appl. Electrochem.*, 1974, **4**, 259.
[93] P. Selånger, *J. Appl. Electrochem.*, 1974, **4**, 263.
[94] P. Selånger, *J. Appl. Electrochem.*, 1975, **5**, 255.

and in the activity of the reaction surface, was set up for the galvanostatic anodic high-rate transients. The calculated overpotentials were compared with experimental values, showing a satisfactory correspondence. It was shown that the failure of low-porosity electrodes that have great surface activity is generally caused by the blockage of the pores that occurs at low rates of discharge. At high rates, the depth of discharge is limited by transport of mass. For a porous cadmium electrode with porosity of 0.6, the transition from 'pore-blockage' failure to mass-transfer failure occurred at values of between 100 and 200 mA cm^{-2} for the discharge rate. The recovery in potential after a period of discharge follows from an extension to the equations for the porous electrode. This recovery is due to the diffusion of electrolyte solution in the porous matrix. A composite process of 'discharge and rest' has been simulated, and the results were compared with an experimental process. It was found that the recovery of the potential could be predicted by means of numerical mass-transfer models for known initial concentrations and profiles inside porous electrodes. The general influence of porosity on energy-storage capacity is examined in a charge–porosity diagram. Selånger[92] considers that this may be an efficient means of storing information on electrode systems. In a following paper[93] Selånger presents a simple charge–porosity diagram for the rapid estimation of charge capacity in battery electrodes. The basis of the technique rests on changes in porosity being dependent upon molar volumes of reactants and products. This specific sensitivity to changes in porosity for each system is a useful tool for the assessment of mass- and current-transport performance of electrodes that give insoluble electrode products. The characteristic lives for the cadmium, zinc, and iron electrodes in alkali are compared.

The natural technological goal of the work on porous battery electrodes is the optimization of the behaviour of the electrodes. Accordingly, in a further paper, Selånger[94] has attempted the optimization of the current efficiency. The analysis of the system was considered *via* a three-problem representation. These were to determine the charge capacity which gives the best utilization of the active material for a given polarization current and conditions of final potential; to determine the smallest capacity for a given current load and a given limit on the termination of the reaction and the final polarization; and finally to determine the capacity which gives the best utilization of the active material for the given conditions. The overpotentials of the electrode were calculated as in the previous work.[91] The resulting numerical solutions to the equations showed that the best high-rate electrodes (~ 100 mA cm^{-2}) have high porosity, with the half-width thickness in the interval of 0.1—0.18 cm for a final overpotential of 200 mV. There appears to have been no reported extension to this work, the suitability of the technique to the identification of the ultimate performance notwithstanding.

Tye and co-workers[95,96] have presented experimental work comparing the electrochemical responses of planar and sintered (porous) cadmium electrodes in KOH solutions. The cyclic voltammetric method was used. The relatively poor behaviour of sintered electrodes was considered to be intimately connected with the discharging (oxidation) process, as shown by the determination of concentrations of cadmium in the electrolyte before and after charging. Polarization-interruption

[95] R. Barnard, K. Edmondson, J. A. Lee, and F. L. Tye, *J. Appl. Electrochem.*, 1976, **6**, 107.
[96] R. Barnard, G. S. Edwards, J. A. Lee, and F. L. Tye, *J. Appl. Electrochem.*, 1976, **6**, 431.

experiments showed that a critical potential was reached at which passivation occurred, subsequent to which a plateau region was observed for sintered electrodes. Evidence was presented for the formation of a Ni–Cd alloy by interaction of the nickel plaque with the impregnated cadmium. A comparison of voltammetric data for sintered and planar electrodes showed that the distribution of the cadmium through the porous sintered electrodes improved the charging properties of these electrodes. The involvement of forms of cadmium of different electrochemical activity was demonstrated. Optical and scanning electron microscopy have been used to study the development and subsequent repositioning of phase cadmium and $Cd(OH)_2$ in sintered (porous) electrodes, due to cycles of charge and discharge.[96] Barnard et al.[96] reported the expected growth of both materials with increasing number of cycles and decreasing rate of charging. The sizes of $Cd(OH)_2$ crystallites were difficult to measure because the electrode always contained metallic cadmium, even in the discharged state. High rates of reaction promoted a pronounced aggregation and redistribution of active material towards the edge of the electrode. The result of this was a considerable decrease in the available effective volume of the pores, leading ultimately to blocking of the pores. The isolation of active cadmium by highly textured, unchargeable hexagonal platelets β-$Cd(OH)_2$ resulted in the effective loss of about 50% of the active material after about 100 cycles, at high rates of both charge and discharge; only the finely divided cadmium metal in the interior of the electrode continued to function. Deposits of more uniform size were produced by lowered rates of charging (reduction); however, even these deposits contained a high proportion of large particles of cadmium, which would clearly discharge much less efficiently than those produced at the higher rates of charging.

With porous cadmium electrodes (produced in either a sintered plaque, pocket, or plastic-bonded format), it is clear that the electrochemistry and chemistry are in a sense much more important than the porous structure. Czech workers[97] have discussed the bonding of porous electrodes with plastic materials. The binder used was polyethylene, and a process was described in which the mass of the electrode was first homogenized, then applied to the current-collector, and finally sintered at ~ 450 K and $\sim 5.5 \times 10^6$ Pa. Such electrodes were reported to possess superior output and very satisfactory cycling life in comparison with pocket-type electrodes. On the other hand, electrochemically impregnated sintered porous electrodes seem to be the preferred type for a number of aerospace applications.[98] The question of which type ultimately proves to be the best in service cannot be taken as completely settled. The claims for the plastic-bonded battery electrodes have been advanced by Jindra et al.[99] This group of workers describe plastic-bonded cadmium electrodes that were prepared by rolling a mixture of the active material with a plastic binder onto a current-collector. These electrodes are able to sustain a current density in excess of that of pocket-type electrodes; they possess outstanding resistance against deep and rapid discharge and they can deliver over 1000 cycles of charge/discharge. A content of 5% of poly(tetrafluoroethylene) in the mixture is sufficient to bind the

[97] M. Cenek, O. Kouřil, J. Šandera, A. Touškavá, and M. Calábek, in 'Power Sources', ed. D. H. Collins, Academic Press, London, 1977, Vol. 6, p. 215.
[98] J. D. Harkness, D. F. Pickett, and J. C. Want, *28th Power Sources Symp. 1978, Proc.*, p. 123.
[99] J. Jindra, J. Mrha, K. Micka, Z. Zábranský, V. Koudelka, and J. Malik, *J. Power Sources*, 1979, **4**, 227.

electrode. It is claimed that the toxicity problem that is encountered is very much reduced with this production process.

Micka et al.[100] have derived a theory for the porous cadmium electrode, based on the macrohomogeneous model and on rigorous transport equations. This extended, but was in contrast with, the cylindrical-pore model of Dunning et al.[29] Micka et al. used the same model for cadmium as was used in a previous treatment[101] of the negative electrode of the lead–acid battery. They now used a new system of basic equations, avoiding consideration of the individual ionic activities and the inner potential of the electrolyte. The behaviour of experimental cadmium electrodes was as predicted by the model at low current densities, where no passivation of cadmium occurred. The behaviour of the experimental system was dominated by the passivation mechanism. Discharge at current densities much greater than $0.01\,A\,cm^{-2}$ led to a gradual loss of capacity through passivation. In discharge experiments it was found that thinner electrodes reached the critical end-point polarization earlier than thicker ones, because of their stronger tendency to passivation. This led to the paradoxical situation where utilization of the active material increased with electrode thickness. This result suggests an optimum thickness for maximum utilization, since transport processes become limiting in very thick electrodes. Theoretically derived current distributions were in good agreement with the experimental observations of Bro and Kang,[34] and were qualitatively similar to the theoretical curves of Dunning et al.[29] Theory[100] also matched experiment[34] in the distribution of $Cd(OH)_2$ in the pores, expressed in terms of percent conversion of cadmium.

The crucial factor in the efficient utilization of the active mass of porous electrodes is the uniform distribution of electrode processes.[95,96] Russian workers[102] consider that this is largely determined by the resistance of the electrolyte in the pores. This was estimated, using the four-electrode technique,[102] and it was shown that this varied continuously during the process of discharging.

Carbon.—Armstrong et al.[76] have proposed a method (see also Section 2) whereby information about the structure and composition of porous electrode materials may be obtained from impedance data. Their procedure was applied to porous graphite, and found to be more suitable for comparative rather than absolute determinations of total surface area.

Blaedel and Wang[103] have described a rotated porous carbon disc electrode for analytical applications. The electrode, prepared from reticulated vitreous carbon, was characterized by using the oxidation of ferrocyanide ion. Well-defined polarization curves were obtained at different speeds of rotation. It was concluded that the work demonstrated an improved sensitivity, obtained with a rotated electrode of large surface area.

Copper.—Alkire et al.[104] have made available a procedure for preparing structures

---

[100] K. Micka, I. Roušar, and J. Jindra, Electrochim. Acta, 1978, **23**, 1031.
[101] K. Micka and I. Roušar, Electrochim. Acta, 1976, **21**, 599.
[102] Yu. M. Novak and D. K. Grachev, Elektrokhimiya, 1980, **16**, 57.
[103] W. J. Blaedel and J. Wang, Anal. Chem., 1980, **52**, 76.
[104] R. C. Alkire, E. A. Grens, and C. W. Tobias, J. Electrochem. Soc., 1969, **116**, 809.

that include uniform pores, based on sintering fine wires that are stacked in a parallel close-packed configuration. The procedure was illustrated for copper, but is applicable to any material that can be drawn down to a fine wire. This method produced linear, porous structures with a highly uniform and well-defined porosity. Such a structure can be of value in dissolution studies, and lends itself to the investigation of theories based on well-characterized configurations of pores. Interest has been concentrated, however, on more realistic electrode structures, generally prepared from powders, but sometimes using particle-bed models. Such electrodes are structurally more complex, but simpler to relate to practical electrode systems.

Alkire et al.[22] have proposed a theoretical analysis which takes into consideration the temporal variation of the structure of the electrode. This was one of the first attempts to model this aspect of behaviour, and a simple case of anodic dissolution was selected. The theory is dealt with in an earlier section of this Report. Calculations illustrating the behaviour of the model were presented for the acid–copper system, but no experimental work was undertaken.

Ng et al.[105] have studied the deposition of copper in cylindrical pores, such as might be found in printed-circuit boards, using a pulsed plating technique. A computer-based mathematical model was developed to solve the diffusion equations for such pores. This accounted for the observed difference in plating efficiency inside and outside of the pores, and could be used to predict optimum plating parameters. Whereas steady-state plating produced deposits of poor quality and uneven thickness, the use of a pulsed square-wave potential (typically, on for 1 second and off for 2 seconds) produced sound, even deposits. By taking a sufficiently large number of sample points and small time intervals, an all-numeric method could be devised, with the advantages that no differential equations needed to be solved and that it was not necessary to specify the precise nature of the perturbation on the system. However, this approach resulted in considerably increased demands on computer time and memory size.

Iron.—The porous iron electrode is of considerable interest, owing to the use of such electrodes in the Ni–Fe cell. Although this cell is being displaced by Ni–Cd and Ni–Zn cells in contemporary battery technology, the availability of iron ensures that research into the improvement of the electrode continues. Novakovski et al.[106] found the electrical conductance and the passivation behaviour of iron electrodes to be improved by additions of arsenic, antimony, nickel, cadmium, and lead. Pretreatments for obtaining electrodes with a high energy-storage capacity were described.

Bryant, Liu, and Buzzelli[107] have described an iron–air cell which employs conventional porous electrodes, prepared by using powder-metallurgical techniques. Power deliveries of the order of 100 W kg^{-1} were mentioned. No details of the porous electrode were given, and indeed, with the general electrode employed in batteries, French workers[108] have emphasized that only in very rare cases do

[105] H. K. Ng, A. C. C. Tseung, and D. B. Hibbert, *J. Electrochem. Soc.*, 1980, **127**, 1034.
[106] A. M. Novakovski, S. A. Grushkina, and R. L. Kozlova, *Zh. Prikl. Khim.*, 1973, **46**, 2183.
[107] W. A. Bryant, C. J. Liu, and E. S. Buzzelli, *28th Power Sources Symp. 1978, Proc.*, p. 152.
[108] J. Epelboin, M. Keddam, Z. Stoynov, and H. Takenouti, *Int. Soc. Electrochem., 28th Meeting*, 1977, p. 24.

steady states exist. These authors consider that impedance is the best method for study, and preliminary results have been discussed. Bryant[109] has also shown the importance of physical structure to the capacity of porous iron electrodes. Measurements of the discharge capacity were made and correlated with the porosity of the sinter. Using the Kuczynski relationship,[110] which relates the half-thickness of the neck that is formed during sintering with the duration of the sintering, it was shown that the available capacity for discharge of the iron electrode is controlled by either the extent of growth of inter-particle necks or the amount of surface area that is retained during sintering. In a somewhat later paper, Bryant[111] gave more specific details. Maximum capacities of ~ 0.4 A h kg^{-1} have been found for electrodes with densities within the range 21.5—23.5% of the theoretical density of iron. Relatively high capacity was also obtained by a low charge density (25 mA cm^{-2}). Moreover the capacity of a low-density electrode ($\sim 19.6\%$) was found to be much less sensitive to current density of the charge than that of the high-density electrodes. Interestingly, charge and discharge potentials were very dependent on the density of the electrode.

Andersson and Öjefors[112] have used a slow potentiodynamic method to charge and to discharge a porous iron electrode fully in alkali. The effects of changing the thickness of the electrode, the temperature, and the composition of the electrolyte were investigated. It was shown that the charging efficiency of the electrodes was favoured at high temperatures and at high concentrations of KOH. The latter parameter promotes the direct oxidation of Fe to FeOOH. Concentrations of K_2CO_3 up to very high values of 250 g dm^{-3} do not appreciably influence the reaction of the porous electrode. The porosity of the electrode also affects the efficiency of operation of the electrode. The effect of supporting iron nets within the electrode is explicable if the shielding effect that they have on the electrode is considered.

Lead.—The electrochemistry of porous lead is almost exclusively confined to the application of that electrode in the lead-acid cell, *i.e.* the behaviour in sulphuric acid. This topic has been reviewed recently.[113] The lead electrode has been the model for the development of much of the theory of porous electrodes and (together with the porous lead dioxide electrode) occupies a rather special position; the reader is referred to the above review (149 references) for a detailed account of the porous lead electrode to 1980.

As with many porous metal electrodes, the second law of thermodynamics implies a minimization of surface energy, and hence a tendency for a porous electrode structure to become less porous. With the lead electrode in sulphuric acid, this occurs during the reduction from lead sulphate to the metal. It has become customary to include 'expanders' to counteract this process, and the effective materials are $BaSO_4$ and derivatives of lignin. A recent paper describes experiments to investigate the effects of $BaSO_4$ and lignosulphonate on the redox processes at a

[109] W. A. Bryant, *J. Electrochem. Soc.*, 1979, **126**, 1899.
[110] G. C. Kuczynski, *Trans. Am. Inst. Min., Metall. Pet. Eng.*, 1949, **185**, 169.
[111] W. A. Bryant, *Electrochim. Acta*, 1979, **24**, 1057.
[112] B. Andersson and L. Öjefors, *J. Electrochem. Soc.*, 1976, **123**, 824.
[113] N. A. Hampson and J. B. Lakeman, *J. Power Sources*, 1981, **6**, 101.

porous lead electrode in H_2SO_4 solution.[114] It was shown that $BaSO_4$ provided nucleation centres for the growth of $PbSO_4$ and that lignosulphonate facilitated the nucleation of lead on a $PbSO_4$ phase in the reduction process. In the latter process the $BaSO_4$ exerted a certain synergistic effect with the lignosulphonate. Moreover, with the presence of the expanders, the nominal density of growth points that are associated with the porous lead electrode actually increased on cycling.

The fabrication of polymer-bonded porous lead electrodes without expanders has been described by Weininger and Secor.[115] These were prepared by pasting a mixture of carbon black, PbO, and a solution of neoprene in toluene on to a substrate of a lead–calcium alloy. Sodium benzoate can be added, which leaves a more porous structure after it has been baked out. Electrodes were cycled at the four-hour rate and their behaviour was compared with that of conventional porous lead negatives. Scanning electron microscopy was used to follow the structural changes throughout the life of a cycle. It was found that the electrode performance of the polymer-bonded electrodes was superior to that of conventionally pasted electrodes. This conclusion was based upon an enhanced cycle life, better charge acceptance, and a lower rate of evolution of hydrogen at the charging potential. The SEM observations provided further evidence for a dissolution–precipitation mechanism for the electrode reaction. The incorporation of organic materials into commercial porous electrodes is likely to be of considerable importance to modern battery technology.

Micka and Roušar[116] have formulated a set of equations which describe the transport of mass and electricity in the pores of the negative electrode of the lead–acid battery. The theory is based on exact transport equations and makes the assumption that the electrode potential is governed by the Nernst equation. Although changes in the specific volumes of both phases are taken into account, since the current-controlling reaction for both oxidation and reduction is electrocrystallization, this treatment is unlikely to be a successful reflection of the state of affairs in the lead cell, although the model cited is likely to apply to other systems.

The oxidation (discharge) of porous lead electrodes in sulphuric acid solutions has been followed, using a rotating-disc technique.[117] By this method it was possible to separate out the current that was generated at the front regions of the electrode, which could be affected by the hydrodynamics of the solution. It was found that the major component of the oxidation current arose from the inner porous structure. Potential-step studies of the same system[118] showed that transient responses have the same general features; however, the precise form of the transients indicated that the porous structure was exerting a considerable influence on the current–time relationships.[119] Whereas, for a planar electrode, a two-dimensional process of instantaneous nucleation and growth engenders an initial current that is proportional to elapsed time, the porous electrode gave a half-power dependence, in keeping with the theory for porous electrodes.

[114] N. A. Hampson and J. B. Lakeman, *J. Electroanal. Chem. Interfacial Electrochem.*, 1981, **119**, 3.
[115] J. L. Weininger and F. W. Secor, *J. Electrochem. Soc.*, 1974, **121**, 1541.
[116] K. Micka and J. Roušar, *Electrochim. Acta*, 1974, **19**, 499.
[117] N. A. Hampson and J. B. Lakeman, *J. Power Sources*, 1979, **4**, 21.
[118] N. A. Hampson and J. B. Lakeman, *J. Appl. Electrochem.*, 1979, **9**, 403.
[119] N. A. Hampson and J. B. Lakeman, *J. Electroanal. Chem. Interfacial Electrochem.*, 1980, **107**, 177.

The reduction of lead sulphate that has been grown anodically on porous electrodes has been studied, using a potential-step technique.[120] It was observed that both 'planar' and porous electrodes behave in a similar manner, which indicates that the resulting surfaces produced from both planar and porous electrodes are porous. This conclusion was based on the observation of a current which increased as (time)$^{\frac{1}{2}}$. The mechanism for the formation of metallic lead from the lead sulphate is instantaneous nucleation and two-dimensional growth, with subsequent limitation of the current due to overlap and depletion of the lead sulphate. The complete current transients were not explicable by any established theory.

The behaviour of the porous lead electrode in the microform[121,122] and as commercial-format plates[123] has been investigated at temperatures down to $-30\ °C$, in order to discover the electrochemical effects that occur at low temperatures. It has been convincingly demonstrated how ice crystals that are formed in porous lead electrodes play an important role in high-rate–low-temperature behaviour. The behaviours reported are complex, and it is only rarely that the response to a step function fits any of the theoretical models. A number of reasons have been advanced for the gross deviations that are observed; however, much more work is urgently required if the behaviour is to be understood. At a low temperature ($\sim -30\ °C$), the output of the lead–acid battery is quite poor, and limitation is due to the behaviour of the negative plate; if improvements could be made through a better understanding of the electrochemistry, the commercial rewards would be significant.

Matthews[124] has reported recent work on high-energy-density and mechanically rechargeable lead–acid batteries for electric vehicles. Cyclic voltammetric studies of lead and PbO_2 in various sulphate electrolytes have led to improved discharge capacities. Work on slurry electrodes has shown the possibility of a lead–acid battery that is capable of very rapid mechanical recharging. Tests of batteries in electric vehicles have demonstrated high charge and energy efficiencies for regenerative braking, and that the performance of the battery is unaffected by circulation of electrolyte.

Wales et al.[125] have attempted to relate the microstructure and performance of the electrodes of lead–acid accumulators in high-current cycling regimes (up to 2000 A m^{-2}), such as might be encountered in practice. Commercial negative plates were cycled to deep discharge at high current densities and then examined by light and electron microscopy, and by image analysis. Electrochemical performance was monitored by measuring the capacity and electrode potential of the plate as a function of current density and the number of cycles. The decrease in size of crystals of $PbSO_4$ in the discharged plates roughly corresponded with the decrease in capacity of the plate that occurred during cycling. This suggested that passivation is the major cause of loss of capacity, since it is accepted that passivation becomes more critical as the size of the $PbSO_4$ crystals decreases. The observation that through-thickness profiles of $PbSO_4$ were uniform supports this conclusion.

[120] N. A. Hampson and J. B. Lakeman, *J. Electroanal. Chem. Interfacial Electrochem.*, 1980, **108**, 347.
[121] N. A. Hampson and J. B. Lakeman, *J. Electroanal. Chem. Interfacial Electrochem.*, 1980, **112**, 355.
[122] N. A. Hampson and J. B. Lakeman, *J. Appl. Electrochem.*, 1981, **11**, 361.
[123] N. A. Hampson, J. B. Lakeman, K. S. Sodhi, and J. G. Smith, *Surf. Technol.*, 1980, **11**, 377.
[124] D. B. Matthews, *J. Electroanal. Chem. Interfacial Electrochem.*, 1981, **118**, 157.
[125] C. P. Wales, S. M. Caulder, and A. C. Simon, *J. Electrochem. Soc.*, 1981, **128**, 236.

However, the form of the polarization curves of the electrode indicated that limited rates of diffusion of the electrolyte and reduced contact between crystals of the active material (as a consequence of the porosity that is produced by cycling) could also act to reduce the capacity of a plate.

Lead Dioxide.—Lead dioxide forms the porous active material of the positive electrode of the lead–acid cell. Under normal ambient conditions it is this electrode which limits the discharge capacity of the cell, so that, not unnaturally, the behaviour of the porous PbO_2 has been the subject of much work. Conventionally, porous PbO_2 is produced either by direct oxidation of lead in sulphuric acid (with the addition of a 'forming agent', which acts by promoting the 'pitting' attack of the lead that is beneath the layer of lead sulphate) or by oxidizing a cured paste of commercial (leady or grey) lead oxides and dilute sulphuric acid. The porosity of the final PbO_2 is about 50% void.[126] With continued cycling, the cohesive structure of the porous mass tends to break down, and fine particles are lost from the electrode. Although the use of organic binders for secondary use is ruled out by the high positive potential of the electrode, they have been described as successful for PbO_2 primary cells. Paulson[127] has described latex-bonded electrodes, prepared from PbO_2 powder, which behaved well in a reserve battery with a negative lead electrode and 48% fluoroboric acid as the electrolyte. Owing to the difficulties of the high potential (see above), the binding of PbO_2 with organic compounds has not been commercially successful to date; with the advent of poly(fluorohydrocarbons) and isotactic hydrocarbon polymers, however, the field seems ripe for re-investigation.

The modelling of the porous PbO_2 electrode is of great importance and has been tackled by many workers.[3] In recent Czech work,[128] partial differential equations have been developed that describe the transport of mass and electricity in the pores of the PbO_2 electrode. The equations were formulated by using exact transport equations and with the assumption that the solid porous matrix had a high (metallic) conductivity. The changes in volume that occur in both phases during charge/discharge were incorporated in the treatment. Potentials were calculated, *via* the change in concentration, for the field equations for the rates of reaction (current), using the recalculated values of Harned and Hamer.[129] Computer-generated numerical solutions were presented for the cases where the concentration of the electrolyte was sufficiently high for the shortage of H_2SO_4 not to be a reaction-limiting process. The relationships between discharge time, current density, and porosity were estimated as relatively simple product functions, which did not agree very well with cited data for discharging.[128]

What must be the most complete treatment of the lead dioxide electrode to date has been given in a series of three papers by Simonsson.[44,45,53] In the first of these,[44] the distribution of current in the porous electrode was calculated, using the macrohomogeneous model and a simple equation for the conductivity of the porous matrix and the electrolyte phase to obtain a value for the current density in the electrolyte solution that is within the pore as a function of the gradient of electrode

---

[126] P. Casson, N. A. Hampson, and K. Peters, *J. Electroanal. Chem. Interfacial Electrochem.*, 1978, **87**, 213.
[127] J. W. Paulson, *28th Power Sources Symp. 1978, Proc.*, p. 99.
[128] K. Micka and J. Roušar, *Electrochim. Acta*, 1973, **18**, 629.
[129] H. S. Harned and W. J. Hamer, *J. Am. Chem. Soc.*, 1935, **57**, 27.

potential. The relationship between the current density and the electrode potential as a function of the concentration is all that is required to obtain the initial distribution of the current. The experimental estimation of the conductivities and of the polarization curve of the porous electrode permits the distribution of the current to be established, enabling the solution of the differential equations, which are both trivial and exact. Moreover, by the careful analysis of the profiles of the concentration of lead sulphate within the electrode in such a way as to avoid galvanic effects, the theoretical profiles were shown to agree with the experimental ones to well within experimental limitations.

Simonsson developed the macrohomogeneous model of the discharging porous lead dioxide electrode.[45] Two main structural effects, due to the precipitation of the lead sulphate that was produced, were considered to determine the local degree of discharge. These structural processes are the plugging of the pores and the gradual insulation of the active surface of the electrode by the reaction product. The equations were solved by numerical methods and indicated that, at high current densities, the discharge capacity is limited by both the structural effects and the transport of sulphuric acid into the discharged part of the electrode. At the termination of the useful discharge, lead sulphate blocks the outer regions of the electrode. The concentration of electrolyte solution at this blocked region is at a minimum at the end of discharge.

In a final paper, Simonsson[53] uses the pseudo-steady-state approach[22,29] to consider the discharge at low rates. In this method, a concentration profile is assumed to be established rapidly, which corresponds to the rate of discharge. The method fails at high rates of discharging, and it is likely that the limit of this method is ~ 10 minutes.[24] It was shown that the distribution of current at low rates of discharge becomes non-uniform during discharging because of changes in concentration and decreasing porosity. The maximum of the distribution of current starts at the front of the electrode and moves into the interior during discharging, owing to the increasing resistance of the 'discharged' areas of the electrode. Provided that the porosity is high enough, the discharging spreads uniformly through the electrode, so that, at the end of a discharge that has taken place at a low rate, the electrode has been uniformly discharged throughout. At low porosities ($< 50\%$), it was shown that the inner regions of a thick electrode may not always be fully utilized, owing to the plugging of fine pores by $PbSO_4$.

Czech workers[130] have also applied the macrohomogeneous model to the porous PbO_2 electrode, reacting in $5M-H_2SO_4$. Changes in volume during the discharge were expressed mathematically for both the solid and liquid phases, and the equations coupled with the electrode kinetics and the exact transport equations for the concentrated electrolytes. Distributions of concentration and porosity were calculated, from which discharge times (to the knee of the galvanostatic polarization curve) were established at various current rates. In a later paper, Micka and Roušar[101] combine this calculation with a similar one made for the porous lead negative electrode of a battery to form the complete estimation of discharge for the lead–acid battery. The authors make the interesting conclusion that the concentration of electrolyte does not indicate the degree of discharge of the

---

[130] K. Micka and I. Roušar, *Collect. Czech. Chem. Commun.*, 1975, **40**, 921.

lead–acid cell. Another important conclusion is that the maximum capacity of the cell occurs at an optimum separation of the plates. The conclusions emphasize the difficulty of the problem of optimization in batteries generally.

Wiesener and Reinhardt[131] investigated the importance of the conditions of preparation on a number of physical properties of lead positive electrodes of an accumulator, with a view to improving performance and stability. They found that the content of α-PbO_2 in the electrode decreased, with an increase of that of the β-form, during cycling, reaching a steady state after 100 cycles. In a study of the mechanisms of discharge of these two forms of PbO_2, the same authors[132] were able to account for the different behaviour of the two oxide types in terms of the overpotential for crystallization of $PbSO_4$ on PbO_2.

The use of Peukert's equation[133] is traditional for calculating the effect of the thickness of the electrode (t_d) on the discharge current (i_d). The form of the equation is a simple transcendental one, $t_d = k/i_d^n$, in which k and n are constants, and it has been used as a good approximation.[101] Recently, the general application of this equation has been examined for the case of porous PbO_2 electrodes.[134] It was found that the value of n was very dependent upon the applied current density, and this fact must be borne in mind when comparing data from different workers. A more serious current-sensitive characteristic was the overpotential of the electrode, which was observed to change from the familiar 30 mV decade^{-1} at the lower current densities to 120 mV decade^{-1} at very high current densities (e.g. as needed for engine-starting duty). This transition was considered to represent a change from activation/diffusion control to control by the ohmic resistance of the pore and of the electrolyte therein. The maximum degree of discharge in the lead dioxide electrode, X_{max}, is an important characteristic of the electrode, and it is a crucial choice in most of the modelling investigations.[45,53] Simonsson terms X_{max} a 'fitting parameter', in that it is adjusted outside the program. However, he has shown that X_{max} varies with current density, the thickness of the electrode, and the cycle number. A further shortcoming of the models of the porous lead dioxide electrode has been the use of the Erdey-Gruz–Volmer-type equation generally to include the potential dependence of the activation reaction. This is only satisfactory if this process is current-controlling in the potential domain that is of interest. There exist a number of papers to show that the major current-controlling process is that of nucleation and growth. This process results in a current characteristic which increases (after starting from relatively small values) and, after passing through a maximum, begins to fall as growing centres overlap. So far, the application of the electrocrystallization process to porous electrodes has not been made; however, it is obvious that gently rising transients must result from the ultimate model. This behaviour has been observed, and is a well-known characteristic. Recent measurements have shown that the distinctive rising transients occur in simple porous disc electrodes[135] and in commercial systems.[136] These shortcomings in the operational electrochemistry of

---

[131] K. Wiesener and P. Reinhardt, *Z. Phys. Chem. (Leipzig)*, 1975, **256**, 285.
[132] P. Reinhardt and K. Wiesener, *Z. Phys. Chem. (Leipzig)*, 1976, **257**, 412.
[133] W. Peukert, *Elektrochem. Z.*, 1897, **18**, 287.
[134] J. M. Asher, N. A. Hampson, S. Kelly, and G. S. Holmes, *Surf. Technol.*, 1980, **10**, 371.
[135] J. M. Asher, N. A. Hampson, and G. S. Holmes, *Surf. Technol.*, 1980, **11**, 17.
[136] S. Kelly, N. A. Hampson, and K. Peters, *J. Electrochem. Soc.*, 1977, **124**, 1656.

the model porous lead dioxide electrode have resulted in overvoltage curves which fall too quickly with time.[45]

A further experimental contribution to the modelling of the porous electrode has been made by Whyatt and Hampson.[137] Circular porous disc electrodes of PbO_2 in lead annuli were electroformed from conventional pastes and reduced in 5M-H_2SO_4. The porous electrode was sectioned and the advancing boundary of $PbSO_4$ was examined by SEM. The results showed that the discharge begins at the front of the electrode and proceeds inwards (with a sharp boundary), normal to the surface, with an apparent layer-by-layer conversion of the porous mass. If the porous electrode is sufficiently thin, the reaction proceeds right through to the centre; if not, a region of what is effectively unreacted PbO_2 remains deep in the electrode. The authors concluded that the simple Simonsson model was not completely satisfactory, and that the assumption of a constant current density over the whole face required revision.

A further complicating factor of porous PbO_2 electrodes has been identified by Micka *et al.*[138] as their structure. The investigation, made with commercial electrodes, concerned the density of the PbO_2 and the tortuosity factor of the pores. The results showed that charged and cycled porous electrodes contained a certain amount of PbO_2 of a lower density, possibly amorphous. This material is preferentially reduced; what is more, the changes in volume that occur in a porous PbO_2 electrode were irreversible.

It is interesting that not only is the physical representation of the porous PbO_2 electrode a problem of considerable interest and importance, but the chemical behaviour of porous PbO_2 electrodes is still not fully understood. The simplest porous lead dioxide electrode is that due to the original process of Planté, in which the porous deposit is prepared from the base (pure) lead support by corrosion in sulphuric acid. The addition of a forming agent to promote this process is well established. Conventionally, perchlorate ion has been used for many years; recently, the projected use of 'Planté' cells for the storage of solar power has stimulated interest in this area. The optimization of the amount of additional aggressive perchlorate ion has been reported.[139] The optimization was carried out by using the potentiostatic pulse technique, and it was concluded (for the oxidation of Pb to PbO_2 in 0.5M-H_2SO_4) that, if the electrode is not to be rendered passive, the formation process must change from one of instantaneous to progressive nucleation. The process of formation of PbO_2 was a solid-state one, promoted by an optimum concentration of perchlorate ion (30—50 mmol dm^{-3}). Increasing the concentration of perchlorate ion above the optimum produces decreasing rates of formation, because the perchlorate ion, in displacing the sulphate ion from the lead surface, inhibits the regeneration reaction. As the concentration of ClO_4^- is increased, a limit is ultimately exceeded, and the system cannot be rendered completely passive.[140] The effect of the perchlorate ion is to promote the attack on the lead by displacing the passivating sulphate ion. It has been shown[141] that a range

---

[137] P. R. Whyatt and N. A. Hampson, *Surf. Technol.*, 1979, **9**, 351.
[138] K. Micka, M. Svatá, and V. Koudelka, *J. Power Sources*, 1979, **4**, 43.
[139] N. A. Hampson, C. Lazarides, and M. Henderson, *J. Power Sources*, 1981/82, **7**, 181.
[140] C. Lazarides, N. A. Hampson, and G. M. Bulman, *J. Power Sources*, 1981, **6**, 83.
[141] C. Lazarides, N. A. Hampson, and M. Henderson, *J. Appl. Electrochem.*, 1981, **11**, 605.

of concentrations of perchlorate ion exist (30—40 mmol dm^{-3}) for a given concentration of H_2SO_4 (0.5 mol dm^{-3}) where the attack is an 'advancing front'. The process is analogous to the electrodeposition of a metal from solution, where the layer of PbO_2 steadily advances into the metal. By a combination of the chemistry and the electrochemical processes of phase formation, a reaction sequence can be written which satisfactorily accounts for the entire mechanism of formation of porous PbO_2 electrodes by the Planté method.[141]

An alternative method has been proposed[142] for the preparation of Planté porous positive electrodes in fully assembled cells. This involves the use of nitric acid as the forming agent and depends on the electrochemical destruction of the nitrate-forming agent. The final destiny of the nitrogen is NH_4^+ ion, and it is clear that the presence of NH_4^+ is not detrimental to the behaviour of the cell. If the perchlorate ion cannot be removed from the porous PbO_2 electrode, the result is very rapid corrosion of the lead frame of the porous lead electrode.

Much work has been done recently on the chemistry of the porous lead dioxide electrode in connection with the effect of the interphase between the porous PbO_2 structure (prepared by the oxidation of Pb^{II} oxides/sulphates) and the support (grid). Conventionally, the support materials were (the very castable) lead–antimony alloys; however, industrial progress has demanded a reduction in the content of antimony since this metal was not only expensive but introduced a material with a low hydrogen overpotential into the system whose ultimate destination was the negative (lead) electrode, causing excessive local reaction. The removal of antimony improved this aspect but affected the bonding and the interaction between the porous matrix and the support. So far, no alloy has yet been perfected which combines the desirable properties of freedom from antimony with the mechanical stability and electrochemical reactivity of an alloy that contains 7% of antimony. Various alloys have been tried out, and considerable work has been done in order to identify this 'antimony effect'.

The reduction of porous PbO_2 has been studied by linear-sweep voltammetric techniques,[143] and the rate of the process shown to be completely controlled by solid-state processes of phase growth. The reduction peak was broadened by the porosity, and this was interpreted in terms of the reaction being driven more deeply into the electrode as the front of the electrode became progressively more resistive. The oxidation of porous $PbSO_4/PbO_2$ on a lead phase in $5M$-H_2SO_4 has been studied in detail, using the potentiostatic pulse technique. It was found[144] that the form of the current transient depended upon the relative amounts of $PbSO_4$ and PbO_2. For small amounts of $PbSO_4$, a simple falling transient is obtained. For larger amounts, a complicated rising transient was observed which could be deconvoluted with the aid of a computer. The decomposition indicated the occurrence of two separate two-dimensional processes of nucleation and growth, followed by a final progressive three-dimensional process of nucleation and growth. The conclusion was drawn that the first peak corresponds to the inner region of the electrode (the self-corrosion layer), the second to those regions beneath the outer electrode surface that are at distances greater than the penetration depth,

---

[142] W. H. Edwards and N. A. Hampson, *J. Appl. Electrochem.*, 1979, **7**, 381.
[143] P. Casson, N. A. Hampson, and K. Peters, *J. Electroanal. Chem. Interfacial Electrochem.*, 1978, **87**, 213.
[144] P. Casson, N. A. Hampson, and K. Peters, *J. Electroanal. Chem. Interfacial Electrochem.*, 1978, **92**, 191.

and the final peak to the front of the electrode (to the limit of the penetration depth for the process of reduction). The electrode behaves like three systems in parallel rather than as a simple, uniform, porous matrix. In effect, during the initial period of re-oxidation, the front of the 'reduced PbO_2 electrode' behaves like a porous medium[114] rather than a porous electrode. The re-oxidation behaviour of the electrode depended upon the extent to which the porous electrode had been cycled; it was shown that cycling the porous system resulted in the fine-grained $PbSO_4$ that had been produced in early cycles being gradually transformed into large block-type crystals. These structural changes, which occur to an extent that is linked with the extent of the redox process, have been followed by SEM.[145] The existence of the three well-defined regions of electrode structure was confirmed.

Pavlov and Ruevski[146] have encountered a problem in the production of PbO_2 positive plates which they call thermopassivation. This occurs if drying occurs at temperatures $>70\,°C$, the resulting loss of energy being determined from galvanostatic discharge curves for thermopassivated and unpassivated plates. The dependence of the thermopassivation effect on a variety of parameters was determined. It was concluded that thermopassivation might be due to semiconductor properties of the non-stoicheiometric PbO_n ($n<2$) corrosion layer of the plate, and to the contacts between the active mass and this corrosion layer as well as with the electrolyte.

The effect of porosity on the a.c. response of porous electrodes has been reviewed by de Levie.[5] This aspect of behaviour of a porous electrode has become very important recently, owing to the developing interest in impedance methods, especially in connection with primary and secondary storage cells. In view of the very complex structure of the porous PbO_2 electrode, a simple behaviour is hardly likely to be found. Casson, Hampson, and Willars[147] have compared the a.c. response of massive (electrodeposited β-PbO_2) and porous electrodes. The influence of the porous structure was profound, for while the massive electrode behaved in an interpretable manner, the behaviour of the porous electrode was complicated. The most significant feature was an inductive region at high frequencies which is not to be interpreted on any theory of simple flooded electrodes. Others have noticed this aspect in connection with measurements on complete lead cells.[148,149]

The a.c. impedances of thin porous layers of PbO_2 on lead and its alloys have been examined.[150] Formed potentiostatically on the base metals, the PbO_2 layer is thick and porous, and the a.c. impedance spectrum is more simple than that of the porous electrode (0.1 cm deep). The kinetics were interpreted as due to an electrode of the second kind; however, the Warburg line dihedral was greater than $\pi/4$. This was due to the capacitance of the frontal $PbSO_4$ film. Inductive responses were only observed to any extent with the lead–antimony alloys. The effect of antimony (and bismuth) on the impedance of the porous PbO_2 was considerable, engendering

[145] P. Casson, N. A. Hampson, K. Peters, and P. R. Whyatt, *J. Electroanal. Chem. Interfacial Electrochem.*, 1978, **93**, 1.
[146] D. Pavlov and S. Ruevski, *J. Electrochem. Soc.*, 1979, **126**, 1100.
[147] P. Casson, N. A. Hampson, and M. J. Willars, *J. Electroanal. Chem. Interfacial Electrochem.*, 1979, **97**, 21.
[148] F. Gutmann, *J. Electrochem. Soc.*, 1965, **112**, 94.
[149] M. Keddam, Z. Strognov, and H. Takenouti, *J. Appl. Electrochem.*, 1977, **7**, 539.
[150] N. A. Hampson, S. Kelly, and K. Peters, *J. Appl. Electrochem.*, 1981, **11**, 751.

thickening of the layer and apparently its semiconductor properties. Australian workers[151] have shown that the antimony becomes widely distributed throughout the porous matrix and is probably associated with the PbO_2 phase rather than with the $PbSO_4$.

The importance of the interphase between the porous lead (alloy) support and the PbO_2 is clearly important, and Hampson, Peters, and Casson[152] have described a method for the assessment of the importance of the interphase on the porous PbO_2 electrode. Fourteen different alloys were examined and compared with pure lead, with reference to charge output on reduction, ease of re-oxidation, and adhesion between the electrode support and the porous PbO_2. It was found that an alloy that contains 6% of antimony is far superior to all others in all respects. Bismuth is also of some interest. Some work has been described recently on the effect of reduction potential on the behaviour of porous PbO_2 on a lead–bismuth base[153] in order to compare the alloy with the lead and lead–antimony systems, which exhibit reluctance to accept charge if the reduction potential is too low (700 mV vs $Hg_2SO_4/5M$-H_2SO_4).[154,155] With 0.13% of bismuth, two distinct potential regions were observed, separated by a plateau. The output of charge to the plateau (limit 700 mV) could always be readily replaced by a potentiostatic oxidation at 1200 mV. If the potential was reduced to a value below this limit, the complete re-oxidation could not be effected under these conditions.

With the porous lead electrode, the chemistry and electrochemistry of the material itself, and especially the nature of the support material, are of considerable importance. It seems that improvements to the theory of the porous electrodes in order to describe the behaviour more fully will not be effective until there exists a complete understanding of the electrochemistry of PbO_2.

Lithium.—Lithium–aluminium alloys[156] have been tested as possible electrode materials for secondary batteries of high specific power and high specific energy, for use in off-peak storage of energy and electric vehicles. Such cells would comprise positive electrodes of either iron sulphide or iron pyrites, negative lithium–aluminium electrodes, and a molten salt electrolyte, $e.g.$ LiCl–KCl eutectic, and they would operate at 400—450 °C. The problems attendant upon using pure lithium as an electrode material can be avoided by using a lithium–aluminium alloy. A 48 atom % lithium–aluminium alloy has shown promise in meeting performance requirements for these high-power cells.

Vissers and Anderson[156] have studied the major parameters determining the performance of the porous lithium–aluminium electrode. An interesting aspect of this work was their use of a liquid lithium electrode as both the working electrode and the reference electrode, since this material is unpolarized even at high current densities. Vissers and Anderson found the utilization of lithium in the alloy electrodes to decrease rapidly as the thickness of the electrode was increased from

---

[151] R. Barrett, M. T. Frost, J. A. Hamilton, K. Barris, I. R. Harrowfield, J. F. Moresby, and R. A. J. Rand, *J. Electroanal. Chem. Interfacial Electrochem.*, 1981, **118**, 131.
[152] N. A. Hampson, K. Peters, and P. Casson, *J. Power Sources*, 1981, **6**, 63.
[153] S. Kelly, N. A. Hampson, and K. Peters, *J. Appl. Electrochem.*, 1982, **12**, 81.
[154] S. Kelly, N. A. Hampson, and K. Peters, *J. Appl. Electrochem.*, 1981, **11**, 601.
[155] S. Kelly, N. A. Hampson, and K. Peters, *J. Appl. Electrochem.*, 1981, **11**, 269.
[156] D. R. Vissers and K. E. Anderson, Argonne Nat. Lab. Report ANL 76—8, 1976.

0.32 to 0.64 mm. The utilization of lithium and the cycling characteristics were improved by reducing the pore fraction of the electrode from 0.7 to 0.2. The performance of alloy electrodes was greatly improved by the incorporation of a porous metal current-collector. The performance of the lithium–aluminium electrode appeared to be limited by physical and chemical properties, including electronic conduction and the transport of lithium in the solid.

Dey[157] evaluated a range of different porous carbons for their suitability as cathodes for $Li/SOCl_2$ cells. The carbons possessed a large variation in such properties as particle size, BET surface area, density, and electrical conductivity. The performance of the cell remained unaffected by these physical properties, except in extreme cases (*e.g.* graphite, the particle size of which was larger by about three orders of magnitude).

Tiedemann and Newman[43] and Pollard and Newman[59] have presented theoretical treatments of the operation of porous electrodes, the lithium–aluminium electrode being used in both cases as an illustrative example. Their contributions are considered in the theoretical section, earlier in this Report.

Mercuric Oxide.—The constant-power test is a well-known method of evaluating battery quality, and Szpak and McWilliams[158] have described such a procedure for testing a Zn–HgO(C) battery. The decline of cell power was associated with penetration of the reaction zone into the electrode structure, the mode of discharge determining the rate of penetration. The electrode effectiveness factor used by Szpak and McWilliams employed Winsel's λ^{-1} factor,[7] and was determined for the porous HgO electrode without resorting to tedious chemical procedures. The decrease in the effectiveness factor was associated with the initial discharge of the battery, followed by its stabilization to a value $3 < \lambda^{-1} < 5$ after 30% of all the active material had reacted.

Molybdenum.—There has been very little interest in the electrochemistry of molybdenum during the review period. However, in a recent communication, Jacobson *et al.*[159] briefly reported the electrochemical behaviour of amorphous molybdenum disulphide that had been prepared by the reaction of lithium sulphide with molybdenum chloride. X-Ray diffraction indicated the amorphous compound to be produced by reaction at ambient temperature, with properties radically different from those of product that was obtained at high reaction temperatures. Crystallization is initiated on heating to 200—300 °C, and is not complete until 800 °C. Initiation of crystallization is followed by a large increase in surface area, and consequently a large capacity for electrochemical reaction with lithium. This low-temperature molybdenum sulphide phase is highly reversible; after 244 charge/discharge cycles in a lithium cell, its capacity exceeded 50% of that for the second discharge.

Nickel.—The major interest in the porous nickel oxide electrode arises from the power-source application. Some technologically interesting contributions have

---
[157] A. N. Dey, *J. Electrochem. Soc.*, 1979, **126**, 2052.
[158] S. Szpak and G. E. McWilliams, *J. Electrochem. Soc.*, 1973, **120**, 635.
[159] A. J. Jacobson, R. R. Chianelli, and M. S. Wittingham, *J. Electrochem. Soc.*, 1979, **126**, 2277.

been made in connection with porous nickel oxide electrodes for application in Ni/H$_2$ cells.[160-162] Seiger and Puglisi[160] have developed a porous nickel electrode that works in conjunction with the hydrogen electrode, using an impregnation technique. An aqueous impregnation process is preferred in which a solution of mixed nickel and cobalt nitrates is cathodically reduced in a nickel plaque. Such electrodes show considerable retention of their dimensions on cycling, and require only a single impregnation step.

Pickett et al.[161] have described an electrochemical process in which, it is claimed, the substantial change in shape (expansion) that is observed with conventional chemical/vacuum impregnation processes is not apparent. Moreover, only very slight corrosion of the electrode occurred, and it could be reduced further if the acidity of the electrolyte was controlled and the plaque was efficiently passivated prior to deposition of the nickel hydroxide from its salt. An interesting phenomenon which often occurs with nickel oxide electrodes is the enhancement of recoverable capacity as a result of cycling. This effect is not of great importance with the Ni–Cd cell, operating in aqueous solution; however, for the Ni/H$_2$ system, the pressure is driven higher at the end of charging, and can lead to failure of the system.

Korovin et al.[163] obtained stationary polarization curves that were similar for both porous and planar nickel electrodes in alkali and in NH$_4$OH. The dissolution behaviour of the anode was similar for both types of electrode. The effects of the addition of NH$_4$OH and of Al^{III} ions were investigated.

Riba and Aussaresses[164] produced a porous nickel–sulphur electrode by compressing powders of nickel and sulphur, and subsequently reducing the sulphur in 1M-KOH. The temporal variation of potential between the nickel of the cathode and the electrolyte layer near the front of the cathode was recorded for galvanostatic discharge. The slope of the linear part of the relation gave the ratio of the tortuosity factor to the conductance of the electrolyte in the pores.

Prema et al.[165] have presented procedures for producing composite nickel powders with inert core materials, such as TiO$_2$, and describe their physical and electrochemical properties. They found that some pure nickel powder had to be mixed in, to improve the conductivity, so that the sintered plaque could be used as a battery plate. Electrodes that were prepared in this way contained up to 50% less nickel and functioned as well as those prepared from pure nickel powder. These authors[166] later found that the addition of tetra-alkylammonium compou .ds to sintered nickel and cadmium electrodes improves the performance of Ni–Cd cells, especially at high discharge rates.

Takamura et al.[167] studied the incorporation of Ni(OH)$_2$ into porous nickel by the electrochemical reduction of Ni(NO$_3$)$_2$. The amount of Ni(OH)$_2$ that was incorporated depended strongly on current density and concentrations of

---

[160] H. N. Seiger and V. J. Puglisi, *27th Power Sources Symp. 1976, Proc.*, p. 115.
[161] D. F. Pickett, U. D. Martin, J. W. Logsdon, and J. F. Leonard, *27th Power Sources Symp. 1976, Proc.*, p. 120.
[162] H. N. Seiger, *28th Power Sources Symp. 1978, Proc.*, p. 120.
[163] N. V. Korovin, V. N. Savel'eva, and Yu. P. Shishkov, *Elektrokhimiya*, 1972, **8**, 855.
[164] J. P. Riba and H. Aussaresses, *J. Chim. Phys. Physicochim. Biol.*, 1972, **69**, 585.
[165] R. Prema, P. V. Vasudeva Rao, and H. V. K. Udupa, *J. Electrochem. Soc. India*, 1973, **22**, 303.
[166] P. V. Vasudeva Rao, T. Vasanthi, H. V. K. Udupa, *J. Power Sources*, 1976, **1**, 81.
[167] T. Takamura, T. Shirogami, and T. Nakamura, *Denki Kagaku*, 1974, **42**, 582.

Ni(NO$_3$)$_2$, and, under certain conditions, the method could produce electrodes of greater capacity than those produced by the conventional technique. The mechanism of impregnation was discussed in terms of the formation of NH$_3$ by reduction of NO$_3^-$.

Yamamoto[168] investigated the behaviour of alkaline storage-battery electrodes of pressed mixtures of 20% Ni(OH)$_2$, 30% graphite, 20% NH$_4$HCO$_3$, and 30% of a 2% solution of polystyrene. These electrodes were characterized in terms of potential (overpotential) in charged and discharged states, and of self-discharge rates. Results were compared with similar electrodes that had been prepared using polyethylene.

The distribution of current along a segment pore has been measured by Russian workers.[169] Measurements were subject to error, owing to the significant values of the current-measuring resistances. It was shown experimentally that the current changes with time and with the depth of the pore. The results are interpreted in terms of non-uniform oxidation along the pore, the extent of the oxidation being greater near the front. An a.c. polarization procedure for the production of a metal/ceramic/nickel oxide electrode was suggested by the authors on the basis of their model.

Rademacher et al.[170] employed various physical methods of examination to show that ageing of porous nickel electrodes was associated with variations in the secondary structure, which in turn depended on the crystallization process of β-Ni(OH)$_2$.

Micka and Svatá[171] have estimated the tortuosity of the pores in the porous nickel oxide battery electrode. The tortuosity was defined in terms of the electrical conductivity of the electrolyte in the pores of the matrix. The method used was based on the earlier work of Euler and Rieder[172] and of Kzenzhek et al.,[173] which depends upon the measurement of field potential (with Luggin probes) of each side of a porous electrode at a fixed current density.

A new term, 'coefficient of transport hindrance', was introduced by Micka and Svatá as the ratio (volume fraction of pores)/tortuosity factor, and it was obtained experimentally as the ratio of conductivity of electrolyte in a pore to that of the free electrolyte. The method involves a regression analysis in order to find a correction factor for errors that were introduced by the tips of the capillary. The method allows the estimation of the coefficient of transport hindrance in the pores, which was shown to be related to the performance of the electrode at high current densities. The tortuosity of the pores can be obtained as the ratio of the porosity to the coefficient of transport hindrance. The authors[171] were able to use their experimental values to speculate on the porous structure of the matrix and the transport properties of the solution within the pores.

Further work on the nickel oxide electrode has been reported by Micka and co-workers.[174] Studies were reported on plastic-bonded electrodes, prepared by the

[168] Y. Yamamoto, *Denki Kagaku*, 1974, **42**, 557.
[169] Yu. D. Kudryavtsev, F. I. Kukoz, and L. N. Fesenko, *Elektrokhimiya*, 1975, **11**, 378.
[170] O. Rademacher, K. Wiesener, and E. Prikryl, *Z. Phys. Chem. (Leipzig)*, 1976, **257**, 354.
[171] K. Micka and M. Svatá, *J. Power Sources*, 1978, **3**, 167.
[172] J. Euler and E. Rieder, *Elektrotech. Z.*, 1964, **85**, 557.
[173] O. S. Ksenzhek, E. A. Kalinovskii, and E. L. Baskin, *Zh. Prikl. Khim.*, 1964, **37**, 1045.
[174] B. Klápště, J. Mrha, K. Micka, J. Jindra, and V. Maraček, *J. Power Sources*, 1979, **4**, 349.

addition of a conducting component (carbon black, graphite, or nickel carbonyl), with poly(tetrafluoroethylene) as a binder, to the active material that is used in pocket-type NiO electrodes. These electrodes have shown the absence of a second step in the discharge curve which is usually observed when a pressed nickel oxide electrode is discharged in the range from -100 to -600 mV (vs HgO/Hg). A sintered nickel electrode does not give this stepped discharge. The experiments showed that the parameters of plastic-bonded porous nickel oxide electrodes can be controlled to a large extent by the type and quantity of the conductive component, which influences the degree of participation of electro-active particles in the electrode reaction. PTFE binder particles improve the rolling properties of the electrode mix. The extent and quality of the ohmic contact among individual particles control the presence of a stepped discharge curve.

Crespy et al.[175] have improved the energy density and high-rate discharge behaviour of alkaline batteries by developing a nickel positive electrode with a highly porous foam structure. The electrochemical characteristics of these electrodes were superior to those of conventionally sintered electrodes.

Shirogami et al.[176] prepared alkaline battery electrodes by hot-pressing a powder mixture of γ-NiOOH, carbon, and polystyrene on to nickel-screen current-collectors. The use of nickel in this form, instead of the usual $Ni(OH)_2$, increased the capacity of the electrode. Mrha et al.[177] found carbon black to be the most suitable conducting addition for PTFE-bonded NiO electrodes, for high current-carrying capacity and long-term cycling. The cycle life of the electrode depended on the interface between the active layer and the collector. At loads of 3—100 mA cm^{-2}, the type of conducting admixture and current-collector, as well as its surface treatment, strongly influenced the current-carrying capability of the electrode.

Micka et al.[178] have shown the active layer of plastic-bonded nickel oxide electrodes to undergo expansion during discharging and contraction during charging, but that the latter process does not fully compensate for the former. The changes in volume could be attributed both to a difference between the molar volumes of $Ni(OH)_2$ and NiOOH and to a loss of cohesion of the particles of active material. Changes in volume could be made reversible by the application of an external pressure. Graphite showed the least anodic corrosion of the carbonaceous conductive additions.

An important contribution to the theory of porous nickel oxide electrodes was presented by Micka and Roušar[179] in 1980. A model of the electrochemical behaviour of porous electrodes of hydrated nickel oxide was constructed from the known electrochemistry of hydrated nickel oxides and from transport equations derived for a continuous macrohomogeneous model. Two cases were considered as important technological designs, i.e. the current-collector in the interior (sintered type) and on the surface (pocket type) of the electrode. The basic equations for the major part of the calculation had been given previously.[100] The mathematical treatment enables the prediction of the discharge characteristics, which are

---

[175] J. G. Crespy, R. Schumitt, M. A. Gutjahr, and H. Saenffrier, *Power Sources*, 1978, **7**, 219.
[176] T. Shirogami, T. Nakamura, H. Niki, and K. Sasaki, *Denki Kagaku*, 1979, **47**, 440.
[177] J. Mrha, I. Krejči, Z. Zábranský, V. Koudelka, and J. Malik, *J. Power Sources*, 1979, **4**, 239.
[178] K. Micka, J. Mrha, and B. Klápště, *J. Power Sources*, 1980, **5**, 207.
[179] K. Micka and J. Roušar, *Electrochim. Acta*, 1980, **25**, 1085.

compared with the experimental data. The comparison is surprisingly good, considering the uncertainties involved. An interesting fact to emerge from the work is that the composition of the fully charged active nickel material is between $NiO_{1.5}$ and $NiO_{1.8}$.

Paskiewicz[180] has reviewed the various methods used for the impregnation of porous nickel plates in alkaline batteries. Chemical methods of impregnation produced electrodes with higher volumetric capacities but shorter lives; methods of impregnation that involved thermal decomposition reversed this situation. Starter batteries, designed for very fast discharge, had greater capacities when they were constructed with finer nickel powders.

Paskiewicz[181] also investigated the influence of cathodic polarization during the precipitation of $Ni(OH)_2$ in porous sintered nickel electrodes. The simple dependence of rate of precipitation on temperature reveals the important part that is played by the diffusion of Ni^{2+} ions in conjunction with applied polarization. Choice of polarization conditions which avoid the evolution of hydrogen led to low electrode capacities and erratic performances. Greater electrode capacities were obtained by a cyclic treatment, alternating polarization with immersion in KOH.

Platinum.—Platinum is of interest predominantly for its use as a catalytic material in gas electrodes for electrochemical conversion and fuel cells. These aspects of its use are not within the scope of this Report. However, two communications concerning flooded porous platinum electrodes have been found. Shustov and Kalinovskii[182] recorded polarization curves for porous platinum anodes, 0.5—1 mm thick, in sulphuric acid solutions, and they observed minimum polarization in the porosity range 40—50%.

Podlovchenko and Gladysheva[183] studied the electrochemical sintering of platinum electrodeposits, under potentiostatic conditions, in a variety of aqueous solutions. The dependence of the sintering process on electrode potential could be related to changes in the reversible work function of the surface. The authors suggest that changes in the surface during sintering are chiefly due to migration of vacancies, and to surface self-diffusion of platinum atoms, these processes leading to a decrease in microporosity.

Ruthenium.—Burke and Murphy[184] have attempted to establish an electrochemical technique for estimating the real surface area of RuO_2-coated titanium foils, based on a measurement of voltammetric charge. These authors encountered errors in measurement of areas by the BET technique which they attributed to the blocking of pores by adsorbed H_2O and OH species, whose removal required temperatures high enough to cause sintering. The high values for interfacial charge and capacitance were attributed to a combination of ionic double-layer effects and surface redox processes. For this type of microporous electrode, surface area is determined largely by oxide loading and the annealing temperature. Burke and

---

[180] M. Paszkiewicz, *J. Appl. Electrochem.*, 1981, **11**, 135.
[181] M. Paszkiewicz, *J. Appl. Electrochem.*, 1981, **11**, 443.
[182] V. A. Shustov and E. A. Kalinovskii, *Elektrokhimiya*, 1974, **10**, 436.
[183] B. I. Podlovchenko and T. D. Gladysheva, *J. Electroanal. Chem. Interfacial Electrochem.*, 1979, **103**, 375.
[184] L. D. Burke and O. J. Murphy, *J. Electroanal. Chem. Interfacial Electrochem.*, 1979, **96**, 19.

Murphy were able to offer the basis of a convenient, though approximate, electrochemical technique for evaluating the real surface area.

Silicon Carbide.—Some interest has recently been shown in the use of sintered silicon carbide as both an electrode and as a container material for use in sodium–sulphur cells.[185] Sintered silicon carbide is chemically inert in a range of sulphur-based melts, under a variety of electrochemical conditions. The formability and mechanical properties of this material were shown to be adequate for its use in the first generation of sodium–sulphur cells. Silicon carbide can be used as either electrically conducting or non-conducting cell components, in contact with both the catholyte and oxidizing atmospheres.

Silver.—Mao et al.,[186] writing in 1970, briefly recapitulate the various methods for producing porous silver electrodes and note their common disadvantage in not controlling porosity. They describe the fabrication and electrochemical behaviour of silver electrodes of controlled porosities. The basic procedure involves sintering a mix of silver powder, binder, and a pore-forming agent such as wood flour or charcoal. In this way, these workers were able to fabricate gridless silver electrodes that were in many cases superior to conventional grid-based electrodes, regarding their physical properties and simple discharge behaviour. Complex electrodes were fabricated of multiple layers of different porosity, and these performed significantly better than simple electrodes, of uniform porosity.

In a series of communications, Wales[187—189] made a detailed study of the microstructural effects of working silver sinter electrodes in 35% KOH. While all electrodes were charged at a 20-hour rate, two discharge rates were used, i.e. a slow (20 hour) rate and a fast (1 hour) rate. Wales[187] found that Ag_2O forms on AgO clumps at both rates of discharge. The faster discharge rate left a small proportion of unchanged AgO remaining at the end of the discharge. When all the AgO was coated with Ag_2O, the discharge potential fell to the Ag/Ag_2O plateau, at which level metallic silver began to form. At the fast discharge rate, silver first formed on the surface and then spread to the interior. At the slow rate, silver formed throughout the electrode.

For rapidly discharged electrodes,[188] the first discharge produced very fine particles, which slowly increased in size with cycling. A slow-discharge cycling regime produced large particles, whose size then remained unchanged, after the first discharge. Consequently, electrodes that were cycled using the fast discharge rate lost capacity more slowly than those that were cycled with slow discharge. Eventually (within 30 cycles), cycling with fast discharge led to clumping of silver particles, with the formation of large void spaces and a decreasing proportion of small silver particles. This effect was most pronounced in the interior of the electrodes, indicating that conclusions drawn after examining the surface of an electrode should be made with caution.

Wales[189] observed that, during charging, the lower oxide (Ag_2O) formed as readily

---

[185] R. R. Dubin and S. Prochazka, *J. Electrochem. Soc.*, 1979, **126**, 2156.
[186] G. W. Mao, D. S. Polcyn, and R. E. Tieder, *J. Electrochem. Soc.*, 1970, **117**, 1319.
[187] C. P. Wales, *J. Electrochem. Soc.*, 1969, **116**, 729.
[188] C. P. Wales, *J. Electrochem Soc.*, 1969, **116**, 1633.
[189] C. P. Wales, *J. Electrochem. Soc.*, 1971, **118**, 7.

in the interior of the electrode as on the surface, but with an uneven distribution. Most particles of silver were coated with Ag_2O before the Ag/Ag_2O potential plateau was half completed, yet few of these particles were completely oxidized to Ag_2O at the end of this plateau. In a cycled electrode, $\sim 60\%$ of the silver had been oxidized to Ag_2O at the potential peak that separates the Ag/Ag_2O and the Ag_2O/AgO plateaux. The lower oxide did not form large solid masses, but was itself porous, and, being less dense than silver, reduced the overall porosity of the electrode.

Dunaeva et al.[190] have studied the structure and capacity of porous silver electrodes as a function of discharge current. With increasing current density, the fraction of fine porosity increased, as did the proportion of open pores, with a consequent increase in the specific area of the electrode. Thus the observed decrease in capacity of the electrode during cycling at low discharge currents was due to a decrease in the area of the electrode. However, the observation that the behaviour is similar with high discharge currents was ascribed by the authors to severe structural changes in the electrode, surface material being dispersed to the counter-electrodes. The same authors[191] also found that admixture of 0.6% of ZrO_2 with the silver dust, when used as an accumulator electrode, increased the real surface area as well as the electrochemical capacity.

Gagnon and Austin[192] studied the cathode performance of porous Ag/Ag_2O electrodes at low temperatures (down to $-60\,°F$). The galvanostatic discharge curves at low temperatures displayed six regions of polarization. The first region is due to an immediate ohmic polarization; the second to double-layer charging. In the third region, some active material is discharged, followed by a polarization peak, which has been attributed to nucleation overpotential. A fifth (plateau) region was thought to correspond to the activated charge-transfer reaction; finally, the increasing polarization, associated with more complete discharge, was ascribed to layers of discharged material that were formed at the face of the electrode and to mass-transfer of the dissolved Ag_2O. The authors note that it is important, in the design of practical electrodes, to determine thickness and porosity so as to avoid entering the terminal region at the outer face while the interior remains slightly discharged.

Bennion and co-workers have made a significant contribution to the understanding of porous electrodes. The development of their model for porous electrodes with sparingly soluble reactants, between the years 1971 and 1976, is considered in an earlier section of this Report. The Ag/AgCl electrode couple was used as a test for the model, and that experimental work is reviewed here. The electrode was generally in a well-characterized particle-bed arrangement, first introduced by Katan et al.[193]

In a purely electrochemical study of the Ag/AgCl electrode, Katan et al.[194] distinguished two radically different reaction paths, involving transfer of charge either at the metal/salt interface or at the metal/solution interface. They considered a situation comparable with a practical battery electrode, where the AgCl layer is

---

[190] T. I. Dunaeva, G. P. Ereiskaya, and M. F. Skalozubov, *Elektrokhimiya*, 1971, **7**, 56.
[191] G. P. Ereiskaya, T. I. Dunaeva, and M. F. Skalozubov, *Zh. Prikl. Khim.*, 1971, **44**, 1272.
[192] E. G. Gagnon and L. G. Austin, *J. Electrochem. Soc.*, 1971, **118**, 497.
[193] T. Katan, S. Szpak, and E. A. Grens, *J. Electrochem. Soc.*, 1965, **112**, 1166.
[194] T. Katan, S. Szpak, and D. N. Bennion, *J. Electrochem. Soc.*, 1973, **120**, 883.

<1 μm thick, in contrast with previous work[195,196] on thick films, where reduction occurs at the Ag/AgCl interface. The effective diffusion coefficients of Ag^+ ions were calculated (by direct application of the Levich equation[197]) as 1.4×10^{-5} and 0.31×10^{-5} cm^2 s^{-1} in 2M- and 4M-KCl, respectively, both being saturated with AgCl. The reduction of films of AgCl on silver could be driven very fast, about two orders of magnitude faster than the reduction of AgCl in solution, at speeds of up to 2000 r.p.m. of a rotating disc. This suggested high current densities to be inherently possible in the porous electrodes by virtue of the short diffusion paths. The linearity of the i/η relationship, and its independence of the state of charge, suggested the occurrence of transport of the liquid phase, possibly followed by a rate-controlling process of surface diffusion. In the case where the area of AgCl is small, its dissolution then becomes rate-controlling.

Katan et al.[198] extended this study[194] to a morphological investigation of the formation and growth of films in Ag/AgCl particle-bed electrodes, in order to provide data to fit into the model of Katan et al.[31] Their previous work[194] had shown that the reduction of AgCl involves diffusion in solution and at the surface to growth sites on the surface of the silver, and that dissolution of silver becomes rate-limiting when the AgCl crystallites become small. No clear connection between the electrochemistry and morphological changes was established; however, SEM observations indicated that the stages of anodic oxidation were (i) dissolution of silver as a soluble complexed species, then (ii) transport in solution, followed by (iii) deposition and growth with a characteristic spreading morphology, the final stage being the formation of a uniform coating that is ~ 3500 Å thick. The silver was found to dissolve at preferred sites (thought to be emergent dislocations) to produce an array of pits ahead of the advancing front of the film of silver chloride. The cathodic reduction of the AgCl that was thus formed proceeded by a reversal of the anodic process.

The contribution of Szpak et al.[199] took these studies[194,198] further, to assume a more restrictive model, where the rate-determining step changes during the discharging of the electrode, and they tested this against observations of a Ag/AgCl particle-bed electrode. Szpak et al.[199] paid attention to the dimensionless parameter $\kappa_1 = (R_1/Z_1)^{\frac{1}{2}}$, which is the square-root of the ratio of the resistance of the solution to the impedance of the electrode reaction (which Winsel' has referred to as 'reduced pore length' and which Bro and Kang[34] have termed the 'electrochemical Thiele parameter'). The use of this parameter is, however, limited to electrodes under a steady-state regime. It cannot, therefore, be applied to a battery electrode in the course of discharging, where the time-dependent reaction profile is related to the state of discharge. In previous experimental studies[34,158] the Thiele parameter had been found to vary with the state of discharge of an electrode.

The porous electrode comprised silver spheres, of diameter 30—40 μm, that were loosely packed together upon a silver backing plate. Measurements of the amount of charge transferred were as predicted by the model, for depths greater than 3 mm below the surface of the electrode. The experimental curves showed a break at a

[195] W. Jaenicke, R. P. Tischer, and H. Gerischer, *Z. Elektrochem.*, 1955, **59**, 448.
[196] G. W. D. Briggs and H. R. Thirsk, *Trans. Faraday Soc.*, 1971, **48**, 1952.
[197] V. G. Levich, 'Physicochemical Hydrodynamics', Prentice Hall, Englewood Cliffs, N.J., 1962.
[198] T. Katan, S. Szpak, and D. N. Bennion, *J. Electrochem. Soc.*, 1974, **121**, 757.
[199] S. Szpak, A. Nedoluha, and T. Katan, *J. Electrochem. Soc.*, 1975, **122**, 1054.

depth of $x \approx 3$ mm, so that, within a shallow surface layer, significantly less charge had been transferred than was predicted by the model. An examination of the charged silver particles by SEM indicated two different electrochemical processes to be operating above and below the critical depth of ~ 3 mm. These observations could be accounted for by the assumption that, in the (shallow) principal reaction zone, the silver chloride redissolved as a complex, subsequently recrystallizing from the liquid phase in the chloride-deficient interior of the electrode.

Katan, Gu, and Bennion[31] tested the theory developed by Bennion and co-workers[26,29,30] (see Section 2) to describe the performance of porous secondary electrodes with sparingly soluble reactants. SEM and microprobe analyses of fully charged particle-bed electrodes revealed a layer of AgCl sealing the frontal surface of the electrode. To a depth of ~ 0.02 cm, the silver particles were enveloped in AgCl. The final inability of the electrode to accept further charge is then due to two kinds of blockage, i.e. blockage of the pores and blockage of the walls of the pores. Discharging of the electrodes produced a new morphology of the silver in this shallow reaction zone, now comprising very fine particles, of diameter <2 μm. The observed location of the reaction profile, and measured polarization/time plots, were in good agreement with the theory of Gu et al.[30] Thus, with reasonable values for physical parameters, the theory could correctly predict the onset of failure to accept charge, the location of the reaction profile, and the shifts in rate-governing processes that determine blockages during cycling.

Szpak and Katan[56] utilized a segmented electrode construction to study reaction profiles for the Ag/AgCl system. These authors note the value of such an approach in gaining access to reaction profiles in working electrodes, compared with the study of macroscopic electrodes. They were able to draw on technology that was developed by the microelectronics industry in the construction of their single-pore segmented electrode. This arrangement permitted the charging and discharging processes to be monitored as they occurred. Observations on both processes were wholly consistent with the formation of a reaction layer that progressed along the length of the segmented pore. During charging, measurements indicated a change in reaction path, independent of the magnitude of the charging current, resulting from a change in electrochemical mechanisms, rather than an increase in the resistance of the solution. The charging process was dominated by choking of the second kind, owing to an increase in reaction resistance. This is distinct from choking of the first kind, which has elsewhere[199] been defined as the shift of the current profile towards the electrolyte due to an increase in the resistance of the solution. As the charging current was increased, choking of the second kind still occurred, but at greater depths within the pore.

These observations, made by Szpak and Katan[56] on a single segmented-pore electrode, were in good qualitative agreement with data obtained by Szpak et al.[199] by using a silver particle-bed electrode. Szpak and Katan[56] concluded that studies of segmented selectrodes could be of value in interpreting the behaviour of more complex geometrical structures.

Titanium Carbide.—Freid and Lilin[200] have drawn attention to inconsistencies in

---

[200] M. K. Freid and S. A. Lilin, *Elektrokhimiya*, 1979, **15**, 1464.

the literature on the electrochemical behaviour of titanium carbide, and noted that its resistance to corrosion depended strongly on the method of preparation and on the condition of the surface. However, aerially oxidized titanium carbide, regardless of the method of preparation, assumed a constant potential in a range of concentrations of both HCl and H_2SO_4. During cathodic polarization, the surface oxides were reduced and a titanium hydride layer was formed. The reduction of surface oxide varied according to the method of preparation. The authors concluded that titanium carbide is a material of promising stability in corrosive media such as HCl.

Zinc.—Zinc was the first (and is the most commonly used) material for negative electrodes in electrochemical energy conversion, for both primary and secondary cells. Zinc offers a very attractive range of properties, including low cost, no toxicity, ease of fabrication, low density, low electronegativity (high cell voltage), and high exchange-current density. The chemistry and electrochemistry of solid zinc are complex, and are the subject of a great deal of research, though not considered in this Report. The behaviour of porous zinc electrodes is still more complex, and models of practical electrodes are still very much under development.

McBreen and Cairns[201] have reviewed the literature on zinc electrodes up to 1977 to give an account of the history of the development of the zinc cell, the solution chemistry in battery electrolytes, electrode processes of zinc, cycling behaviour, and the status of practical zinc cells.

Myers and Marchello[202] utilized a galvanostatic pulsed current technique to measure the capacitance of flat and porous zinc electrodes, and hence to estimate their wet surface area. Measurements were made near to the evolution potential for hydrogen, at 1.76 V (vs S.C.E.), so as to eliminate errors due to the dissolution reaction of zinc. Values of roughness (defined as the ratio of double-layer area to geometric area) were in the range 40—200 for specimens of electroplated porous zinc. Average diameters of pores were calculated to be 0.05—0.10 cm. Some exchange currents were determined for various zinc electrodes and compared with the exchange currents for standard metals (silver, platinum, *etc.*), determined by the same method.

Breiter[203] has studied the anodic dissolution and the passivation of vertical porous zinc electrodes in 6M-KOH+0.25M-ZnO. Potentiostatic current–potential curves were recorded for a rate of 1 mV s^{-1}, and the physical structures of the electrodes were examined. A comparison of results for smooth and porous zinc indicated that only a small fraction of the interior of the porous electrode participated in electrochemical processes. The lowermost parts of vertical electrodes of both smooth and porous zinc were found to be much less reactive than other parts of the electrodes. This was related to the effect of the hydrodynamic flow pattern on the thickness of the diffusion layer. Repeated cycling of porous electrodes resulted in the reduction of particle size, though this effect could be diminished by amalgamation.

---

[201] J. McBreen and E. J. Cairns, *Adv. Electrochem. Electrochem. Eng.*, 1978, **11**, 273.
[202] R. A. Myers and J. M. Marchello, *J. Electrochem. Soc.*, 1969, **116**, 790.
[203] M. W. Breiter, *Electrochim. Acta*, 1970, **15**, 1297.

Elsdale et al.[204] determined times for galvanostatic transition between active and passive states for both horizontal and vertical porous zinc electrodes in KOH. Results suggested that the anodic reaction of zinc proceeds first by the formation of a soluble zincate, which then decomposes to give an oxide film within the pores of the electrode. This oxide film makes a large ohmic contribution to the total overpotential of the electrode. It was suggested that this ohmic overpotential was the major factor in determining the useful discharge life of a porous zinc electrode. These authors found that the transition from the active to the passive conditions was not abrupt, as in the case of plane electrodes,[205] but gradual, corresponding to the electrode reaction being driven deeper into the electrode as the discharging pores become blocked with oxide.

McBreen,[206] writing in 1972, identified changes in shape, arising from the redistribution of electrode material, as the predominant factor in limiting the life of secondary zinc batteries. The typical pattern of movement is transfer of material from the edges to the centre of the electrode. McBreen briefly summarizes the numerous constructional and operating parameters that had been correlated with changes in the shape of the electrode, concluding that no proven mechanism had been reported. His contribution was an analysis of the patterns of distribution of potential and of current over the surface of a cycling zinc electrode, using a sectioned cadmium counter-electrode.

McBreen[206] distinguished between the theoretically predicted primary distribution of current, with no polarization of the electrode, and the actual secondary distribution of current, determined by the pattern of the polarization of the electrode. Variations in this secondary distribution of the current, caused by changes in polarizability of the electrode, resulted in exhaustion of the reducible zinc species at the edges of the plate in the early stages of cycling. The initiation and progress of changes of shape were explained on the basis of concentration cells that corroded zinc off the edges of the plate, to deposit it at the centre. Moreover, current-distribution effects at the beginning of discharge cycles favoured the dissolution of zinc at the periphery of the electrode, with its consequent transfer as zincate to the centre of the electrode. McBreen was thus able to account for a number of diverse observations on porous zinc electrodes by considering the aspect of secondary distribution of current and its consequences for changes in shape.

Gregory et al.[207] studied the rates of self-discharge of porous zinc electrodes in alkaline electrolytes, under a variety of conditions, by monitoring the rate of evolution of hydrogen. The self-discharge reaction was inhibited by the addition of lead or mercury, both of which increased the overpotential for hydrogen. Rates of evolution of hydrogen on various materials showed minima between 2M- and 5M-KOH. Rates of self-discharge were lower in NaOH than in KOH, and saturation of both electrolytes with zincate also reduced the rates of self-discharge in both cases. Amalgamation of the copper grid material significantly reduced the rates of self-discharge, while the same treatment of a silver grid had no such effect.

---

[204] R. N. Elsdale, N. A. Hampson, P. C. Jones, and A. N. Strachan, *J. Appl. Electrochem.*, 1971, **1**, 213.
[205] N. A. Hampson and M. J. Tarbox, *J. Electrochem. Soc.*, 1963, **110**, 95.
[206] J. McBreen, *J. Electrochem. Soc.*, 1972, **119**, 1620.
[207] D. P. Gregory, P. C. Jones, and D. P. Redfearn, *J. Electrochem. Soc.*, 1972, **119**, 1288.

By 1972, various studies[203,204,208] had been made of the anodic behaviour of porous zinc in terms of current density, temperature, porosity of the electrode, and composition of the solution, but the distribution of current within the electrode had not then been considered. Nagy and Bockris[209] made an examination (by SEM) of the morphology of the zinc oxide that was formed within a galvanostatically discharged porous zinc electrode. The distribution of current in the electrode was determined by chemical analysis of micro-slices of the electrode. SEM showed the oxide film to have a porous 'carpet-like' structure, of long acicular crystals with occasional side-arms, which suggested a dissolution–precipitation mechanism of formation. The pattern of distribution of current in the porous electrode and its dependence on current density was explained on the basis of the duplex-film model, based on studies of solid zinc,[210,211] in which a thin, resistive, compact film formed beneath the porous oxide.

Coates et al.[212] investigated the relationship between the structure of a porous electrode and its electrochemical discharge behaviour. It was found that samples with larger surface areas did not give proportionally improved anodic performance. It was concluded, in agreement with Breiter,[203] that, unless very low rates of reaction were used, the inner regions of the porous electrode contributed little to the total current passing. Specific rates of polarization for porous zinc displayed no general relationship with current density, unlike plane electrodes.[205] Coates et al.[212] suggested a qualitative explanation whereby the potential that was required to support the flow of current was increased, owing to the blocking of the more accessible areas of the electrode, driving the discharge deeper within the interior of the electrode. The specific rate of polarization, a parameter representing the early stages of discharge, was found to correlate better with the true (BET) surface area than with the estimate of electrochemically accessible surface area that was obtained from the methylene-blue-adsorption method. This indicated that the true surface area determined the current–voltage relation when the diffusion layer was thin; subsequently, the important area was that which was superficially more accessible.

By 1976 the phenomenon of changes in shape of cycled porous zinc electrodes was experimentally well established.[206] Choi et al.[213] offered an alternative mechanism to that put forward by McBreen,[206] proposing that a change in the shape of the zinc in a zinc/silver oxide cell resulted from convective flows of electrolyte, driven primarily by membrane pumping during cycling. Their analysis postulates that zinc is redistributed by the coupling of changes in concentration of zincate with convective flow, and this is driven primarily by membrane pumping effects, assisted to a minor extent by changes in the volume fractions that arise because chemical reactions occur. The electrolyte, flowing upward toward the reservoir within a vertical porous zinc electrode during charging, is depleted of zincate; during discharging, the downward-flowing electrolyte is supersaturated with zincate. The net result is a pumping of zinc from the upper to the lower portions of the electrode.

[208] T. P. Dirkse, J. Electrochem. Soc., 1955, 102, 497.
[209] Z. Nagy and J. O'M. Bockris, J. Electrochem. Soc., 1972, 119, 1129.
[210] R. W. Powers and M. W. Breiter, J. Electrochem. Soc., 1969, 116, 719.
[211] R. W. Powers, J. Electrochem. Soc., 1971, 118, 685.
[212] G. Coates, N. A. Hampson, A. Marshall, and D. F. Porter, J. Appl. Electrochem., 1974, 4, 75.
[213] K. W. Choi, D. N. Bennion, and J. Newman, J. Electrochem. Soc., 1976, 123, 1616.

The model suggested that changes in shape could be eliminated if this tidal convective flow in the plane of the zinc electrode could be suppressed. Choi et al.[214] tested their convective-flow model[213] by experiments with modified silver/zinc oxide secondary cells, in one of which the rates of flow of electrolyte could be measured and the other designed so as to suppress convective flow. Observations on these cells indicated that redistribution of material in these cells is determined primarily by convective flow of electrolyte within the zinc electrode, and concordant with predictions based on considerations of osmotic and electro-osmotic pressure. Measurements of flow rates and distributions of ZnO were in agreement with those predicted by the model. Limitation of convective flow was stated virtually to suppress changes in shape of the zinc electrode. However, the authors noted that this is not a practicable solution to the problem of changes in shape. Convective flow is an important factor in determining the capacity of the cell and its capability to cope with a high rate of charging or discharging. Furthermore, steps taken to suppress convective flow may have detrimental side-effects on the control of the evolution of gases.

Turner and Hutchison[215] took up the suggestion by Nagy and Bockris[209] that the accumulation of oxide in a porous zinc electrode could effect a redistribution of the current density, in an investigation of the galvanostatic and pulse discharge of porous zinc electrodes. They derived a theoretical analysis of the current-density distribution (c.d.d.) as a function of porosity of the anode, depth, and resistance of the electrode/electrolyte interface. The efficiency of porous zinc anodes was found to be critically related to the c.d.d. Kinetic measurements showed that the presence of superficial metallic lead on the zinc particles increased the resistance of the electrode/electrolyte interface, thus producing a more uniform c.d.d., and a consequent improvement in the galvanostatic behaviour of porous anodes. Turner and Hutchison suggested a mechanism to account for differences in efficiency between the pulsed and steady-state regimes.

Cenek et al.[216] have reported the promising performance of plastic-bonded zinc electrodes for alkaline accumulators. In this approach, a plastic skeleton, together with a current-collector, forms the stable carrier structure for the active electrode mass. Plastic-bonded zinc electrodes, typically prepared from 75% ZnO, 20% polyethylene, and 5% PTFE, gave increased service life in comparison with pocket-type electrodes, showed no growth of dendrites, and displayed high resistance to changes in shape during cycling. The performance of the electrode could be further improved by adding potassium fluoride to the KOH electrolyte.

Poa and Wu[217] have studied changes in the shape of a zinc electrode in commercial-type zinc/silver oxide secondary cells. They examined various factors that influence changes in shape and tested different methods for alleviating the phenomenon. The observations indicated that the extent of change in the shape was increased with higher rates of charging and discharging of the cell. Addition of ZnO to the electrolyte reduced the loss of material by anodic dissolution and increased the cycle life of the cell. Orientation of the cell in the earth's gravitational field had no

---

[214] K. W. Choi, D. C. Hamby, D. N. Bennion, and J. Newman, *J. Electrochem. Soc.*, 1976, **123**, 1628.
[215] J. Turner and P. F. Hutchison, *Power Sources*, 1977, **6**, 335.
[216] M. Cenek, O. Kouřil, J. Šandera, A. Toušková, and M. Calábek, *Power Sources*, 1977, **6**, 215.
[217] S-P. Poa and C. H. Wu, *J. Appl. Electrochem.*, 1978, **8**, 427.

effect on cycling performance; neither was the method of preparation of the electrode (by electrodeposition or from a slurry paste) significant. The incorporation of $Fe(OH)_2$ around the edges of the separators had the effect of reducing the overpotential of hydrogen, reducing the zinc-plating efficiency, and so delaying the onset of changes in shape. A similar improvement was obtained by increasing the thickness of the separators at the edges.

Poa and Wu[218,219] went on to investigate the relationship between current distribution and changes in shape in zinc electrodes of zinc/silver oxide batteries, using a sectioned-electrode technique. During discharging, the distribution of current started off non-uniform and became uniform. During charging, the reverse process occurred, until the potential of the cell reached a critical value, when the current distribution rapidly became uniform. Poa and Wu ascribed this last event to the initiation of oxidation of Ag_2O to AgO. The non-uniformity of current distribution during both charging and discharging was explained in terms of variations in polarizability of the electrode affecting the secondary distribution of current. The initiation and progress of changes in shape was directly related to the uniformity of distribution of the current on the zinc electrode. The incorporation of Fe_2O_3 at the edges of separators improved the distributions of current and of potential and inhibited changes in shape of the zinc electrode during cycling of the cell.

Poa and Lee[220] performed an experimental programme to establish optimum parameters for the fabrication of slurry paste zinc electrodes for alkaline zinc/silver oxide primary cells. The parameters investigated were composition of the active material, thickness of the electrode, and applied compression. The factorial experimental programme established the optimum values for these parameters as 2—4% HgO, 0.5—2% PVA, and 94—97% ZnO, an electrode thickness of 0.265—0.35 mm, and a compression of 500—1500 p.s.i.

Choi et al.[214] had found that limiting the convective flow in their experimental cell suppressed changes in shape of the zinc electrode. However, qualitative observations[213,221] on this cell suggested that distributions of current and of potential were not uniform over the zinc electrode, as implied by the convective-flow theory. Hamby and Wirkkala[221] constructed a limited-convection alkaline zinc cell, incorporating reference electrodes, so as to test the convective-flow theory. Non-uniformity of overpotentials was directly established, and non-uniformity of current was indicated by qualitative observations of the distribution of zinc metal after charging. The experimental results still indicated that the rate of lateral change of shape was significantly reduced by limiting the convective flow. Hamby and Wirkkala[221] were unable to correlate their measured overpotentials with changes in shape of their zinc electrodes. They found that non-vented zinc electrodes, in which convective flow was suppressed, developed anodic overpotentials which increased with cycling, corresponding to a loss of electrode capacity. Hamby and Wirkkala discussed possible reasons for these and the overpotential observations, but were unable to come to any definite conclusions.

[218] S-P. Poa and C. H. Wu, *J. Chin. Inst. Chem. Eng.*, 1977, **8**, 171.
[219] S-P. Poa and C. H. Wu, *J. Appl. Electrochem.*, 1978, **8**, 491.
[220] S-P. Poa and S. J. Lee, *J. Appl. Electrochem.*, 1979, **9**, 307.
[221] D. C. Hamby and J. Wirkkala, *J. Electrochem. Soc.*, 1978, **125**, 1020.

In a further study, Hamby et al.[222] constructed a silver oxide/zinc cell with which flow rates, distribution of material, cumulative variations in concentration, and overpotentials were compared with the predictions of the convective-flow model.[213] Changes in concentration within a single cycle were as predicted by the Choi model,[213] though these had not been reported by them. However, Hamby et al.[222] failed to observe the large cumulative changes in concentration that were predicted by the model for a cycling regime. They attributed this to mismatch between the actual and modelled systems, despite having designed the experimental cell to conform with the model.[213,214] Calculated and measured values of concentration overpotential showed poor correlation. Hamby et al.[222] concluded that concentrations in the cell were not simple one-dimensional functions, as predicted by the convective-flow theory. Various physical possibilities were suggested as causes of non-uniform distribution of the current.

Katan, Savory, and Perkins[223] constructed an experimental analogue cell to model a single pore in a zinc electrode (14 μm thick, 0.6 cm deep, and 1 cm wide), so as to gain information concerning morphological changes and reaction profiles in porous zinc electrodes. The studies of the pore analogue portray the dissolution–precipitation mode of formation of ZnO as the predominant process, with a primary penetration depth of 0.09 cm, corresponding to a calculated 'effective' exchange current density of 1.6×10^{-3} A cm^{-2}. Sites of dissolution of zinc formed with an initial density of 6×10^5 cm^{-2} under experimental conditions, forming pits which grew and merged, though less than 2% of the original surface initially participated in the oxidation step. Examination of the precipitated ZnO by SEM showed it to have a needle-like microstructure, with about 80% porosity, and with a specific surface area of $\sim 8 \times 10^4$ cm^2 cm^{-3}. The onset of what Katan et al.[223] termed 'passivation' was associated with the formation of a white precipitate, filling the entire pore, with loss of OH$^-$ and consequent concentration polarization. The performance of the anode deteriorated before this event, there being gradual consolidation and thickening of the ZnO film. Growth of zinc dendrites on the cathode was found to be related to the presence of bubbles of hydrogen, producing areas of locally increased current density. Under such conditions, the formation of dendrites appeared to be related to the occurrence of blockage of pores by deposited zinc crystals and bubbles of hydrogen.

Szpak and Gabriel[224] have made a detailed study, using SEM, of the operation of the dissolution–precipitation mechanism of formation of ZnO in an experimental porous zinc analogue electrode (of the type used by Katan et al.[223]) in zincate-saturated KOH. They found the precipitation mechanism in solution to be consistent with the morphological observations; however, a strong coupling between kinetics of growth and local transport of mass determined the shape of individual crystallites and the details of the structure of duplex films. The formation and growth of stable nuclei occurred within that part of the diffusion layer adjacent to the electrode. Deformation of the ZnO film was ascribed to densification processes that occur during growth of the film. Szpak and Gabriel[224] considered the

---

[222] D. C. Hamby, N. J. Hoover, J. Wirkkala, and D. Zahnle, *J. Electrochem. Soc.*, 1979, **126**, 2110.
[223] T. Katan, J. R. Savory, and J. Perkins, *J. Electrochem. Soc.*, 1979, **126**, 1835.
[224] S. Szpak and C. J. Gabriel, *J. Electrochem. Soc.*, 1979, **126**, 1914.

effects of electric field to be negligible at moderate current densities, but possibly to become significant for morphology of the film at higher current densities.

Thornton and Carlson[225] distinguished three related processes that occur at the zinc electrode and that are collectively responsible for the limited cycle life of zinc/nickel oxide batteries; the zinc tends, first, to concentrate in the centre of the electrode (shape change); secondly, to densify, and thirdly, to grow dendrites during charging, resulting in internal shorting. These three effects are all the result of the high solubility of zinc hydroxide in NaOH and KOH. These authors investigated several mixed aqueous electrolytes, made by partially replacing the KOH with salts such as KF or K_3PO_4 in order to reduce the high solubility of zinc hydroxide. Major additions of fluoride, phosphate, and borate salts to KOH reduced the solubility of zinc hydroxide, whilst reducing the conductivity of the electrolyte by less than a factor of two. Thornton and Carlson[225] suggested that the use of these electrolytes should suppress changes in shape and the growth of dendrites in zinc electrodes.

Kocherginski et al.[226] have presented designs for nickel–zinc traction cells that are capable of rapid charging and have extended cycle life. These benefits are only obtained by using soluble zinc anodes, which must be completely dissolved after discharging and then re-formed, under vibration, during charging. Also, two electrolytes must be used, one being pure alkali and the other alkali that is saturated with zincate.

Sunu and Bennion[227] have presented a model to describe the transient behaviour of zinc in the porous electrode. The discharge product (ZnO) is highly soluble in KOH, resulting in a ternary mixture of potassium hydroxide, potassium zincate, and water, which can be supersaturated with zincate. Sunu and Bennion therefore derived their model on the basis of the theory of ternary electrolytes. The results were used to predict modes of failure and limitations of performance of the zinc electrode.

The model predicted highly non-uniform reaction distributions and very thin reaction zones, which accentuated the failure of the electrode that arises from interior depletion of electrolyte. The zinc at the rear of the electrode thus serves simply as an inert matrix, and does not contribute to the capacity of the cell. Operation of the model cell without a membrane resulted in loss of most of the zincate discharge product into the counter-electrode compartment. Introduction of a membrane restricted this loss of zincate, but also severely limited the utilization of zinc by depleting the concentration of hydroxide ions within the zinc electrode compartment. Sunu and Bennion[227] interpreted such failure of discharging in terms of limited mass-transfer of soluble species to and from the surface of the dissolving zinc. The occurrence of this phenomenon on planar electrodes has been termed passivation. However, Sunu and Bennion distinguished three inter-related subsidiary mechanisms for the porous electrode, i.e. passivation, plugging of pores, and depletion of the electrolyte, and proposed precise definitions for each.

The model indicated that the difference between distributions of anodic and cathodic current, together with variations in saturation of the electrolyte, caused the

---

[225] R. F. Thornton and E. J. Carlson, *J. Electrochem. Soc.*, 1980, **127**, 1448.
[226] M. D. Kocherginski, L. P. Esayan, K. A. Belyankina, and L. L. Penchukova, *Elektrokhimiya*, 1980, **16**, 1110.
[227] W. G. Sunu and D. N. Bennion, *J. Electrochem. Soc.*, 1980, **127**, 2007.

redistribution of solid zinc. This mechanism is similar to that described by McBreen.[206] If convective flow due to membrane pumping (the model of Choi et al.[213]) was also allowed, then more serious redistribution was predicted. Various possible modes of failure were predicted, either during charging or discharging, or as a result of repeated cycling.

Sunu and Bennion[228] constructed an experimental cell to observe and verify these predicted mechanisms of failure, particularly those occurring during discharging. Distributions of zinc and ZnO, and corresponding overpotentials during galvanostatic discharge, compared well with theoretical predictions. The behaviour of zinc electrodes during discharging was interpreted according to the modes of failure proposed earlier:[227] depletion of the electrolyte, plugging of pores, and passivation. Observed reaction profiles were highly non-uniform, resulting in a reaction zone equivalent to $\sim 20\%$ of the thickness (1 mm) of the electrode. The utilization of zinc depended strongly on the applied current density, the initial porosity, the amount of electrolyte solution, and the type of membrane. At high rates of discharging, utilization of zinc was severely limited by depletion of the electrolyte within the compartment of the zinc electrode. At low rates of discharging, utilization of zinc was increased, and discharging continued until a decrease in the pore size in the reaction zone limited further discharging. The highly non-uniform distribution of reaction favoured failure by plugging of the pores.

Yamazaki and Yao[229] have derived a migration model for the distribution of current and of reaction in a soluble porous electrode, and calculated reaction profiles in a zinc/zincate electrode for a typical set of electrode parameters. The model predicted considerable differences in distribution of the current during charging and discharging. Two physical causes were accounted responsible for these differences. The first was a change in the resistivity of the electrolyte in the pore, which was itself brought about by a chemical reaction between the electrolyte in the pore and the product species of the electrochemical reaction on the wall of the pore. The second physical cause was a change in resistance of the interfacial reaction at the junction of the electrolyte and the surface of the pore matrix. The model suggested that the structure of the pores of an alkaline zinc electrode would change with cycling, even with small currents, when no passivation occurred. Pore size was a critical parameter in the model. Its reduction would bring an increase in the real surface area of the electrode, with a corresponding fall in the impedance of the electrode; but, this also reduced the depth of the reaction zone, and hence the utilization of the electrode.

Yamazaki and Yao[230] tested their migration model,[229] using a sectioned porous zinc/zincate electrode that was constructed using photolithographic techniques. Measured reaction profiles revealed differences in distributions of reactions between charging and discharging processes, which is in good agreement with predictions of the migration model, though theory and experiment were beginning to diverge at high current densities. High discharging currents resulted in the formation of a passivating film, which masked active areas of the wall of the pore. Passivation started at the mouth of the pore and extended inwards. The use of a sectioned

---

[228] W. G. Sunu and D. N. Bennion, *J. Electrochem. Soc.*, 1980, **127**, 2017.
[229] Y. Yamazaki and N. P. Yao, *J. Electrochem. Soc.*, 1981, **128**, 1655.
[230] Y. Yamazaki and N. P. Yao, *J. Electrochem. Soc.*, 1981, **128**, 1658.

electrode enabled Yamazaki and Yao to follow the temporal progression of the reaction profile. It was found that reaction currents at all penetration depths initially decreased and then recovered slightly to steady values. These temporal variations were most pronounced towards the front of the electrode.

Since the introduction of the zinc/silver oxide battery, it has been the universal practice to incorporate metal oxides in the pasted zinc electrodes. Whilst the common addition of 1—4% of HgO suppresses the evolution of hydrogen very well, it also accelerates changes in shape and losses of capacity. McBreen and Gannon[231] examined the effect of incorporating various oxides of metals that have a high overvoltage for hydrogen [e.g. HgO, Tl_2O_3, PbO, CdO, In_2O_3, $In(OH)_3$, and Ga_2O_3] in pasted zinc electrodes. Some of these additives [e.g. PbO and $In(OH)_3$] increased the density and the polarizability of the nickel electrode. Others (e.g. Ga_2O_3 and CdO) produced finely divided deposits, of low polarizability. These observations were attributed to a substrate effect, with the additives being reduced on the electrode prior to deposition of zinc. The beneficial effects of some additives [e.g. mixtures of $In(OH)_3$ and PbO] were ascribed to increased polarizability of the electrode, which improved the distribution of current and decreased the rates of change of shape.

Wagner and Himey[232] surveyed potential substitutes for mercury in alkaline zinc batteries and found PbO, Tl_2O_3, and CdO to be valuable. These additions belong to a family of compounds that exhibit exchange-current densities for the evolution of hydrogen that are equal to or less than that of zinc. The best additive was thallium (0.5—5%) for minimizing the rate of evolution of hydrogen. As regards minimizing changes in shape, the best additives were lead (2—5%) and cadmium (0.5%). The best formulations for optimum performance of a zinc anode were mixtures comprising 1% thallium and 2—5% lead, and also 1% thallium and 0.5% cadmium.

In a study of long-life silver oxide/zinc cells, Charkey[233] identified five possible routes towards improving and maintaining the capacity of a battery. The distribution of current in the zinc negatives was improved by using solid foil current-collectors. Incorporation of Teflon in the zinc electrode was found to improve its mechanical integrity, while potassium titanate minimized the distribution of electrolyte. The performance of a cell was improved by using an inorganic separator that was resistant to penetration by zinc and to oxidative attack by the silver.

4 Discussion and Conclusions

The mathematical theories of the behaviour of porous electrodes have been developed to an advanced level. The contrasting approaches of the single-pore and macrohomogeneous models are clearly not mutually exclusive, but complementary. Both treatments have yielded much valuable insight into mechanisms of the behaviour of porous electrodes, and in turn have stimulated more precise experiments. These two basic approaches have, in the past decade or so, been elaborated and applied to a variety of practical electrode systems, often with

---

[231] J. McBreen and E. Gannon, *Electrochim. Acta*, 1981, **26**, 1439.
[232] O. Wagner and A. Himey, *28th Power Sources Symp. 1978, Proc.*, p. 135.
[233] A. Charkey, *26th Power Sources Symp. 1974, Proc.*, p. 87.

Electrochemistry of Flooded, Static Porous Electrodes

considerable success, as in the case of lead dioxide. However, our understanding of porous electrodes is still deficient in two important respects. First, we require a much more detailed knowledge of the actual structure of an electrode. It is unlikely to be homogeneous, and the structure will probably vary with depth from the front of the electrode. Such a structure can result if electrodes are prepared from base metal, such as occurs in Planté processes, where porous electrodes are prepared from a lead base. Even porous lead dioxide electrodes that are prepared from the oxidation of Pb^{II} salts show unequivocal electrometric evidence of complex layered structures. The structure of the porous zinc electrode is also non-uniform with depth. This situation becomes more complex when these electrodes are operated (cycled), with consequent progressive structural changes. Thus the common porous electrodes of industry (battery electrodes, *etc.*) exhibit a varying behaviour, according to their service life and their history. This aspect requires urgent investigation, for the application of theoretical models to structurally changing systems remains unsatisfactory.

The second area of deficiency concerns our knowledge of the processes of phase formation which occur at the inner and outer surfaces of porous electrodes. Two examples will serve. In the lead–acid cell the processes of phase formation control the current that flows from the electrodes during a significant amount of the operating time (both discharging and charging). Up to the present, no attempt has been made to incorporate such processes in any mathematical description of the porous electrode. For the porous zinc electrode, operating in fairly strong alkali, the process involves what is almost certainly the direct production of solid ZnO in the electrode pores. This is again likely to be current-controlled by the phase-formation processes for significant periods. Such omissions from the porous model materially affect the result of calculations, and in a number of cases are likely to account for many of the observed inconsistencies.

A further point concerning the porous electrode is the inability to control the hydrodynamics. A practical point which arises from this is that in many of the systems that are of interest in electrochemical engineering there exists a density gradient from the bottom of the electrode to the top, as well as from the front of the electrode to the rear. This affects the results of many calculations (especially on lead cells) when using the macrohomogeneous approach.

A final important feature which is not always recognized is the chemical interaction between the support of the porous electrode and the electrode itself. It is not always feasible to construct the support of the electrode from the same material as the electrode itself, and the need to provide a support for the porous matrix that can also remove the current can introduce chemical, electrochemical, and semiconductor effects. The use of lead–antimony alloys in lead–acid cells, for example, introduces extremely significant interactions.

2
Electrode Processes in Molten Salts

BY J. ROBINSON

1 Introduction

The last review of molten-salt electrochemistry to appear in these Specialist Periodical Reports covered work published in the two years up until the end of 1972.[1] On that occasion the authors dealt comprehensively with this topic, but, as they stated in their introduction, the field was really becoming too large for such encyclopaedic coverage to be attempted in the future. In recognition of this fact, this Report is much more modest in its scope, restricting itself to studies of electrode processes. Thus topics which fit into the general category of ionics (such as conductivity measurements and studies of transport processes) will not be covered. No attempt either will be made to review technological investigations of areas such as corrosion, batteries, and the electrorefining and electrowinning of metals except where results are of fundamental interest. Within these terms, a comprehensive coverage of work appearing in the literature of 1980 and 1981 has been attempted, though no doubt there are omissions; to those authors whose work has been overlooked, the Reporter offers his apologies.

Since the last Report, several reviews and books have been published which should be of general interest to students of molten-salt electrochemistry. In particular, Plambeck[2] has produced a book which reviews publications prior to 1976 on electrode processes in a wide range of molten salts and mixtures of molten salts (including, amongst others, halides, chloroaluminates, carbonates, thiocyanates, and nitrates) as well as the properties of the melts themselves; it deserves a place on the bookshelf of every molten-salt electrochemist. More recently, Inman and Lovering[3] have edited a book on ionic liquids which contains several chapters of interest to electrochemists.

Before proceeding with this Report, a couple of general observations are in order. When the last Report appeared, the range of electrochemical techniques being used in molten salts was rather limited, chronopotentiometry and linear-sweep voltammetry perhaps being the most widely applied, owing to the difficulties inherent in working at elevated temperatures. Since then, researchers appear to have overcome these problems, and it is now probably true to say that almost any technique that has been applied in aqueous solutions has also been used in molten salts. For example, rotating-disc studies have been made at temperatures in excess of

---

[1] D. Inman, J. E. Bowling, D. G. Lovering, and S. H. White, in 'Electrochemistry', ed. H. R. Thirsk (Specialist Periodical Report), The Chemical Society, London, 1974, Vol. 4, p. 78.
[2] J. A. Plambeck, in 'Encyclopaedia of Electrochemistry of the Elements', ed. A. J. Bard, Vol. 10 'Fused Salt Systems', Dekker, New York, 1976.
[3] 'Ionic Liquids', ed. D. Inman and D. G. Lovering, Plenum Press, New York, 1981.

1000 K, and *in situ* spectroelectrochemical studies are now being successfully undertaken in chloroaluminate melts. The other major change is the increased range of fused salts being studied, and in particular the development of chloroaluminate systems that are molten at room temperature, which is opening many new areas of investigation.

2 Metal Halide Systems

Despite the increasing variety of molten salts that are now being studied electrochemically, systems based on mixtures of alkali-metal halides remain the most popular, and it is such systems that will be considered in this section, as well as some other halide melts.

Metal-deposition Processes.—*Chloride Melts.* For many years now, great efforts have been expended with the intention of developing baths suitable for the electrodeposition of refractory metals, and halide melts do appear to be amongst the more promising electrolyte solutions for such a purpose. Perhaps the major difficulty that has thus far been encountered is that these metals generally exhibit several stable oxidation states in solution; frequently, these oxidation states interconvert, by, for example, disproportionation. As a consequence of this, there will be many different species present that may be electroreduced, and satisfactory electrodeposition can only be achieved with very fine control of the composition of the solution.

With this in mind, Chassaing *et al.*[4] have made a fairly thorough investigation of the chemistry of cationic titanium species in halide melts, in particular of Ti^{II} and Ti^{III}. The equilibria that have been found to describe this system are:

$$2\,Ti^{III} + Ti \rightleftharpoons 3\,Ti^{II} \tag{1}$$

$$2\,Ti^{III} \rightleftharpoons Ti^{IV} + Ti^{II} \tag{2}$$

$$4\,Ti^{III} \rightleftharpoons Ti + 3\,Ti^{IV} \tag{3}$$

$$2\,Ti^{II} \rightleftharpoons Ti + Ti^{IV} \tag{4}$$

In CsCl melts, Ti^{III} was found to be very stable, and little Ti, Ti^{II}, or Ti^{IV} was formed, whereas the stability was significantly reduced in LiCl, with up to twenty times as much Ti^{II} being generated. When the alkali-metal cation that was present was larger than K^+, Ti^{II} was found to disproportionate, in accord with equation (4), with the consequent formation of large amounts of powdery titanium metal. Having established the chemistry of these species in solution, this group of workers proceeded to investigate the electroreduction of Ti^{III} in LiCl, CsCl, and LiCl–KCl melts at 700 °C, using linear-sweep voltammetry at a nickel electrode. In all melts, the reduction was found to be a two-step process, the first being the one-electron reduction to Ti^{II} ($E_{\frac{1}{2}} = -1.6$ V *vs* Cl^-/Cl_2) and the second a two-electron step, with Ti^0 as the product ($E_{\frac{1}{2}} = -2.1$ V). Whilst satisfactory electrodeposits were formed in LiCl and in LiCl–KCl, these were only obtained in the case of CsCl if small amounts of metallic titanium were added. This presumably hindered the disproportionation of

---

[4] E. Chassaing, F. Basile, and G. Lorthioir, *J. Appl. Electrochem.*, 1981, **11**, 187.

Ti^{II}. It is interesting to note that these results differ from those obtained for reduction of Ti^{III} in the lower-temperature CsCl–LiCl eutectic melt (400 °C), where Ti is formed in a single three-electron step. As a complement to the above study, the same authors have also reported the results of an investigation of the same system by Raman spectroscopy, and discuss the solvation of the ions.[5] Electrodeposition of titanium has also been considered by Rao et al.,[6] who describe a pilot plant for electroreduction of fused TiCl$_4$, and by Vytlacil and Neckel,[7] who used an electrolyte of 14% K$_2$TiF$_6$ and 86% NaCl.

Chassaing et al.[8] have investigated the electrochemical reduction of Zr^{IV} species in chloride melts. They used cyclic voltammetry at vitreous carbon and tungsten electrodes to study this system, and have shown the behaviour to be highly dependent upon the nature of the melt. In LiCl–KCl (at 520 °C), the reduction is a two-step process; first the chloro-complex of Zr^{IV} is reduced to a Zr^{II} species, which is subsequently reduced to the metal, whereas in CsCl (at 700 °C) the principal process is the direct reduction of the Zr^{IV} complex to the metal. This latter system is also slightly complicated by the presence of dissolved ZrCl$_4$ gas in the melt, as a result of the higher temperature, which is reduced at a different potential to the complex. The behaviour in molten NaCl (at 820 °C) was found to be both complex and time-dependent. The authors explain the observed variations in terms of both the temperature and the stabilizing effects that the cations afford to the complex anions. Thus Cs$^+$ is thought to stabilize ZrCl$_6^{2-}$, thereby preventing the reduction of Zr^{IV} to Zr^{II} from being observed, whereas Na$^+$ does not afford the Zr^{IV} complex such stability.

The remaining member of the Group IVa metals, i.e. hafnium, has only been very briefly studied; Puzanova and Smirnov used potentiometric measurements at a hafnium electrode to investigate the system in molten KCl[9] and the NaCl–KCl eutectic,[10] but they do not appear to have reached any significant conclusions as to whether the metal may be satisfactorily electrodeposited.

Deposition studies have not, of course, been restricted to the refractory metals. Lantelme and Chevalet[11] have used chronoamperometry and chronopotentiometry to investigate the reduction of Cu^I in fused LiCl–KCl (at 410 °C) at both vitreous carbon and graphite electrodes. As with many such studies of deposition in molten salts, the results are shown to be consistent with a mechanism of instantaneous nucleation and subsequent diffusion-controlled three-dimensional growth. Using the same melt as in the above study, Uchida et al.[12] have investigated the electroless deposition of precious metals on glassy carbon electrodes. A chronopotentiometric stripping technique was used to estimate the extent of the electroless deposition, and it was found that deposits, up to several atomic layers thick, of those metals with deposition potentials more positive than -0.5 V vs Pt/Pt^{II} (1M), e.g. Au, Pt, Ir, Ru, Rh, and Pd, were built up on a substrate of glassy

---

[5] E. Chassaing, F. Basile, and G. Lorthioir, J. Appl. Electrochem., 1981, **11**, 193.
[6] C. S. Rao, T. K. Mukherjee, C. V. Sundaram, C. S. Subramanian, and K. Swaminathan, Trans. Indian Inst. Met., 1980, **33**, 275.
[7] R. Vytlacil and A. Neckel, Metall (Berlin), 1980, **34**, 538.
[8] F. Basile, E. Chassaing, and G. Lorthioir, J. Appl. Electrochem., 1981, **11**, 645.
[9] T. A. Puzanova and M. V. Smirnov, Elektrokhimiya, 1980, **16**, 1397.
[10] M. V. Smirnov and T. A. Puzanova, Elektrokhimiya, 1980, **16**, 1563.
[11] F. Lantelme and J. Chevalet, J. Electroanal. Chem. Interfacial Electrochem., 1981, **121**, 311.
[12] I. Uchida, J. Niikura, and S. Toshima, J. Electroanal. Chem. Interfacial Electrochem., 1980, **107**, 115.

carbon. An electroless deposition process of this type, of course, requires a participating anodic reaction. It is suggested that this involves either bulk O^{2-} impurities, adsorbed water, or surface functional groups on the glassy carbon. Furihata et al.[13] have suggested that water impurities are also important if one wishes to investigate the electroreduction of Mg^{2+} in the LiCl–KCl melt. In earlier studies by various workers it had been found to be impossible to detect a reduction wave for this process. Furihata and co-workers have now, however, been able to detect a well-defined wave at a platinum electrode $[E_{\frac{1}{2}} = -2.53 \text{ V } vs \text{ Pt/Pt}^{II}$ (1M)], and they observed that the addition of even very small traces of water results in this process being totally masked.

Deposition of aluminium from dilute solutions of $AlCl_3$ in NaCl–KCl melts has been studied by Delimarski et al.[14] They report that the half-wave potential for this process lies between -0.7 and -0.8 V vs a lead reference electrode, and that there is also some evidence for the formation of univalent aluminium. With respect to this latter point, it should be noted that it has been suggested in the past that Al^I is formed in chloroaluminate melts during deposition of aluminium, but this is not now thought to be the case, the observations being attributable to impurities. Sergeev and Fedorov[15] have also used KCl–NaCl melts (at 973 K) to investigate the formation of uranium–aluminium alloys by the deposition of uranium into molten aluminium cathodes. Skyllas-Kazacos and Welch[16] have turned their attentions towards studying the electrochemistry of PbS in the $PbCl_2$–NaCl eutectic (at 430 °C) in an effort to explain the rather low current efficiencies that are observed for the electrowinning of lead from this melt. They concluded that, upon electrolysis, S^{2-} ions are oxidized at the anode to produce insoluble sulphur, which reacts with further sulphide to form soluble subsulphide species. It is the reduction, at the cathode, of these subsulphides, according to reaction (5), that is thought to be the

$$S_{n+1}^{2n-} + 2e^- \rightarrow (n+1)S^{2-} \tag{5}$$

major cause of the decreased current efficiency, whilst a secondary cause is the reduction of sulphur itself, which becomes increasingly soluble as the concentration of PbS is increased. Zarubitskii et al.[17] have also been concerned with electrowinning, this time from either $ZnCl_2$ (31.5 mole%)–KCl (68.5 mole%) or $ZnCl_2$ (46%)–KCl (34.5%)–NaCl (19.5%). They found that Bi^{3+}, Sb^{3+}, Ag^+, Cu^{2+}, and Sn^{2+} could all be reversibly deposited from either melt, though the latter appeared to be more suitable for electrorefining or electrowinning, as the reductions occurred at less negative potentials.

Some systems of interest to the electronics industry have also been investigated. Tyagai et al.[18] have cathodically polarized Si(111) single crystals in KCl–NaCl–LiCl eutectic melts and have grown thin films of solid solutions of lithium and potassium in silicon. At potentials more negative than about -2.7 V vs Ag/Ag^I, the liquid alkali

---

[13] S. Furihata, K. Akashi, and S. Kurosawa, Electrochim. Acta, 1981, **26**, 1107.
[14] Yu. K. Delimarski, V. F. Makogan, and O. P. Gritsenko, Ukr. Khim. Zh. (Russ. Ed.), 1980, **46**, 115.
[15] V. L. Sergeev and V. A. Fedorov, Elektrokhimiya, 1981, **17**, 1267.
[16] M. Skyllas-Kazacos and B. J. Welch, Electrochim. Acta, 1980, **25**, 179.
[17] O. G. Zarubitskii, A. A. Omelchuk, and V. T. Melekhin, Ukr. Khim. Zh. (Russ. Ed.), 1980, **46**, 233.
[18] V. A. Tyagai, I. A. Stepanova, B. N. Kabanov, I. I. Astakhov, E. V. Panov, and Yu. M. Shirshov, Elektrokhimiya, 1980, **16**, 483.

metal itself was deposited and the surface of the silicon was etched. Markov and Ilieva[19] have also been interested in the formation of thin films. They have been able to electrodeposit films of CdTe (with a cubic zinc blende structure) on copper substrates, by the electrolysis of solutions of $CdCl_2$ and TeO_2 in LiCl–KCl melts. At high temperatures and high concentrations of $CdCl_2$ it was found that these films will even form electrolessly, whilst low temperatures and high concentrations of TeO_2, on the other hand, favour the formation of Cu_2Te.

Fluoride Melts. The only electrochemical studies in fluoride melts that appear to have been reported during the period under review are in fact studies of deposition of metals; the area of greatest interest appears to be the deposition of semiconductor materials, which has recently been reviewed by Feigelson.[20] Within this general area, electrowinning of silicon has received the greatest attention, and in particular it is widely thought that this approach should be capable of producing thin films of solar-grade silicon on cheap substrates, thus reducing the cost of silicon-based solar cells. Rao and co-workers[21,22] have shown that such films, with promising properties, can indeed be deposited from a solution of 8—14 mole % K_2SiF_4 in a LiF–KF melt (at 745 °C) on either silver or graphite electrodes. The authors point out that care has to be taken in controlling the experimental conditions if the formation of unsuitable spongy deposits is to be avoided. The deposition of molten silicon from a $BaO–SiO_2–BaF_2$ melt at ca 1450 °C has been studied by De Mattei et al.[23] Silicon of high purity (99.97 w/o) can be formed and, owing to the high temperature and the consequent molten nature of the deposit, very high rates of deposition are possible. The general field of electrowinning of silicon from fluoride melts has been reviewed by Elwell.[24]

The deposition of two other metals, niobium and gadolinium, from fluoride media has been studied. Cohen[25] has investigated the effects of periodic current reversal on the electrodeposition of niobium from a melt consisting of K_2NbF_7 (6 mole %), KHF_2 (10 mole %), and the balance LiF–KF (47 and 37 mole %) at 750 °C, and found that better deposits were indeed formed by this procedure. Deposition of niobium has also been studied by Konstantinov et al.,[26] from a KCl–KF–K_2NbF_7 system with and without added Nb_2O_5. In the absence of added oxide, the reduction, as determined by linear-sweep voltammetry and quasi-steady-state electrolysis, has been shown to be a two-step process:

$$Nb^V + e^- \rightleftharpoons Nb^{IV} \qquad (6)$$

$$Nb^{IV} + 4e^- \rightleftharpoons Nb \qquad (7)$$

These conclusions agree with an earlier study, made in the KCl–NaCl–K_2NbF_7 system. As Nb_2O_5 is added, the reduction process becomes much more complex, with up to four peaks being evident on the sweep voltammogram. This complexity,

---

[19] I. Markov and M. Ilieva, *Thin Solid Films*, 1980, **74**, 109.
[20] R. S. Feigelson, *Adv. Chem. Ser.*, 1980, **186**, 243.
[21] G. M. Rao, D. Elwell, and R. S. Feigelson, *J. Electrochem. Soc.*, 1980, **127**, 1940.
[22] G. M. Rao, D. Elwell, and R. S. Feigelson, *J. Electrochem. Soc.*, 1981, **128**, 1708.
[23] R. C. De Mattei, D. Elwell, and R. S. Feigelson, *J. Electrochem. Soc.*, 1981, **128**, 1712.
[24] D. Elwell, *Sol. Energy Mater.*, 1981, **5**, 205.
[25] U. Cohen, *J. Electrochem. Soc.*, 1981, **128**, 731.
[26] V. I. Konstantinov, E. G. Polyakov, and P. T. Stangril, *Electrochim. Acta*, 1981, **26**, 445.

which varies as the ratio of Nb_2O_5 to K_2NbF_7 varies, is thought to be due to the formation of oxyfluoro-complexes, and the possible nature of such species is discussed. Deposition of gadolinium has been investigated by Zwilling.[27,28] He has shown that gadolinium of high purity, in the form of crystals several centimetres long (rather than the more usual spongy deposits), can be obtained by the electrolysis of either LiCl–LiF–GdF_3 (80:18:2, at 750 °C) or LiF–GdF_3 (75:25, at 870 °C). Replacement of ten parts of GdF_3 by ten parts of BaF_2 in the latter melt changes the morphology of the deposit; block-type deposits are formed rather than the (001) needles.

Iodide Melts. There only appears to have been one study in this type of medium, *i.e.* that of Sato *et al.*,[29] who have investigated the electrochemistry of solutions of CrI_3 (5—40 mole %) in the NaI–KI melt, in the temperature range 600—800 °C. In particular, they have recorded polarization curves for chromium electrodes and also investigated the evolution of iodine on tungsten anodes. Co-deposition of sodium with chromium occurred far more readily than would be predicted from the decomposition voltage of NaI.

Inorganic Electrochemistry in Chloride Melts.—This section is concerned with the study of the electrochemistry of inorganic species except systems involving the deposition of a metal.

It has been realized for a considerable time that molten salts are a very suitable medium in which to process nuclear fuels. To this end, the UO_2^{2+}–UO_2^+–UO_2 system in LiCl–KCl and other chloride melts has been studied by several groups of workers. Komarov and Nekrasova[30] have made potentiometric measurements on this system in fused CsCl, RbCl, and the LiCl–KCl eutectic. They concluded that the anodic dissolution of UO_2 results in the introduction of both hexa- and penta-valent uranium into the melt. Uchida and co-workers[31,32] have made rather more extensive and detailed studies, using a variety of electrochemical techniques, including cyclic and pulse voltammetry. In particular, they were interested in the electrochemical reduction of UO_2^{2+} in LiCl–KCl melts. On both tin oxide and glassy carbon electrodes, they observed that this species underwent a well-defined, reversible, one-electron reduction $[E_{\frac{1}{2}} = -0.487$ V *vs* Pt/Pt^{II} (1M)], followed by a further one-electron step to give a deposit of UO_2. The potential of this second process depended on the nature of the electrode material. From their results they determined that the equilibrium potentials for the processes described by equations (8) and (9) were -0.268 V and -0.049 V, respectively [*vs* Pt/Pt^{II} (1M)].

$$UO_2^{2+} + 2e^- \rightleftharpoons UO_2 \qquad (8)$$

$$UO_2^+ + 1e^- \rightleftharpoons UO_2 \qquad (9)$$

From these values, the equilibrium constant for the reaction described by equation

---

[27] G. Zwilling, *Electrochim. Acta*, 1981, **26**, 637.
[28] G. Zwilling, *Electrochim. Acta*, 1981, **26**, 643.
[29] T. Sato, S. Igarashi, and T. Tachikawa, *Denki Kagaku Oyobi Kogyo Butsui Kagaku*, 1980, **48**, 195.
[30] V. E. Komarov and P. P. Nekrasova, *Elektrokhimiya*, 1981, **17**, 1263.
[31] I. Uchida, J. Niikura, and S. Toshima, *J. Inorg. Nucl. Chem.*, 1981, **43**, 549.
[32] I. Uchida, J. Niikura, and S. Toshima, *J. Electroanal. Chem. Interfacial Electrochem.*, 1981, **124**, 165.

(10) was calculated to be about 10^{-3} mol dm^{-3}, confirming the instability of the UO_2^+ ion.

$$UO_2^{2+} + UO_2 \rightleftharpoons 2UO_2^+ \qquad (10)$$

The authors also present a potential *versus* pO^{2-} diagram for this system. Komarov and Mityaev[33] have anodized uranium into LiCl–KCl (60:40) melts containing various amounts of fluoride ion at a range of temperatures. A mixture of U^{3+}, UF_4^-, and UF_3 was formed, and empirical expressions for the temperature dependences of the apparent stability constants of these species are given.

Inorganic species are, of course, of interest for use as either anode or cathode materials in molten-salt-based primary or secondary batteries; however, there only appear to have been a few studies in halide melts that are directly relevant to this application. Kam and Johnson[34] have used cyclic voltammetry to study the electrochemistry of iron and lithium sulphides in fused LiCl–KCl, and in particular they were interested in the sulphur–sulphide couple. The FeS electrode in LiCl–KCl,[35] as well as in a LiF–LiCl melt,[36] has also been investigated in considerable detail by Tomczuk *et al.*, using a combination of phase studies, e.m.f. measurements, and slow cyclic voltammetry. The sweep rates in this latter technique were very low (typically 0.01 to 0.02 mV s^{-1}), so as to be comparable to the discharge rates of cells; thus the voltammogram peaks correspond to complete conversion of material, and no peaks arising from transient species are observed. The results are complex, and are discussed in terms of possible electrochemical and chemical reactions.

Lithium–aluminium anodes are now receiving wide interest, in view of their possible use in batteries, and Fung and Inman[37] have used a variety of electrochemical techniques to identify the processes that occur when lithium is deposited on aluminium cathodes from a LiCl–KCl melt. Their results indicate that initial formation of a monolayer of lithium on aluminium was followed by undervoltage deposition at potentials that are positive to the formation of the β-phase, which required a nucleation overvoltage. Eventually, the γ-phase, which was formed below the equilibrium potential, and other lithium-rich phases formed. At potentials more negative than -0.4 V (*vs* α/β LiAl reference electrode), potassium was evolved, owing to the replacement reaction:

$$Li + KCl \rightarrow K + LiCl \qquad (11)$$

Unless extreme care is taken in preparing the melt, trace levels of oxide ion impurities will invariably be present in molten salts. A knowledge of the behaviour of these ions is therefore very important. The use of the Pt/O_2 electrode (isolated from a melt by a solid electrolyte of zirconia that is stabilized with CaO, MgO, or Y_2O_3) as a specific oxide-ion electrode is well established. Recently, Stern[38] has shown that these electrodes may be used to titrate oxide ions coulometrically

---

[33] V. E. Komarov and V. S. Mityaev, *Elektrokhimiya*, 1981, **17**, 1160.
[34] K. W. Kam and K. E. Johnson, *J. Electroanal. Chem. Interfacial Electrochem.*, 1980, **115**, 53.
[35] Z. Tomczuk, S. K. Preto, and M. F. Roche, *J. Electrochem. Soc.*, 1981, **128**, 760.
[36] Z. Tomczuk, M. F. Roche, and D. R. Vissen, *J. Electrochem. Soc.*, 1981, **128**, 225.
[37] Y. S. Fung and D. Inman, 158th Meeting Electrochem. Soc., 1980, Extended Abstr. No. 643, p. 1607.
[38] K. H. Stern, *J. Electrochem. Soc.*, 1980, **127**, 2375.

into, and out of, fused NaCl, as described by equation (12).

$$\tfrac{1}{2} O_2 + 2 Na^+ + 2 e^- \rightarrow Na_2O \qquad (12)$$

This property could, of course, be particularly useful in purifying melts, and the lower limit of oxide ion that can be achieved is restricted only by the solubility of the zirconia itself. It should, however, be noted that the electrode is highly polarized by the passage of current, and therefore the electrode that is used to add or remove oxide ions cannot also be used as an indicating electrode for the amount of O^{2-} ion present. This application of these electrodes is not unique to chloride melts, and their use in molten Na_2SO_4 has subsequently been demonstrated.[39] Deanhardt and Stern[40] have used these electrodes, in conjunction with cyclic voltammetry and chronopotentiometry at gold electrodes, to study the electrochemistry of oxide ions in fused NaCl (at 830 °C). Their results show that Na_2O-containing melts take up molecular oxygen reversibly to form peroxide ions. Cyclic voltammograms of oxide-containing solutions that have been exposed to oxygen show two quasi-reversible two-electron redox couples. These are thought to be due to the oxidation of oxide and of peroxide ions respectively. Thus the overall mechanism may be explained by the reactions:

$$2 O^{2-} + O_2 \rightleftharpoons 2 O_2^{2-} \qquad (13)$$

$$2 O^{2-} \rightleftharpoons O_2^{2-} + 2 e^- \qquad (14)$$

$$O_2^{2-} \rightleftharpoons O_2 + 2 e^- \qquad (15)$$

No evidence was found for the formation of superoxide ions, though the generation of small amounts of this species cannot be discounted.

Tkalenko et al.[41] have investigated the reduction of the tungstate ion, ultimately to tungsten metal, in KCl–NaCl melts and, in particular, they have considered the effect of the addition of Li^+, Ca^{2+}, and Mg^{2+} ions on the reduction processes. Finally, from a more industrial standpoint, Ito et al.[42,43] have considered the reprocessing of the NH_4Cl (to yield NH_3, H_2, and Cl_2) that is formed in the Solvay process. Both aqueous and non-aqueous processes were considered, but the most promising was electrolysis in LiCl–KCl melts at a zinc cathode. The added NH_4Cl is reduced by the zinc to ammonia, hydrogen, and zinc chloride, which is electrolysed to return the zinc and evolve chlorine. The performance of a 100-ampere pilot plant is discussed.

3 Aluminium Halide Based Melts

Melts containing an aluminium halide as a major constituent were not considered in the previous section of this review because they form a group of systems with distinctly different properties. These melts are being very widely used, and whilst the bulk of work continues to be concerned with alkali-metal halide–aluminium halide

---

[39] M. L. Deanhardt and K. H. Stern, J. Phys. Chem., 1980, **84**, 2831.
[40] M. L. Deanhardt and K. H. Stern, J. Electrochem. Soc., 1980, **127**, 2600.
[41] D. A. Tkalenko, N. A. Chmilenko, and S. V. Sezhin, Elektrokhimiya, 1980, **16**, 1160.
[42] Y. Ito, T. Ohmori, S. Nakamatsu, and S. Yoshizawa, J. Appl. Electrochem., 1980, **10**, 419.
[43] Y. Ito, T. Ohmori, S. Nakamatsu, and S. Yoshizawa, J. Appl. Electrochem., 1980, **10**, 427.

systems (particularly NaCl–AlCl$_3$ mixtures), increasing interest is being shown in the development of low-temperature, and, particularly, of room-temperature melts, as will be shown later.

Alkali-metal Chloroaluminates.—*Deposition of Metals.* Welch and Osteryoung[44] have reviewed studies of the deposition of metals in sodium tetrachloroaluminates, as well as in the AlCl$_3$–n-butylpyridinium chloride melt. This review is concerned with both aluminium plating and the deposition of other metallic species such as Fe, Cu, Ni, Cr, Cd, Pb, and Zn, though little previously unpublished data is presented. Deposition of aluminium on to iron and steel from AlCl$_3$–NaCl–KCl melts has been studied by Grjotheim and Matiasovsky.[45] They found that the best deposits were formed from melts containing 76—80 weight % of AlCl$_3$, when compact polycrystalline layers were formed, whereas lower amounts of AlCl$_3$ resulted in grainy deposits. This finding is unfortunate, as it is at these high concentrations of AlCl$_3$ that the vapour pressure of Al$_2$Cl$_6$ over the melt is at its greatest, thus greatly complicating any potential commercial process. In the same study, these authors have also investigated co-deposition of aluminium with manganese, tin, and lead, and the effect of organic additives. In respect of this latter point, tetra-alkylammonium halides were found to give the best results.

It was stated in the previous section that Li–Al alloys are now being widely considered for possible use in batteries. Carpio and King[46] have made a study (using cyclic voltammetry, chronoamperometry, and coulometry) of the formation of such alloys in LiAlCl$_4$ melts. Their initial experiments were made with tungsten electrodes, but it was soon found that a thin layer of aluminium was always deposited prior to any formation of an alloy, and therefore all subsequent studies were made with aluminium electrodes. Contrary to the behaviour found in other melts (*e.g.* LiCl–KCl, where Li–Al alloys are formed by an implantation mechanism) β-LiAl was found to be formed by direct co-deposition in LiAlCl$_4$, although, not surprisingly, this alloy corroded in the melt, with lithium being replaced by aluminium. Long-term deposition resulted in the formation of a uniform, dense, non-dendritic deposit of a lithium–aluminium alloy in which the content of lithium was significantly below 50%. Whilst the electrochemical properties of this alloy with a low lithium content are inferior to those of β-LiAl, they can potentially be prepared very much more cheaply, and therefore might be useful in battery systems.

Deposition of cobalt on platinum electrodes from NaAlCl$_4$ melts (at 175 °C) has been studied by Tumanova *et al.*[47] These workers were particularly interested in the effects of organic additives such as urea, β-naphthylamine, and aluminium stearate.

The Chlorine-evolution Reaction. Evolution of chlorine from chloroaluminates has been studied by several research groups, the most detailed work having been undertaken by Uchida and co-workers. They have investigated both the evolution and the dissolution reactions in basic AlCl$_3$–NaCl melts (pCl = 1.1 at 175 °C) at

---

[44] B. J. Welch and R. A. Osteryoung, *J. Electroanal. Chem. Interfacial Electrochem.*, 1981, **118**, 455.
[45] K. Grjotheim and K. Matiasovsky, *Acta Chem. Scand., Ser. A*, 1980, **36**, 666.
[46] R. A. Carpio and L. A. King, *J. Electrochem. Soc.*, 1981, **128**, 1510.
[47] N. Kh. Tumanova, M. V. Prikhod'ko, V. L. Kislenko, and V. B. Krinitskii, *Teor. Eksp. Khim.*, 1980, **16**, 834.

glassy carbon,[48] RuO_2,[49] and SnO_2 electrodes.[50] In the cases of the first two electrode materials, they have made the first detailed mechanistic studies of the evolution process. At the composition of the melt that was used, the overall reaction can be represented as shown in equation (16). The discharge of $AlCl_4^-$ ions occurs about 0.6 V positive of this oxidation, and it can consequently be neglected.

$$2\,Cl^- \rightleftharpoons Cl_2 + 2\,e^- \qquad (16)$$

By analogy with the evolution of hydrogen, possible reactions involved in this process are shown in equations (17)—(19), which are the so-called electron-transfer, ion–atom recombination, and atom–atom recombination steps, respectively. Several overall mechanisms are possible, and the following are considered: (a), with reaction (17) fast and reaction (18) rate-determining; (b), with reaction (17) fast and reaction (19) rate-determining; and (c), with reaction (17) rate-determining and

$$Cl^- \rightleftharpoons Cl_{ads} + e^- \qquad (17)$$

$$Cl_{ads} + Cl^- \rightleftharpoons Cl_2 + e^- \qquad (18)$$

$$Cl_{ads} + Cl_{ads} \rightleftharpoons Cl_2 \qquad (19)$$

reaction (18) fast. Their results showed that, on a glassy carbon electrode, the exchange current density was about 9.2 $\mu A\,cm^{-2}$ and the reaction followed a type (a) mechanism, with the chlorine obeying an activated Temkin adsorption isotherm. The RuO_2 electrode was found to be more catalytic than glassy carbon. The exchange current density increased to about 130 $\mu A\,cm^{-2}$ and the mechanism changed to type (b). This type of behaviour is identical to that observed for the hydrogen-evolution reaction in aqueous systems; platinum follows a type (b), Tafel mechanism whilst less catalytic metals obey the Heyrovsky reaction. In view of this similarity, it might be fruitful to extend this present study to the other metal electrodes.

The same workers[51] have also investigated the photoelectrochemistry of TiO_2 electrodes in the potential region of chlorine evolution. In the dark, the electrode was non-catalytic, and no current was detectable even at a potential of 3 V vs aluminium; however, upon illumination with a mercury lamp there is a measureable photocurrent. It is suggested by the authors that the light induces holes in the TiO_2, which then oxidize the Cl^- ions in the melt. The potential for onset of this photoelectrochemical effect was found to be highly dependent upon the basicity of the melt, indicating that the flat band potential of the TiO_2 had a dependence on pCl. This latter effect was ascribed to specific adsorption of Cl^- ions on the TiO_2.

Evolution of chlorine has also been studied by Van Huong et al.,[52] using impedance techniques, in the $NaCl$–KCl–$AlCl_3$ melt, at high temperatures (790—880 °C), whilst Panov et al.[53] have used the same method at a glassy carbon

---

[48] H. Urushibata, I. Uchida, and S. Toshima, *J. Electroanal. Chem. Interfacial Electrochem.*, 1981, **117**, 43.
[49] I. Uchida, H. Urushibata, and S. Toshima, *J. Electrochem. Soc.*, 1981, **128**, 2351.
[50] I. Uchida, H. Urushibata, and S. Toshima, *J. Electrochem. Soc.*, 1980, **127**, 757.
[51] I. Uchida, H. Urushibata, H. Akahoshi, and S. Toshima, *J. Electrochem. Soc.*, 1980, **127**, 995.
[52] B. Van Huong, Yu. V. Borisoglebskii, and V. M. Vetyukov, *Elektrokhimiya*, 1980, **16**, 1529.
[53] Z. V. Panov, G. V. Molodid, and Yu. K. Delimarski, *Elektrokhimiya*, 1980, **16**, 1193.

electrode in both $AlCl_3$–NaCl–KCl (at 150 °C) and LiCl–NaCl–KCl (at 400 °C). In agreement with Uchida et al., they found that, in the chloroaluminate melt, the mechanism was of type (a) whilst in the alkali-metal halide system it was of type (c).

Inorganic Electrochemistry. Meuris et al.[54] have investigated the electrochemical behaviour of uranium species in acidic $AlCl_3$–NaCl melts (at 175 °C). This continues some earlier work that they had undertaken in basic media. Using cyclic and pulse voltammetries, at both tungsten and glassy carbon electrodes, they observed two one-electron redox couples, which they assigned to U^V/U^{IV} and U^{IV}/U^{III}. From the peak separation in cyclic voltammetry and the shape of the pulse voltammogram, the second of these processes was shown to be reversible. This also appeared to be true of the U^V/U^{IV} couple, though since it occurred close to the anodic limit of the melt, the voltammograms were distorted, and consequently this conclusion is tentative. The redox potentials for both processes were found to be dependent on pCl, and from this the complexation of the uranium ions was elucidated. They concluded that the reduction of U^{IV} to U^{III} may be written as shown in equation (20), where, in the most basic melts studied (pCl = 6), x equals three, but it tends to one as the acidity is increased. The U^V/U^{IV} couple, on the other hand, is represented by equation (21), with $(y-x)$ equal to three for values of pCl in the range 6.1—7, and its value decreases with increasing acidity.

$$UCl_x^{(4-x)+} + e^- \rightleftharpoons U^{3+} + xCl^- \tag{20}$$

$$UCl_y^{(5-y)+} + e^- \rightleftharpoons UCl_x^{(4-x)+} + (y-x)Cl^- \tag{21}$$

In view of the probable presence of oxide impurities in the melt, the formation of oxychloro-complexes should not, however, be discounted.

Basile et al.[55] have investigated the effect of the composition of the melt on the electrochemical reduction of Zr^{IV}. The overall conclusions that they reached were that, in acidic melts, the high concentration of $Al_2Cl_7^-$ ions favoured the direct reduction of Zr^{IV} to Zr^{II} (it has previously been recognized that these anions stabilize low oxidation states of metals), whilst for approximately neutral melts the reduction of Zr^{IV} to Zr^{III} was observed. At intermediate compositions, both processes were seen. It is hoped by the authors that these results will be helpful in developing electrolyte solutions that are suitable for electrodeposition of zirconium.

Spectroelectrochemistry. One of the most significant advances that has been made in electrochemistry during the past decade has undoubtedly been the coupling of electrochemistry and various spectroscopic techniques. This has proved invaluable in many areas, including mechanistic studies and the identification of reaction intermediates. Until recently, however, these spectroelectrochemical methods have not been applied by molten-salt chemists, largely because of the greater technical difficulties that are presented when working in these solvents, particularly those associated with the higher temperatures involved. Mamantov and co-workers have, however, now demonstrated that both u.v.—visible absorption spectroscopy[56] and

---

[54] F. Meuris, L. Heerman, and W. D'Olieslayer, *J. Electrochem. Soc.*, 1980, **127**, 1294.
[55] F. Basile, E. Chassaing, and G. Lorthioir, *Ann. Chim. (Paris)*, 1980, **5**, 553.
[56] G. Mamantov, V. E. Norvell, and L. N. Klatt, *J. Electrochem. Soc.*, 1980, **127**, 1768.

e.s.r. spectroscopy[57] can be applied *in situ* in the electrochemical cell to chloroaluminate melts, and there appears to be no reason why these techniques should not be applied to other systems, particularly those with relatively low melting points, provided that suitable window materials are available.

The instrumentation used for u.v.—visible spectroscopy consists of a computer-controlled rapid scanning spectrometer and a specially designed cell in which the light beam passes through a thin layer within which is incorporated a mini-grid working electrode (of platinum, tungsten, or glassy carbon). This system was initially used to study the processes occurring during the reduction of Nb^V and also during the oxidation of sulphur. More recently it has been used to study the oxidation of sulphur in acidic melts in more detail, and the e.s.r. technique, which uses a similar cell except that it is now designed to fit into the cavity of the e.s.r. spectrometer, has also been applied to the sulphur system. This system is of particular importance in view of the possible use of S^{IV} as a cathode material in secondary batteries.[58]

The electrochemical oxidation of sulphur to SCl_3^+ in acidic chloroaluminates is very complex, but, based on electrochemical experiments and some *ex situ* spectroscopic studies, the following possible mechanism for the first oxidation step (at 1.3—1.7 V) had been proposed:

$$S_8 \rightleftharpoons S_8^+ + e^- \quad (22)$$

$$2 S_8^+ \rightleftharpoons S_{16}^{2+} \quad (23)$$

$$S_8^+ \rightleftharpoons S_8^{2+} + e^- \quad (24)$$

$$S_8 \rightleftharpoons S_8^{2+} + 2 e^- \quad (25)$$

$$S_8^{2+} + S_8 \rightleftharpoons S_{16}^{2+} \quad (26)$$

Whilst the subsequent *in situ* spectroscopic studies were inconclusive, evidence was found for the presence of both S_{19}^{2+} and S_5^+ in this potential range, and therefore the mechanism must be more complex than that indicated in equations (22)—(26). Other possible reaction pathways are discussed. Similarly, routes to S_4^{2+} and S^{2+} during the second and third oxidation steps, for which spectroscopic evidence was also found, are discussed.

The oxidation of I_2 has also been re-investigated by Mamantov *et al.*,[59] using spectroelectrochemical techniques. All the electrochemical results strongly point to the final product of oxidation of I_2 in both acidic and basic melts being a species with an oxidation state of (+1). Ultraviolet—visible spectroscopy shows that this species is ICl, but it also points to the presence of some I_2^+ in acidic melts, as does resonance Raman spectroscopy. This can be explained if the previously reported reaction scheme is modified to allow I_2^+ to be formed as a reaction intermediate, as shown in equations (27) and (28).

$$I_2 \rightleftharpoons I_2^+ + e^- \quad (27)$$

---

[57] V. E. Norvell, K. Tanemoto, G. Mamantov, and L. N. Klatt, *J. Electrochem. Soc.*, 1981, **128**, 1254.
[58] G. Mamantov, R. Marassi, M. Matsunaga, Y. Ogata, J. P. Wiaux, and E. J. Fraser, *J. Electrochem. Soc.*, 1980, **127**, 2319.
[59] K. Tanemoto, G. Mamantov, R. Marassi, and G. M. Begun, *J. Inorg. Nucl. Chem.*, 1981, **43**, 1779.

$$I_2{}^+ \xrightarrow{\text{solvent}} 2\,ICl + e^- \tag{28}$$

Cryolite Melts.—The major area of interest of researchers using cryolite melts is, of course, deposition of aluminium. Duruz et al.[60] have used cyclic voltammetry and ion microprobe analysis of deposits etc. to determine which electrode material is most suitable for such studies. Gold, platinum, nickel, and tungsten were all considered, and, in view of its low solubility in aluminium, the latter was throught to be most suitable. For studying the deposition of metals that are more noble than aluminium and sodium it is suggested that gold should be used, in view of its large electrochemical window. Del Campo and co-workers[61] have made a fairly detailed study of the deposition of aluminium, using a platinum rotating-disc electrode (RDE) upon which some aluminium had already been deposited. In view of the above results, platinum is probably not the best choice of electrode material. Platinum and aluminium are known to form a series of intermetallics; however, the polarization curves do not appear to be complicated by the presence of such species. The use of the RDE enables purely kinetic currents to be determined, and, on the basis of their data, the authors have considered various mechanisms. That most consistent with the results is:

$$2\,AlF_6{}^{3-} + 6\,Na^+ \rightleftharpoons 2\,Al^{3+} + 6\,Na^+F^- + 6\,F^- \tag{29}$$

$$2\,Al^{3+} + 6\,e^- \rightleftharpoons 2\,Al \tag{30}$$

$$6\,F^- + 3\,AlF_4{}^- \rightleftharpoons 3\,AlF_6{}^{3-} \tag{31}$$

Therefore the overall reaction may be written as:

$$6\,Na^+ + 3\,AlF_4{}^- + 6\,e^- \rightleftharpoons 2\,Al + 6\,NaF + AlF_6{}^{3-} \tag{32}$$

In particular, they discount mechanisms involving discharge of Na^+, the direct discharge of $AlF_4{}^-$, and schemes involving $AlOF_2{}^-$, as have been suggested by other workers in the field.

The anodic reaction at a carbon electrode in cryolite-based media, which is of obvious importance to the aluminium industry, has been investigated by Lantelme and co-workers. They have used chronopotentiometry,[62] cyclic voltammetry,[63] and analysis of gases liberated at the anode[64] to investigate this reaction, and in particular the effects of dissolved aluminium and the addition of different types of alumina. Their results generally support earlier observations that dissolved aluminium aids evolution of gases, by affecting the wetting properties of carbon, and that the primary anode gas is CO_2.

Room-temperature Chloroaluminate Melts.—The possibility of using $AlCl_3$-based molten salts as media in which to perform organic reactions, both homogeneous and electrochemical, has been investigated on several occasions in the past. One of the main driving forces for this has been the anhydrous nature of

---

[60] J. J. Duruz, G. Stehle, and D. Landholts, *Electrochim. Acta*, 1981, **26**, 771.
[61] J. J. Del Campo, J. P. Millet, and M. Rolin, *Electrochim. Acta*, 1981, **26**, 59.
[62] F. Lantelme, D. Damianacos, and M. Chemla, *J. Electrochem. Soc.*, 1980, **127**, 498.
[63] D. Damianacos, F. Lantelme, and M. Chemla, *C.R. Hebd. Seances Acad. Sci.*, Ser. C, 1980, **290**, 149.
[64] D. Damianacos, F. Lantelme, M. Chemla, and M. Vogler, *Electrochim. Acta*, 1981, **26**, 917.

these solvents and, in the case of acidic melts, the almost total absence of nucleophiles. The melting point of alkali-metal chloroaluminates does, however, preclude their general applicability, although a number of electrochemical investigations of substances with low volatility have been reported. In an effort to overcome this problem, Osteryoung and co-workers, in particular, have, over the past few years, been developing chloroaluminate melts that are based on mixtures of $AlCl_3$ and alkyl-pyridinium halides and which are molten at or near room temperature. The studies undertaken prior to the period covered by this Report have been reviewed by Chum and Osteryoung,[65] but, in view of their novelty, it is perhaps worth briefly outlining some of the properties of these systems.

The melts are transparent, somewhat viscous liquids at or near room temperature, with acid–base properties very similar to the higher-temperature alkali-metal chloroaluminates, though exhibiting somewhat more extreme conditions (*i.e.*, the pCl range is much greater) and have an electrochemical window of approximately 2 V. As a consequence of their high viscosity, these melts have a relatively low conductivity for a molten salt, though the value is significantly higher than for typical non-aqueous solvent/electrolyte systems. The viscosity may, however, be reduced, and the conductivity increased, by the addition of up to 50 volume per cent of benzene or toluene with little change in the properties. As with other chloroaluminates, the most commonly used reference electrode is an aluminium wire, generally immersed in a 2:1 $AlCl_3$:pyridinium halide melt.

Robinson and Osteryoung[66] have used potentiometric techniques to confirm that the addition of benzene to $AlCl_3$–n-butylpyridinium chloride (BPC) had no chemical effect other than dilution, and they showed that the equilibrium constant for the acid–base reaction described by equation (33) in the presence of 50 volume per cent benzene was about 2.2×10^{-13}; a very similar value to that obtained for the pure melt.

$$2\,AlCl_4^- \rightleftharpoons Al_2Cl_7^- + Cl^- \tag{33}$$

In the same paper they went on to consider the electrodeposition of aluminium, both from the pure melt and from melt–benzene mixtures, at tungsten, platinum, and glassy carbon electrodes. From chronoamperometry, the deposition at all three electrodes was shown to be by instantaneous nucleation, followed by diffusion-controlled three-dimensional growth, the overpotential being greatest for carbon and least for tungsten. Cyclic voltammetric studies indicated the formation of an underpotential deposit on both platinum and tungsten, though not on glassy carbon. In all melts, the deposited aluminium was shown to corrode slowly, though this was probably attributable to traces of impurities in the melt.

In acidic melts, the cathodic limit on, amongst others, platinum, tungsten, and glassy carbon electrodes is always found to be the deposition of aluminium, but in basic melts this is not the case, as the BPC cation is reduced more readily than $AlCl_4^-$ ions. This obviously limits the potential window of basic $AlCl_3$–BPC melts, and it would therefore be advantageous to use a cation that was less easily reduced.

---

[65] H. L. Chum and R. A. Osteryoung, in 'Ionic Liquids', ed. D. Inman and D. G. Lovering, Plenum, New York, 1981.
[66] J. Robinson and R. A. Osteryoung, *J. Electrochem. Soc.*, 1980, **127**, 122.

Gale and Osteryoung[67] have turned their attention to this problem by investigating the factors that might effect the stability of BPC cations. Tafel plots for the primary reduction process in $AlCl_3$–BPC (0.8 :1) at 40 °C reveal the process to be a one-electron step to yield a radical species that rapidly dimerizes, as shown in equations (34) and (35), where R is n-butyl.

$$R-\overset{+}{N}\diagup\diagdown + e^- \rightleftharpoons R-N\diagup\diagdown\cdot \quad (34)$$

$$2\,R-N\diagup\diagdown\cdot \longrightarrow R-N\diagup\diagdown\overset{H}{}\diagup\diagdown N-R \quad (35)$$

DBTBP

The onset of this reduction occurs at -1.1 V vs aluminium, but the Tafel slope changes at about -1.4 V, indicating that further processes or other complicating factors occur. The dimer 1,1'-dibutyl-4,4'-tetrahydrobipyridine (DBTBP), formed in the first reduction, was shown by cyclic voltammetry to be unstable; at high sweep rates, the oxidation of DBTBP could be seen at -0.4 V, but at lower sweep rates, two oxidation processes ($E_{\frac{1}{2}} = -0.56$ and -0.96 V) were seen, and attributed to 1,1'-dibutyl-4,4'-bipyridine that is formed in the decomposition of DBTBP. The authors conclude that, when designing BPC derivatives that will yield melts with larger electrochemical windows, the most important factor to consider is the primary reduction potential of the cation, and they discuss the effects of substitution on this potential. At present, however, it cannot be predicted that such substituted pyridinium cations would yield low-temperature melts with $AlCl_3$. One suitable cation has been prepared by Osteryoung et al.,[68] i.e. p-dimethylamino-N-butylpyridinium ion. This cation is reduced at a potential that is some 0.6 V more negative than that at which the unsubstituted one is reduced, and it forms a melt with $AlCl_3$ for which the equilibrium constant for the acid–base reaction described by equation (33) is about 10^{-10}.

Hussey et al.[69] have also considered some promising salts that yield low-temperature systems with $AlCl_3$, i.e. imidazolium chlorides in which the nitrogen atoms are substituted. Molecular Orbital theory indicated that the values of $E_{\frac{1}{2}}$ for the reduction of these cations should be about one volt more negative than that of n-butylpyridinium. This was indeed found to be the case, and certain of the cations yielded acceptable molten salts; in particular, 1-methyl-3-ethylimidazolium chloride–$AlCl_3$ melts were liquid at room temperature throughout a wide composition range.

The potential dependence of the double-layer capacitance of a mercury electrode in $AlCl_3$–BPC melts has been determined by Gale and Osteryoung,[70] using classical bridge techniques. In acidic and neutral melts, the capacitance–potential plots were

---

[67] R. J. Gale and R. A. Osteryoung, J. Electrochem. Soc., 1980, **127**, 2167.
[68] R. A. Osteryoung, G. Cheek, and H. Linga, 158th Meeting Electrochem. Soc., 1980, Extended Abstr. No. 646, p. 1615.
[69] J. S. Wilkes, J. A. Levisky, C. L. Hussey, and M. Druelinger, 158th Meeting Electrochem. Soc., 1980, Extended Abstr. No. 648, p. 1622.
[70] R. J. Gale and R. A. Osteryoung, Electrochim. Acta, 1980, **25**, 1527.

very similar to those exhibited by other molten salts in being fairly flat, with a shallow minimum in the region of the point of zero charge. In basic melts, however, there was a dramatic increase in the differential capacitance at anodic potentials, due to the specific adsorption of chloride ions. In fact, the excess charge due to the adsorbed chloride ions was very much greater than that observed in aqueous systems under comparable conditions. The authors consider that this is a result of the different structure of the double-layer in a molten salt, in particular the absence of a solvent.

Inorganic Electrochemistry. One continual problem in molten-salt chemistry, and $AlCl_3$-based systems are no exception, is the role of oxide ions, which frequently arise from traces of water in the salts. During an investigation of the electrochemistry of Ti^{IV} in basic $AlCl_3$–BPC melts, Linga *et al.*[71] have observed some interesting behaviour that they attribute to the presence of such oxide ion impurities. In purified melts, Ti^{IV} was seen to undergo a one-electron reduction to Ti^{III} ($E_{\frac{1}{2}} = 0.343$ V) at a potential which was independent of the acidity of the melt. From both cyclic and pulse voltammetries, this process appeared to be reversible; however, it was also observed that there was a second, small reduction step, with a value of -0.77 V for $E_{\frac{1}{2}}$. On the addition of oxide ions (introduced in the form of Li_2CO_3), the current for the second reduction step was found to increase at the expense of that for the first, though the overall current, as determined by pulse voltammetry, for the two processes combined remained constant. It was the conclusion of the authors that this pointed to the presence of two Ti^{IV} species, linked by an equilibrium involving oxide ions. The reduction potential attributed to the oxychloro-complex was also found to be dependent on pCl.

In basic $AlCl_3$–NaCl melts, oxide ions have been shown to react as in equation (36), and it is probable that they behave similarly in basic $AlCl_3$–BPC. From the electrochemical results for reduction of Ti^{IV}, it was proposed that the equilibrium described by equation (37) linked the two Ti^{IV} species, and the equilibrium constant was determined to be 900 at 40 °C in the composition range for $AlCl_3$:BPC of 0.7:1 to 0.9:1. Thus the two reduction steps may be written as in equations (38) and (39).

$$O^{2-} + AlCl_4^- \rightleftharpoons AlOCl_2^- + 2\,Cl^- \tag{36}$$

$$TiCl_6^{2-} + AlOCl_2^- \rightleftharpoons AlCl_4^- + TiOCl_4^{2-} \tag{37}$$

$$TiCl_6^{2-} + e^- \rightleftharpoons TiCl_6^{3-} \tag{38}$$

$$TiOCl_4^{2-} + AlCl_4^- + e^- \rightleftharpoons TiCl_6^{3-} + AlOCl_2^- \tag{39}$$

No evidence was found for oxychloro-complexes of Ti^{III}, presumably because it is a significantly weaker acid than Al^{III}. It is also interesting, and perhaps surprising, to note that the addition of H_2O appears to favour the formation of the chloro-complex and not the oxychloro one. In $AlCl_3$–NaCl systems, water reacts as shown in equation (40), *i.e.* the concentration of oxide ion is increased. It is suggested that,

$$AlCl_4^- + H_2O \rightleftharpoons 2\,HCl + AlOCl_2^- \tag{40}$$

[71] H. Linga, Z. Stojek, and R. A. Osteryoung, *J. Am. Chem. Soc.*, 1981, **103**, 3754.

in the room-temperature systems, the water perhaps reacts further with $AlOCl_2^-$ and $AlCl_4^-$ to form a species such as $Al(OH)_2Cl$, which does not readily react with Ti^{IV}. One of the initial aims of this work was, of course, to attempt to electroplate titanium; this was not achieved, though titanium metal was not corroded by the melt, and it could be anodized into it.

Following on from the above study, Stojek et al.[72] have used this behaviour exhibited by Ti^{IV} as the basis of a method for the determination of oxide ions in basic melts. The equilibrium constant for the reaction between Ti^{IV} and oxide [equation (37)] can be written as:

$$K = [TiOCl_4^{2-}][AlCl_4^-]/[TiCl_6^{2-}][AlOCl_2^-] \qquad (41)$$

but, since $[AlCl_4^-]$ is a constant,

$$K' = K/[AlCl_4^-] \qquad (42)$$

If [Ti] is the total concentration of Ti^{IV}, then

$$[TiCl_6^{2-}] = [Ti] - [TiOCl_4^{2-}] \qquad (43)$$

and therefore

$$K' = [TiOCl_4^{2-}]/([Ti] - [TiOCl_4^{2-}]) \times [AlOCl_2^-] \qquad (44)$$

which, upon rearranging, gives equation (45).

$$\frac{[Ti]}{[TiOCl_4^{2-}]} = \frac{1}{[AlOCl_2^-]}([Ti] - [TiOCl_4^{2-}]) + K'\left(\frac{[AlOCl_2^-]+1}{[AlOCl_2^-]}\right) \qquad (45)$$

The value of [Ti] is known from the amount of Ti^{IV} that has been added, whilst $[TiOCl_4^{2-}]$ can readily be found by pulse voltammetry. Thus, if $[Ti]/[TiOCl_4^{2-}]$ is plotted as a function of $[Ti] - [TiOCl_4^{2-}]$, the oxide content of the melt can be found from the gradient, whilst K', and hence K, can be found from the intercept. Using this procedure, the authors showed that even very carefully purified and prepared melts were about 1 to 2×10^{-3} M in oxide ion. Since such levels of oxide are always likely to be present in these melts, their possible effect must always be considered. This is particularly true of basic melts, but less so of acidic ones, where $Al_2Cl_7^-$ ions are present in large excess.

Whilst this technique does not appear to have been applied to other $AlCl_3$-based melts, it is very possible that, in view of the similarities between these systems, it is more generally applicable. The only other study of inorganic species is that of Hussey and Laher,[73] who have investigated the behaviour of Co^{II} in $AlCl_3$–BPC melts, using both electrochemical and spectroscopic techniques. In acidic melts, Co^{II} was shown to be present as an octahedrally co-ordinated complex, with $AlCl_4^-$ and $Al_2Cl_7^-$ ligands, which could be electroreduced to yield deposits of cobalt. In basic melts, Co^{II} formed tetrahedral $CoCl_4^{2-}$ complexes which appeared to be electro-inactive.

Organic Electrochemistry.—As was stated at the beginning of this section, the major driving force behind the development of mixtures that are molten at room

---

[72] Z. Stojek, H. Linga, and R. A. Osteryoung, *J. Electroanal. Chem. Interfacial Electrochem.*, 1981, **119**, 365.
[73] C. L. Hussey and T. M. Laher, *Inorg. Chem.*, 1981, **20**, 4201.

temperature was the possibility that they could be used as solvents for organic chemical and electrochemical investigations. Osteryoung et al.[68] have studied the dependence of the electrochemistry of anthraquinone on acidity of the medium and have also used i.r. spectroscopy to aid their study. In basic melts, the anthraquinone itself was found to be uncomplexed and to undergo a two-electron reduction to an unstable dianion, co-ordinated by two $AlCl_3$ molecules. As the acidity of the melt increased, the anthraquinone became complexed, first by one and then by two molecules of $AlCl_3$, whilst the dianion became co-ordinated by four such molecules.

Robinson and Osteryoung[74] have investigated the oxidation of a number of aromatic amines; in particular, of species that had previously been studied in molten sodium tetrachloroaluminate. In acidic melts, triphenylamine was found to undergo two one-electron oxidations, first to the stable cation radical and then to the unstable dication. The cation radical was found to be less stable in basic melts, and it readily dimerized. Other amines, e.g. NN-dimethylaniline, behaved in a very similar manner in basic melts but exhibited more complex behaviour in acidic media, owing to the presence of complexes formed between the basic amines and $AlCl_3$. This type of behaviour is generally similar to that observed in $NaCl-AlCl_3$ melts, the major differences being the greater stability of cation radicals in the room-temperature system, presumably due to the lower temperature and the higher acidity of these solvents.

Keszthelyi[75] has studied the oxidation, both electrochemical and homogeneous, of some aromatic species in 2:1 $AlCl_3$–BPC melts. Thianthrene, for example, was found to be spontaneously oxidized to its cation radical on addition to the melt, and this radical was almost indefinitely stable. This oxidation was, however, reversible; i.e., on reducing the acidity of the melt, the cation radical was reduced back to the parent compound, and the cation radical was re-formed on re-acidifying the melt, this process being readily monitored by electrochemical or spectroscopic means. In general, it was found that any aromatic species with a value of $E_{\frac{1}{2}}$ that is more negative than 1.5 V vs the SCE in aqueous solution would, like thianthrene, be spontaneously oxidized by acidic melts. Such behaviour is a little surprising, and cannot at present be explained; nor is the nature of the oxidizing agent known. It is, however, worth noting that similar behaviour is also seen in molten $SbCl_3$, which is another acidic melt.

Organometallic Electrochemistry. Chum et al.[76] have studied the electrochemical behaviour of iron(II) di-imine complexes in 2:1 $AlCl_3$:ethylpyridinium bromide (the first room-temperature system of this type to be used, though now largely superseded by BPC-based melts in view of their superior electrochemical windows). All the complexes were oxidized in a one-electron reversible step, with no concomitant solvent-assisted oxidation of the ligand, as is seen in other solvents. This latter observation is a further example of the favourable properties of this type of non-aqueous system. Correlations were made between the values of $E_{\frac{1}{2}}$ for the oxidation step of the various complexes and polar and steric Taft factors.

---

[74] J. Robinson and R. A. Osteryoung, *J. Am. Chem. Soc.*, 1980, **102**, 4415.
[75] C. P. Keszthelyi, *Electrochim. Acta*, 1981, **26**, 1261.
[76] H. L. Chum, D. Koran, and R. A. Osteryoung, *Inorg. Chem.*, 1981, **20**, 3304.
[77] R. J. Gale, P. Singh, and R. Job, *J. Organomet. Chem.*, 1980, **199**, C44.

Gale and co-workers[77–79] have also undertaken a study of some organometallic species, the metallocenes. Ferrocene itself had previously been shown to undergo a reversible one-electron oxidation at 0.24 V in acidic melts to form the ferrocinium cation, which was itself further oxidized in an irreversible, ill-defined, multi-electron process close to the anodic limit of the melt. The substitution of ten methyl groups into the cyclopentadiene groups to form decamethylferrocene would be expected to shift the oxidation potential negatively by about 530 mV, and this was indeed shown to be the case; decamethyl-ferrocene was oxidized reversibly in basic melts at −0.292 V to form the cation radical. In acidic melts it was spontaneously oxidized to the stable cation radical, which then underwent a one-electron reversible oxidation to the stable dication. This was apparently the first evidence for a stable Fe^{IV} derivative of ferrocene.

Subsequently, the same workers studied other cyclopentadienyl complexes. Cyclic voltammetric, chronoamperometric, and spectroscopic studies revealed that nickelocene and the nickelocenium ion were both unstable in basic melts, but nickelocene underwent a one-electron reversible oxidation at −0.165 V in neutral solution. In more acidic solutions the nickelocene was spontaneously oxidized, and the nickelocenium cation was oxidized reversibly to the stable dication at +0.912 V. This dication was found to be stable for many hours (*cf.* its solutions in acetonitrile, where it is readily decomposed by aqueous impurities). The derivatives of ruthenium and titanium exhibit less interesting electrochemistry. In basic melts they behaved in a manner comparable to that seen in other commonly used non-aqueous solvents, *i.e.* ruthenocene was irreversibly reduced by a one-electron step at +0.57 V (there was some indication of deposition of metal, possibly of ruthenium or a ruthenium–aluminium alloy), whilst the titanocene dication underwent a one-electron, reversible reduction at −0.803 V. In acidic melts, the ruthenium derivative underwent a multi-step oxidation, whilst the titanocene dication was again reduced in a one-electron step ($E = 0.67$ V). Possible complexation reactions of the Ti^{III} complex were discussed.

The major driving force behind these studies of metallocenes has been their possible use in photocells of the type:

n-type $GaAs|Cp_2Fe^+/Cp_2Fe, AlCl_3:BPC|$vitreous carbon

Singh and co-workers[80] have shown that photoelectrochemical devices of this type can be constructed which, with a 1:1 $AlCl_3$–BPC solvent, exhibit reasonable behaviour and an open-circuit voltage of 680 mV. The use of other melt compositions resulted in unstable properties; this was rationalized in terms of the band potentials. The overall behaviour of one of these devices was claimed to be at least as good as that of other systems based on non-aqueous solvents.

4 Nitrate Melts

The major topic of interest in these melts has undoubtedly been the reduction of the nitrate ion, and, in particular, the effect of different cations on this process. It has

---

[78] R. J. Gale and R. Job, *Inorg. Chem.*, 1981, **20**, 40.
[79] R. J. Gale and R. Job, *Inorg. Chem.*, 1981, **20**, 42.
[80] P. Singh, R. Singh, K. Rajeshwar, and J. DuBow, *J. Electrochem. Soc.*, 1981, **128**, 1145.

Electrode Processes in Molten Salts

been studied by several groups of workers and by many techniques,[81–90] but, whilst there is little or no disagreement about the results observed, the interpretation remains a matter of dispute. It is therefore perhaps best, first of all, to outline briefly the effects that various groups of workers have seen.

Using either linear-sweep or cyclic voltammetry at a platinum electrode in fused KNO_3 (at 360 °C), the cathodic limit is about -1.5 V vs Ag/Ag^I. When small amounts of $LiNO_3$, $Ca(NO_3)_2$, $Ba(NO_3)_2$, or $Sr(NO_3)_2$ are added, however, a peak is clearly seen at less cathodic potentials, and, from the nature of the corresponding peak on the reverse sweep of a cyclic voltammogram, the reduction clearly results in the formation of an insoluble (or only sparingly soluble) deposit on the electrode. This is thought to be a metal oxide, and its formation potential is found to be less negative the greater is the ratio of the ionic charge to the ionic radius of the cation that is present. Thus the nitrate ion is reduced less readily in the presence of Li^+ ($Z/r = 1.66$ Å^{-1}) than Ca^{2+} ($Z/r = 1.95$ Å^{-1}). Similar observations have also been made for the reduction of WO_4^{2-} ions.[88] It is the cause of this effect that has led to the disagreements between the various researchers.

Tkalenko and co-workers[84,85] consider that the alkali-metal cations in the primary co-ordination sphere of the nitrate ions affect its ease of reduction. Thus, the greater the depolarizing effect of the cation (high charge to radius ratio), the weaker is the N–O bond, and therefore the more readily the nitrate ion will be reduced. They use the same argument to explain the re-oxidation potential of the sparingly soluble metal oxide film, Li_2O being oxidized more readily than CaO since Li^+ ions are less acidic.

Sternberg and Visan[86] have looked at the process in a more detailed way. They have used cyclic voltammetry to determine the charge ratio for the cathodic and anodic steps in the formation and removal of alkali metal oxide films. It was generally found that the ratio of $Q_{cathodic}$ to Q_{anodic} was 1.5. On the basis of this, they propose that the reduction mechanism may be represented by equations (46) and (47) and the oxidation by equation (48).

$$NO_3^- + 2e^- \rightleftharpoons NO_2^- + O^{2-} \qquad (46)$$

$$O^{2-} + M^{2+} \rightarrow MO \qquad (47)$$

$$MO \rightarrow M^{2+} + \tfrac{1}{2}O_2^- + \tfrac{3}{2}e^- \qquad (48)$$

They point out that the mechanism described by equation (49) is also a possible anodic process, though it is rather unlikely, in view of the low concentration of NO_2^- ions.

$$NO_2^- + MO \rightarrow NO_3^- + M^{2+} + 2e^- \qquad (49)$$

---

[81] Yu. K. Delimarski, D. A. Tkalenko, and N. A. Chmilenko, *Elektrokhimiya*, 1981, **17**, 155.
[82] V. D. Prisyazhnyi, D. A. Tkalenko, and N. A. Chmilenko, *Elektrokhimiya*, 1980, **16**, 138.
[83] N. A. Chmilenko and D. A. Tkalenko, *Elektrokhimiya*, 1980, **16**, 1856.
[84] D. A. Tkalenko, S. V. Sazhin, N. A. Chmilenko, and M. P. Novikov, *Elektrokhimiya*, 1980, **16**, 1199.
[85] D. A. Tkalenko and N. A. Chmilenko, *Elektrokhimiya*, 1981, **17**, 816.
[86] S. Sternberg and T. Visan, *Electrochim. Acta*, 1981, **26**, 75.
[87] M. H. Miles and A. N. Fletcher, *J. Electrochem. Soc.*, 1980, **127**, 1761.
[88] N. A. Chmilenko and D. A. Tkalenko, *Elektrokhimiya*, 1980, **16**, 1368.
[89] S. V. Morozov, G. I. Sukhova, I. A. Kedrinskii, and I. I. Grudyannov, *Elektrokhimiya*, 1981, **17**, 886.
[90] V. I. Shapoval and V. F. Lapshin, *Elektrokhimiya*, 1980, **16**, 235.

The variation in the reduction potential of the NO_3^- ion, they argue, can be simply explained in terms of the free enthalpy of formation of the metal oxide, i.e. the formation of an oxide may be considered to depolarize the cathodic process to less negative potentials as the free enthalpy of formation of the oxide becomes more negative. On the basis of the available data, this latter explanation seems to be the most plausible. It is also worth noting that whilst the alkaline-earth oxides and Li_2O are both insoluble and stable in nitrate melts, Na_2O and K_2O (once formed) readily decompose to peroxide and superoxide respectively.

Apart from the above studies, electrochemical investigations in these melts are rather limited in number. Ramasubramanian[91] has studied the formation of oxide films on zirconium and Zircaloy 2 alloy by polarization in binary and ternary melts, and discusses the transport processes occurring in the films. Turner and Lovering[92] have made a similar study for aluminium. They observed that thin amorphous films of alumina were formed in anhydrous melts, provided that the voltage was kept below the breakdown voltage of the film. In wet solvent, at low voltages (less than 50 V), thick white friable films resulted, whilst amorphous films of alumina were again formed at higher voltages.

The influence of water on electrode processes in molten salts is of interest both theoretically and technologically. One such system that has been studied by Lovering and Oblath[93] is the reduction of Ni^{II} at a mercury electrode. Using d.c. and pulse polarography in the $LiNO_3$–KNO_3 eutectic (at 145 °C), they observed that the addition of small amounts of water (less than 5×10^{-3} mol l^{-1}) caused a positive shift in the value of $E_{\frac{1}{2}}$ for the reduction of Ni^{II}, and the process also became more reversible, until at concentrations of water of about 5×10^{-3} mol l^{-1} it was totally reversible. On the addition of further water, the value of $E_{\frac{1}{2}}$ shifted cathodically; in the wettest solutions, it approached the behaviour observed in aqueous solution. The cathodic shift at high water contents is readily explained; the Ni^{II} becomes increasingly co-ordinated by H_2O, which is much less labile than NO_3^-. The anodic shift at low water contents is less readily interpreted. The authors eliminated double-layer effects, as neither the reduction of Pb^{2+} nor that of Tl^+ showed similar behaviour; liquid junction effects were also discounted. The explanation proposed was that in the anhydrous solvent the Ni^{II} is co-ordinated by bidentate NO_3^- ligands but in the presence of traces of water the co-ordination changes to a combination of unidentate NO_3^- ligands and water. The entropy effects are such that the complex is thereby slightly destabilized, making it more readily reduced. As indicated by the authors, this hypothesis should be tested by parallel spectroscopic studies as well as by investigations of similar systems.

The only remaining electrochemical investigation in these melts is that of Desimoni et al.,[94] who have used a platinum RDE to investigate the platinum-catalysed decomposition of NH_3 in $LiNO_3$. They do not, however, appear to have reached any firm conclusions as to the mechanism of the reaction.

---

[91] N. Ramasubramanian, *J. Electrochem. Soc.*, 1980, **127**, 2566.
[92] A. K. Turner and D. G. Lovering, *Trans. Inst. Met. Finish*, 1980, **58**, 109.
[93] D. G. Lovering and R. M. Oblath, *J. Electrochem. Soc.*, 1980, **127**, 1997.
[94] E. Desimoni, F. Panniccia, and P. G. Zambonin, *Ann. Chim. (Rome)*, 1980, **70**, 351.

5 Carbonate Melts

Virtually all the interest in carbonate melts centres around their potential use in hydrogen- and hydrocarbon-fuelled fuel cells, and most of the recent efforts have concentrated on the cathode process, *i.e.* reduction of oxygen, the mechanism of which is still not fully understood. Earlier studies of this were generally made in either pure Li_2CO_3 or the Na_2CO_3–K_2CO_3 eutectic; however, present workers appear to prefer the Li_2CO_3–K_2CO_3 eutectic. The Lux–Flood acid–base properties of carbonate melts, where CO_2 is the acid and O^{2-} the base, are greatly affected by the composition of the melt; *e.g.*, increasing the concentration of Li^+ greatly increases that of O^{2-}. This is, of course, very important when trying to understand the reduction of oxygen, since oxygen readily reacts with species in the melt, as shown in equations (50) and (51).

$$O_2 + 2CO_3^{2-} \rightleftharpoons 2O_2^{2-} + 2CO_2 \qquad (50)$$

$$O_2 + O_2^{2-} \rightleftharpoons 2O_2^{-} \qquad (51)$$

Appleby and Nicholson[95] have used linear-sweep voltammetry at a gold electrode in the Li_2CO_3–K_2CO_3 eutectic to investigate the mechanism of reduction of oxygen. The behaviour they observed was essentially the same as that seen in the $NaCO_3$–K_2CO_3 system; *i.e.* two reduction processes were observed. The first was assigned to formation of peroxide according to a mechanism involving the reaction described by equation (50), with the CO_3^{-} ions being regenerated according to the reactions defined by equations (52)—(54), and in which equation (52) described the rate-determining step.

$$O_2^{2-} + e^- \rightarrow O^{2-} + (O^-)_{ads} \qquad (52)$$

$$(O^-)_{ads} + e^- \rightarrow O^{2-} \qquad (53)$$

$$2O^{2-} + 2CO_2 \rightarrow 2CO_3^{2-} \qquad (54)$$

The second step was due to formation of superoxide ion, the peroxide ion reacting with oxygen according to equation (51) and being regenerated according to equation (55).

$$O_2^- + e^- \rightarrow O_2^{2-} \qquad (55)$$

White and Bower[96] have also considered the above reduction process, though in this case in the ternary melt Li_2CO_3–Na_2CO_3–K_2CO_3. From the results of their cyclic-voltammetric and steady-state experiments they have estimated the activation energy for the reduction process, and they concluded that it is not diffusion-controlled. Some results in the past have indicated that very much higher currents are observed at silver rather than at gold electrodes. Appleby and Nicholson[97] have investigated this property and concluded that it was due to molecular oxygen dissolved in the silver.

Turning now to the reaction of the fuel cell anode, Vogel *et al.*[98] have made a

---

[95] A. J. Appleby and S. B. Nicholson, *J. Electroanal. Chem. Interfacial Electrochem.*, 1980, **112**, 71.
[96] S. M. White and M. H. Bower, 158th Meeting Electrochem. Soc., 1980, Extended Abstr. No. 654, p. 1632.
[97] A. J. Appleby and S. B. Nicholson, *J. Electrochem. Soc.*, 1980, **127**, 759.
[98] W. M. Vogel, L. J. Bregoli, and S. W. Smith, *J. Electrochem. Soc.*, 1980, **127**, 833.

detailed study of the electrochemical oxidation of H_2 and of CO. The electrode reactions of these molecules in carbonate melts can be represented by equations (56) and (57); however, these are also coupled by the so-called shift reaction [equation (58)], which is the electrically neutral sum of these.

$$H_2 + CO_3^{2-} \rightleftharpoons H_2O + CO_2 + 2e^- \quad (56)$$

$$CO + CO_3^{2-} \rightleftharpoons 2CO_2 + 2e^- \quad (57)$$

$$H_2 + CO_2 \rightleftharpoons CO + H_2O \quad (58)$$

In view of this coupling, it is impossible to vary the partial pressures of the above species independently, and also impossible to study the oxidation of either H_2 or CO independently of the other. By working at very low pressures of CO, the authors were, however, able to obtain data for the oxidation of H_2 at a number of electrodes, with little interference from CO. They concluded that the reaction proceeded with little activation polarization at Ni, Au, Pt, and Pt/Rh electrodes, and that the observed overvoltages were due to the diffusion of either the H_2 or the reaction products, *i.e.* H_2O and O_2. The oxidation of CO, on the other hand, was found to be fairly complex. At low polarization voltages (< 200 mV) it was very slow, and it was thought that the reaction that occurs is the oxidation of dissolved molecular CO. At higher polarizations, other processes occur, probably the oxidation of species formed by the reaction of the CO with the melt. The overall conclusion from this study was that H_2 was to be preferred as a fuel, owing to the low solubility and the low reaction rate of CO. Also with relevance to this anode reaction, Townley *et al.*[99] have investigated the effects of H_2S, which is likely to be found as an impurity in the stream of H_2 gas.

6 Other Molten-salt Systems

In this final section, studies of molten-salt electrochemistry that do not conveniently fit into any of the categories previously considered will be discussed.

Tumanova and co-workers have used a.c. impedance techniques to investigate the double-layer capacitance of platinum electrodes, and in particular its temperature dependence, in various molten salts. In fused $KClO_4$–$LiClO_4$, KNO_2–$LiNO_2$, and KSCN–NaSCN, Tumanova and Delimarski[100] found that the magnitude of the differential double-layer capacitance varied in proportion to the temperature, as has previously been observed for molten halides. In KNO_3–$NaNO_3$, however, it was essentially invariant of temperature. At the present time there appears to be no satisfactory explanation for this difference in behaviour. Tumanova and Cherepanov[101,102] have investigated the adsorption of organic species at platinum electrodes from thiocyanate melts. Urea, phthalimide, and carbamine all behaved as molecular adsorbates whilst benzidine sulphate was shown to be an ionic surfactant in this solvent. Following on from this latter study, Cherepanov and Tumanova[103] have investigated the effects of adsorbates upon the

---

[99] D. Townley, J. Winnicot, and H. S. Huong, *J. Electrochem. Soc.*, 1980, **127**, 1104.
[100] N. K. Tumanova and Yu. K. Delimarski, *Electrochim. Acta*, 1981, **26**, 1737.
[101] N. K. Tumanova and D. S. Cherepanov, *Elektrokhimiya*, 1981, **17**, 413.
[102] N. K. Tumanova and D. S. Cherepanov, *Ukr. Khim. Zh. (Russ. Ed.)*, 1981, **47**, 798.
[103] D. S. Cherepanov and N. K. Tumanova, *Ukr. Khim. Zh. (Russ. Ed.)*, 1981, **47**, 120.

anodic dissolution of metals in thiocyanates. In the absence of additives, iron, zinc, cadmium, and lead were all passivated by sulphide films. The addition of urea or benzidine sulphate prevented passivation of zinc, whilst urea also prevented the passivation of cadmium, and, to a lesser extent, of lead.

Sedea and Fiorani[104] have investigated the electrochemistry of Na_2S_4 and Na_2S in the KSCN–NaSCN melt, with a view to the possible use of these sulphides in batteries. In agreement with other investigators, their cyclic voltammetry of the pure melt indicated that the oxidation of SCN^- ions results initially in the formation of $SCN^{\cdot}$ radicals, which then undergo further chemical processes, whilst reduction results in the formation of sulphide and cyanide ions. Tetrasulphide ions are known to decompose readily in these melts, as shown in equation (59), and the radical anions that are formed were shown to undergo a reversible one-electron oxidation with resultant deposition of polymeric sulphur on the electrode.

$$S_4^{2-} \rightarrow 2\,S_2^{-} \tag{59}$$

It has previously been though that S^{2-} ions, at temperatures in excess of 200 °C, react with the melt. However, at the temperatures used in this study (150 to 190 °C) they were stable, and three poorly defined oxidation processes were observed and assigned as follows:

$$2\,S^{2-} \rightarrow S_2^{2-} + 2\,e^- \tag{60}$$

$$S_2^{2-} \rightarrow S_2^{-} + e^- \tag{61}$$

$$S_2^{-} \rightarrow S_2 + e^- \tag{62}$$

It was also noted that, as time passed, there was a build-up of polymeric sulphur in the solution, which was itself oxidizable to a sulphur (II) species. Cherepanov et al.[105] have used chronopotentiometry to study the reduction of Cd^{2+} in KSCN–NaSCN as well as the reduction of Ag^+ in nitrate media and the reduction of Co^{2+} in halide melts. In particular, they considered the effects of ionic and molecular surfactants such as benzidine sulphate and urea on these deposition processes.

The possibility that they could be used as oxidizing electrolytes in thermal batteries provided the stimulus for the investigation of molten chlorates, perchlorates, and perchlorate–nitrate mixtures by Miles and Fletcher.[106] The cathodic limits in these melts were found to be greatly influenced by the electrode material; e.g., in $LiClO_4$, at 300 °C, a reduction process was observed on nickel electrodes at about -0.5 V vs Ag/Ag^I (0.1M) whereas on platinum the limit was over half a volt more negative. The mechanism given by the authors for the reduction of $LiClO_4$ was that defined by equations (63) and (64), in which the first step was rate-determining, whilst reduction of chlorate was thought to follow a similar scheme, involving an adsorbed (ClO_2) species.

$$LiClO_4 + Li^+ + e^- \rightarrow (ClO_3)_{ads} + Li_2O \tag{63}$$

$$(ClO_3)_{ads} + e^- \rightarrow ClO_3^- \tag{64}$$

It was observed that, at open circuit, bubbles of gas formed on nickel electrodes,

---

[104] M. Andreuzzi-Sedea and M. Fiorani, Ann. Chim. (Rome), 1980, **70**, 341.
[105] D. S. Cherepanov, S. I. Chernukhin, and N. K. Tumanova, Ukr. Khim. Zh. (Russ. Ed.), 1981, **47**, 563.
[106] M. H. Miles and A. N. Fletcher, J. Electrochem.Soc., 1981, **128**, 821.

indicating that decomposition of the melt had occurred. This was thought to be due to the catalytic properties of the nickel oxide layer that spontaneously forms on nickel in this melt, and which also aids the electrochemical reduction of $LiClO_4$. The mechanism for reduction of $LiClO_4$ has similarities with that for the reduction of nitrate ions discussed in a previous section. In common with this process, and for the same reasons, it was found that the presence of cations with a large ratio of charge to radius reduced the cathodic limit of the melt.

Durand et al.[107] have studied the electrochemistry of vanadium cations in molten potassium disulphate at 430 °C. This system is of particular interest, as it is generally though that V^V, dissolved in potassium disulphate, is in fact the catalyst in the oxidation of SO_2 by the V_2O_5/K_2SO_4 system. Using voltammetry at a gold RDE, they found that V^V undergoes a reversible one-electron reduction to V^{IV} ($E^\ominus =$ 0.26 ± 0.1 V vs Hg^I/Hg^{II} at $pSO_4 = 3.2$). From potentiometric and voltammetric measurements it was concluded that V^V and V^{IV} were present as $VO_2(SO_4)_2^{3-}$ and $VO(SO_4)_3^{4-}$, respectively, in basic melts whilst in anodic solution the relevant species were $VO_2SO_4^-$ and $VOSO_4$. The oxidation of SO_2 to SO_3 is discussed in terms of these entities.

Hayashi et al.[108] have used the cell:

$$Na|\beta\text{-alumina}|NaOH(Na_2O)|ZrO_2\text{-}Y_2O_3|O_2(Pt)$$

to study the equilibrium described by equation (65) in molten NaOH.

$$2\,OH^- \rightleftharpoons H_2O + O^{2-} \tag{65}$$

They demonstrated that the stabilized zirconia electrode can be used as an indicator of oxide ion in hydroxide melts, and determined the equilibrium constant for the above reaction as 1×10^{-10}. Zarubitskii[109] has reviewed the literature of electrochemistry in molten hydroxides up to 1975. The bulk of the review is concerned with the acid–base properties of the melt itself, and there is little if anything about electrode processes.

Combes et al.[110] have investigated the properties of molten $SnCl_2$, a little-studied system, and in particular the effect of adding oxide ions. From cyclic-voltammetric studies of the pure melt, at 523 K, they showed that the electrochemical window between deposition of tin and the oxidation of Sn^{II} to Sn^{IV} was about 0.5 V, in agreement with a value calculated from thermodynamic considerations. On the addition of oxide ions, the anodic process changes the formation of SnO_2. From the electrochemical results and available thermodynamic data, a Pourbaix-type diagram for the dependence of concentrations of tin, Sn^{II}, and Sn^{IV} species on pO^{2-} was constructed.

Sorlie and co-workers have used molten $SbCl_3$ as a solvent for aromatic compounds, and have recently investigated the electrochemistry of perylene in this medium,[111] using cells similar to those designed by Mamantov and his group (discussed earlier) that permit simultaneous electrochemical and spectroscopic study. Fused $SbCl_3$ is itself only slightly ionized; however, with the addition of

---

[107] A. Durand, G. Picard, and J. Vedel, *J. Electroanal. Chem. Interfacial Electrochem.*, 1981, **127**, 169.
[108] H. Hayashi, S. Yoshizawa, and Y. Ito, *J. Electroanal. Chem. Interfacial Electrochem.*, 1981, **124**, 229.
[109] O. G. Zarubitskii, *Usp. Khim.*, 1980, **49**, 1014.
[110] R. Combes, M. Cadi, and B. Gindre, *Electrochim. Acta*, 1980, **25**, 867.
[111] M. Sorlie, G. P. Smith, V. E. Norvell, G. Mamantov, and L. N. Klatt, *J. Electrochem. Soc.*, 1981, **128**, 333.

Electrode Processes in Molten Salts

$AlCl_3$, which is a Lewis acid, ionization according to the mechanism in equation (66) is stimulated, whilst the addition of a Lewis base results in the production of free Cl^- ions.

$$AlCl_3 + SbCl_3 \rightleftharpoons AlCl_4^- + SbCl_2^+ \quad (66)$$

For electrochemical studies, glassy carbon was found to be the most suitable electrode material; platinum was also tried, but was found to have a smaller potential window, possibly due to the formation of a Pt–Sb alloy. An antimony wire in a separated compartment containing fused $SbCl_3$ saturated with $Cs_3Sb_2Cl_9$ served as the reference electrode. The behaviour of perylene was found to be similar to that observed for aromatic compounds in chloroaluminate melts, i.e. in basic melts there was a one-electron reversible oxidation ($E_\frac{1}{2}$ =0.26 V) to the stable cation radical, whilst in acidic media this reaction occurred spontaneously, according to the reaction in equation (67), and the cation radical was then reversibly oxidized ($E_\frac{1}{2}$ =0.97 V) to the dication.

$$3\,PeH + 3\,SbCl_2^+ \rightleftharpoons 3\,PeH^{\cdot+} + Sb + 2\,SbCl_3 \quad (67)$$

As was seen in an earlier part of this Report, low-temperature (and in particular room-temperature) systems are being increasingly investigated. A system of this type, based upon triethylammonium chloride and CuCl, has been reported by Silkey and Yoke,[112] and they gave details of its physical properties, such as density, viscosity, and conductivity. Its general usefulness must, however, be questioned in view of its small electrochemical window of about 0.7 V, the limiting reactions being deposition of copper at the cathode and oxidation of Cu^I to Cu^{II} at the anode. Apart from these processes, no other electrode reactions appear to have been investigated, though a galvanic cell with an open-circuit voltage of 0.75 V was constructed; this had a poor performance, being limited by the high viscosity of the melt. Another system with a relatively low melting point, which has been studied by John and Tissot,[113] is the eutectic mixture of CF_3COOK and CF_3COONa (65:35 w/o; m.pt. 116 °C). This system has been shown to have a reasonable conductivity and large electrochemical windows of 5.3, 5.2, and ~ 8 V on Pt, Au, and pyrolytic graphite electrodes, respectively, if some very-low-current-density processes, which are thought to be due to the reduction of species such as CF_3COF, CF_3COOH, and CF_3CHO (the latter two being formed by the reaction of traces of water), are ignored. On all electrodes, the limiting cathodic process was reduction of Na^+ (at -3.2 V vs Ag/Ag^+), whilst the anodic one was oxidation of the trifluoroacetate anion, as shown in equation (68).

$$CF_3COO^- \rightarrow CF_3^{\cdot} + CO_2 + e^- \quad (68)$$

Prolonged oxidation of the melt resulted in the formation of C_2F_6, CF_4, and a little C_3F_8, and possible routes to these species are discussed. Other processes that were studied in this melt included the oxidation of acetate anions, which was ill-defined, and occurred at a potential about 0.5 V less than the value for the trifluoroacetate anion, and the reduction of Pb^{2+}, Cu^{2+}, and Fe^{3+}. All these metals were

---

[112] J. R. Silkey and J. T. Yoke, *J. Electrochem. Soc.*, 1980, **127**, 1091.
[113] O. John and P. Tissot, *J. Appl. Electrochem.*, 1980, **10**, 593.

electrodeposited; in the case of lead there was also an underpotential deposit formed, whilst Fe^{3+} was first reduced to Fe^{2+}. All of these deposition reactions occurred at the anticipated potentials.

Podgornov et al.[114] have investigated the anodic polarization of carbon in molten sodium borosilicate as part of a general study of the oxidation of carbon in oxide melts, whilst Takahashi and Miura[115] have studied a number of metal ions in molten sodium borate. Cadmium, zinc, and nickel were all reduced to the metal without apparently being deposited, whilst lead and thallium were reversibly deposited. A number of redox couples, including Fe^{3+}/Fe^{2+}, Cr^{6+}/Cr^{3+}, and As^{5+}/As^{3+}, were also investigated.

[114] A. D. Podgornov, B. A. Kukhtin, and G. A. Korlev, *Elektrokhimiya*, 1980, **16**, 1566.
[115] K. Takahasi and Y. Miura, *Yogyo Kyokaishi*, 1981, **89**, 107.

3
The Electrochemistry of Transition-metal Complexes

BY C. J. PICKETT

1 Introduction

This Report nominally covers the literature for the years 1980 and 1981. Although the area of electrochemistry concerned with transition-metal complexes has developed rapidly in the past fifteen years or so, there have been no comprehensive surveys recently apart from the thorough article by Denisovitch and Gubin, which deals with organometallic electrochemistry up to about 1975.[1]

The rules of nomenclature defined by IUPAC are used throughout, except where semi-structural formulations are more appropriate or where well-established trivial names are more easily appreciated than their more cumbersome rational counterparts. Numerous papers in the primary inorganic literature contain summary cyclic voltammetric data and comments upon electrochemical reversibility. In the notation of this Report, electrochemically reversible systems are those which satisfy the usual criteria[2] at scan rates < 500 mV s^{-1} and at temperatures of 20 ± 5 °C; partially reversible systems are those which satisfy these criteria at higher rates of scan or lower temperatures, whilst electrochemically irreversible reactions are those in which only one member of a redox couple is observed by cyclic voltammetry.

In the interest of economy of space, precise experimental details of electrochemical reactions are not given unless they are considered to be particularly unusual. The extensive tabulation of redox data is also inappropriate to a review which attempts to survey comprehensively rather than to detail selectively this wide area of electrochemistry. Complexes of Group 1B are not discussed.

2 The Early Transition Metals

The organometallic and co-ordination chemistry of the early transition metals has developed steadily in recent years,[3] with a new emphasis on exploring the chemistry of the heavier elements in Groups 4 and 5. Detailed electrochemical studies of complexes are few, and the behaviour of even the relatively simple M^{IV} halogeno- and organo-metallocenes is now somewhat confused.

Titanium, Zirconium, and Hafnium.—The primary one-electron reduction of [Ti(η^5-C$_5$H$_5$)$_2$Cl$_2$] in non-aqueous electrolytes has been the subject of some controversy between two groups of French workers. The disagreement centres

---

[1] L. I. Denisovitch and S. P. Gubin, *Russ. Chem. Rev.*, 1977, **46**, 27.
[2] A. J. Bond and L. R. Faulkner, 'Electrochemical Methods', Wiley, New York, 1980.
[3] R. R. Schrock and G. W. Parshall, *Chem. Rev.*, 1976, **76**, 243.

around the electrochemical reversibility of this reaction in, primarily, THF and DMF electrolytes at inert cathodes. On the one hand, the reduction is reported to yield the undissociated anion in an electrochemically reversible step;[4,5] on the other hand, fast loss of Cl$^-$ is reported to follow the initial one-electron transfer[6] (see Scheme 1).

$$[\text{Ti}(\eta^5\text{-C}_5\text{H}_5)_2\text{Cl}_2] \xrightleftharpoons[]{+e^-} [\text{Ti}(\eta^5\text{-C}_5\text{H}_5)_2\text{Cl}_2]^- \xrightleftharpoons[-(S)]{(S)} [\text{Ti}(\eta^5\text{-C}_5\text{H}_5)_2\text{Cl}(S)] + \text{Cl}^-$$

[(S) = solvent]

Scheme 1

There have been many electrochemical investigations of the reduction of [Ti(η^5-C$_5$H$_5$)$_2$Cl$_2$] over the past fifteen years or so, and in view of the importance of this compound, both in its own right and to the chemistry of metallocenes of Group 4 in general, a definitive investigation would be worthwhile.

The electrochemistry of various alkyl and aryloxy derivatives of [M(η^5-C$_5$H$_5$)$_2$X$_2$] (M = Ti, Zr, or Hf; X = halide) has been briefly reported.[7—11] After a chemical and e.s.r. study of compounds of the type [M(η^5-C$_5$H$_5$)$_2$R$_2$] (R = alkyl) it was claimed that their reduction gave [M(η^5-C$_5$H$_5$)$_2$R$_2$]$^-$ anions;[12] however, a recent electrochemical investigation suggests that, following transfer of a single electron, a cyclopentadienyl anion is lost, as shown in Scheme 2.[7]

$$[\text{Ti}(\eta^5\text{-C}_5\text{H}_5)_2(\text{CH}_2\text{Ph})_2] \xrightarrow{+e^-} [\text{Ti}(\eta^5\text{-C}_5\text{H}_5)(\text{CH}_2\text{Ph})_2] + [\text{C}_5\text{H}_5]^-$$

Conditions: Pt electrodes, −1.86 V *vs* the S.C.E., THF, [NBu$_4$]$^+$ [PF$_6$]$^-$, at 20 °C

Scheme 2

Alkyl derivatives of zirconium(IV) such as [Zr(η^5-C$_5$H$_5$)$_2$(CH$_2$Ph)$_2$] show irreversible one-electron reduction in THF electrolytes;[8] however, the 'bulky ligand' complexes [Zr(η^5-C$_5$H$_5$)$_2${CH(SiMe$_3$)(C$_6$H$_4$Me-*o*)}$_2$] show reversible one-electron reductions by cyclic voltammetry under similar conditions.[9] Brief studies of bis-cyclopentadienyl metallocycles of titanium, zirconium, and hafnium have been reported.[10,11]

Laviron and co-workers have studied the electrochemical oxidation of 'titanocene(III) monochloride', [Ti(η^5-C$_5$H$_5$)$_2$Cl(L)] (L = THF or PMe$_2$Ph), by a variety of techniques and have interpreted their results according to Scheme 3.[13]

Tuck and co-worker have reported the electrosynthesis of some 26

---

[4] N. El Murr, A. Chaloyard, and J. Tirouflet, *J. Chem. Soc., Chem. Commun.*, 1980, 446.
[5] N. El Murr and A. Chaloyard, *J. Organomet. Chem.*, 1981, **212**, C39.
[6] Y. Mugnier, C. Moise, and E. Laviron, *J. Organomet. Chem.*, 1981, **204**, 61.
[7] A. Chaloyard, A. Dormond, J. Tirouflet, and N. El Murr, *J. Chem. Soc., Chem. Commun.*, 1980, 214.
[8] M. F. Lappert, C. J. Pickett, P. I. Riley, and P. I. W. Yarrow, *J. Chem. Soc., Dalton Trans.*, 1981, 805.
[9] M. F. Lappert and C. L. Raston, *J. Chem. Soc., Chem. Commun.*, 1981, 173.
[10] M. S. Holtman and E. P. Schram, *J. Organomet. Chem.*, 1980, **187**, 147.
[11] M. F. Lappert and C. L. Raston, *J. Chem. Soc., Chem. Commun.*, 1980, 1284.
[12] M. F. Lappert, P. I. Riley, and P. I. W. Yarrow, *J. Chem. Soc., Chem. Commun.*, 1979, 305.
[13] Y. Mugnier, C. Moise, and E. Laviron, *J. Organomet. Chem.*, 1981, **210**, 69.

$$[\text{TiCl}(\eta^5\text{-C}_5\text{H}_5)_2(\text{L})] \underset{+e^-}{\overset{-e^-}{\rightleftharpoons}} [\text{TiCl}(\eta^5\text{-C}_5\text{H}_5)_2(\text{L})]^+$$

$$[\text{TiCl}(\eta^5\text{-C}_5\text{H}_5)_2(\text{L})]^+ \xrightarrow[\text{(fast)}]{[\text{TiCl}(\eta^5\text{-C}_5\text{H}_5)_2(\text{L})]} [\text{Ti}(\eta^5\text{-C}_5\text{H}_5)_2\text{Cl}_2] + [\text{Ti}(\eta^5\text{-C}_5\text{H}_5)_2(\text{L})_2]^+$$

(L = THF or PMe$_2$Ph)

Conditions: Pt electrode, THF electrolyte

Scheme 3

organometallic compounds of zirconium, titanium, and hafnium *via* anodic dissolution of the transition metal,[14] as shown in Scheme 4.

$$\text{Hf (anode)} \xrightarrow{\text{i}} [\text{HfBr}_2\text{Et}_2]\cdot 2\text{MeCN}$$

$$\text{Zr(anode)} \xrightarrow{\text{ii}} [\text{ZrBr}_2(\text{PhCH}_2)_2(\text{bipy})]$$

Reagents: i, MeCN, EtBr, [NEt$_4$]$^+$ [Br]$^-$; ii, MeCN, PhCH$_2$Br, bipyridyl, MeOH, [NEt$_4$]$^+$ [Br]$^-$

Scheme 4

Vanadium, Niobium, and Tantalum.—There have been few recent studies of the electrochemistry of vanadium complexes. Vanadium(III) and vanadyl (V^{IV}=O) acetylacetonato-complexes have been investigated, and some electrochemistry of [(V=O)L] and [VCl$_2$L] complexes [L = NN'-bis(salicylidene)ethylenediamine] has been described.[15,16]

Two studies have dealt with the electrochemical properties of Nb^V=O porphyrinates of the type [NbO(OEP)] and [NbO(TPP)]; the authors report their reversible one-electron reduction to the Nb^{IV}=O complexes.[17,18]

Seven-co-ordinate carbonyl complexes of niobium(I) and tantalum(I), [M(CO)$_2$(dppe)$_2$X] (M = Nb or Ta, X = halide) and [TaX(CO)$_2$(dmpe)$_2$] (X = halide, Me, or H), undergo reversible one-electron oxidation at a platinum anode to give stable tantalum(II) radical cations (Scheme 5).[19,20] It is suggested that the stability of these cations and of related bis(tertiary phosphino-alkane) complexes may be explained in terms of an 'intra-molecular disproportionation' whereby the metal retains a 'stable' eighteen-electron configuration.[19]

---

[14] F. F. Said and D. G. Tuck, *Can. J. Chem.*, 1980, **58**, 1673.
[15] M. A. Nawi and T. L. Riechel, *Inorg. Chem.*, 1981, **20**, 1974.
[16] A. Kaptunkiewicz, *Inorg. Chim. Acta*, 1981, **53**, L77.
[17] R. Guilard, P. Richard, M. El Borai, and E. Laviron, *J. Chem. Soc., Chem. Commun.*, 1980, 516.
[18] Y. Matsuda, S. Yamada, T. Goto, and Y. Murakami, *Bull. Chem. Soc. Jpn.*, 1981, **54**, 452.
[19] A. M. Bond, J. W. Bixler, E. Mocellin, S. Datta, E. J. James, and S. S. Wreford, *Inorg. Chem.*, 1980, **19**, 1760.
[20] F. G. N. Cloke, P. J. Fyne, M. L. H. Green, M. J. Ledouxi, A. Gourdon, and C. K. Prout, *J. Organomet. Chem.*, 1980, **198**, C69.

$$[TaCl(CO)_2(dmpe)_2] \underset{+e^-}{\overset{-e^-}{\rightleftharpoons}} [TaCl(CO)_2(dmpe)_2]^+$$

Conditions: acetone or CH_2Cl_2, $[NEt_4]^+ [ClO_4]^-$

Scheme 5

3 Chromium, Molybdenum, and Tungsten

The co-ordination and organometallic chemistry of metals of Group 6A continues to be one of the most productive areas of inorganic research, and an area which has attracted many detailed electrochemical investigations.

Molybdenum chemistry has received a strong impetus from the recognition that this element is involved in various enzymes which catalyse redox reactions of substrates, *e.g.* aldehyde oxidase and nitrogenase, and consequently there have been many electrochemical studies of complexes which are considered to be relevant to the structure and function of these molybdo-enzymes.

Complexes that Contain M=O and Related Groups.—The molybdo-enzymes sulphite oxidase, xanthine oxidase and dehydrogenase, aldehyde oxidase, and nitrate reductase are closely related,[21] and each is involved in oxygen-atom-transfer reactions, *e.g.* reaction (1).

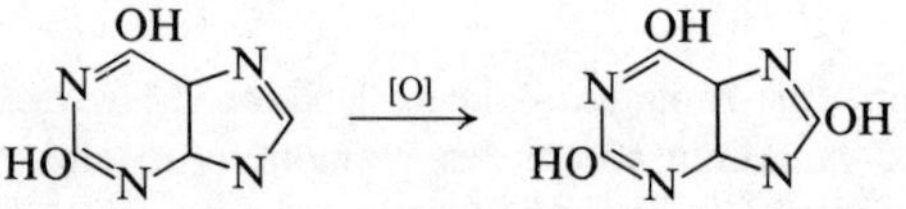

Xanthine → [O] → Uric acid (1)

Recent EXAFS studies suggest that terminal oxo- and thiolato-ligands are co-ordinated to molybdenum in these enzymes, and earlier e.s.r. studies indicate that Mo^{IV}, Mo^V, and Mo^{VI} are involved in their redox transformations.[22]

The reduction of monomeric Mo^{VI} bis-oxo species has been studied in a non-complexing CF_3SO_3H electrolyte. One-electron reduction leads to an unstable Mo^V species which rapidly dimerizes but which can also reduce perchlorate (Scheme 6).[23]

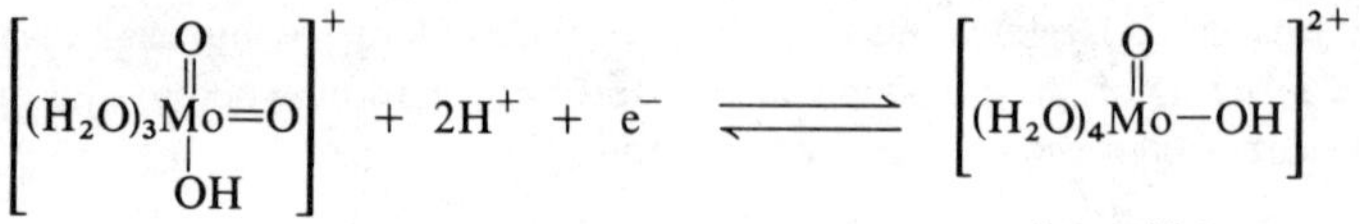

$$\left[(H_2O)_3 \underset{OH}{\overset{O}{\underset{\|}{Mo}}}=O \right]^+ + 2H^+ + e^- \rightleftharpoons \left[(H_2O)_4 \overset{O}{\underset{\|}{Mo}}-OH \right]^{2+}$$
(unstable)

Conditions: 1.9M-CF_3SO_3H, mercury electrode

Scheme 6

---

[21] J. A. Pateman, D. J. Core, B. M. Reva, and D. B. Roberts, *Nature (London)*, 1964, **201**, 58.
[22] S. P. Cramer, H. B. Gray, and K. V. Rajagopalan, *J. Am. Chem. Soc.*, 1979, **101**, 2772.
[23] M. T. Paffett and F. C. Anson, *Inorg. Chem.*, 1981, **20**, 3967.

Catecholato-complexes that possess cis-oxo ligands undergo similar electrochemical reduction reactions, as shown in Scheme 7.[24]

$$[Mo(O)_2(cat)_2]^{2-} + 2H^+ + e^- \rightleftharpoons [Mo(O)(cat)_2(H_2O)]^- + H_2O$$

$$[Mo(O)(cat)_2(H_2O)]^- + 2H^+ + 2e^- \rightleftharpoons [Mo(cat)_2(H_2O)_2]^-$$

Conditions: Mercury electrode, buffered aqueous electrolyte

Scheme 7

Reports of the electrochemistry of other $Mo(O)_2$ complexes that contain thiohydroxamato and Schiffs-base ligands have also appeared.[25–27]

Complexes $[MoO(SR)_4]^-$ (R = alkyl or aryl) and their tungsten and selenium analogues undergo reversible one-electron reductions in non-aqueous electrolytes to give more or less stable Mo^{IV} and W^{IV} complexes. Their one-electron oxidations tend to be irreversible except at low temperature, as shown in Scheme 8.[28, 29]

$$[WO(SPh)_4]^{2-} \underset{[E_{1/2} = -0.92 \text{ V}]}{\xrightarrow[\text{[at 25 °C]}]{+e^-}} [WO(SPh)_4]^- \underset{[E_{1/2} = 0.00 \text{ V}]}{\xrightarrow[\text{[at } -40 \text{ °C]}]{-e^-}} [WO(SPh)_4]^0$$

Conditions: DMF, $[NEt_4]^+$ $[PF_6]^-$, Pt electrode; values of $E_{1/2}$ are versus the S.C.E.

Scheme 8

Other monomeric Mo^V=O species show similar one-electron-transfer processes in DMF electrolytes.[30]

Complexes of Mo=O with tertiary phosphines and with 8-hydroxyquinoline, 8-mercaptoquinoline, o-phenanthroline, or bipyridyl co-ligands have been studied by cyclic voltammetry (see Scheme 9) and the reduction of nitrate by electrogenerated

$$cis\text{-}[Mo^VOCl_3(bipy)] \underset{-e^-}{\overset{+e^-}{\rightleftharpoons}} [Mo^{IV}OCl_3(bipy)]$$
$$[E_{1/2} = -0.285 \text{ V}]$$

Conditions: DMF, $[NEt_4]^+$ $[Cl]^-$; value of $E_{1/2}$ is versus the S.C.E.

Scheme 9

---

[24] F. A. Schultz and L. M. Charney, *Inorg. Chem.*, 1980, **19**, 1527.
[25] C. A. Cliff, G. D. Fallon, B. M. Gatehouse, K. S. Murray, and P. J. Newman, *Inorg. Chem.*, 1980, **19**, 773.
[26] O. A. Rajan and A. Chakravorty, *Inorg. Chem.*, 1981, **20**, 660.
[27] J. Topich, *Inorg. Chem.*, 1981, **20**, 3704.
[28] J. R. Bradbury, G. R. Hanson, I. W. Boyd, S. F. Gheller, A. G. Wedd, K. S. Murray, and A. M. Bond, in *Chem. Uses of Molybdenum*, Proc. Int. Conf., 3rd, ed. H. F. Barry and P. C. H. Mitchell, Climax Molybdenum Co., Ann Arbor, Michigan, 1979, p. 300.
[29] J. R. Bradbury, A. F. Masters, A. C. McDonell, A. A. Brunette, A. M. Bond, and A. G. Wedd, *J. Am. Chem. Soc.*, 1981, **103**, 1959.
[30] J. T. Spence, M. Minelli, C. A. Rice, N. C. Chastean, and M. Schullane, in *Moly. Chem. Bio. Signif.* [*Proc. Int. Symp.*], ed. W. E. Newton and S. Otsuka, Plenum, New York, 1980, p. 263.

monomeric Mo^V species has been discussed.[30] Complexes of $Mo^V{=}O$ with tetrapyrrole ligands, [Mo=O(L)] (H_3L = 2,3,17,18-tetramethyl-7,8,12,13-tetraethylcorrole) and [Mo=O(OMe)(L')] (L' = 5,10,15,20-tetraphenylporphyrin; TPP), undergo metal-based electron-transfer reactions in a CH_2Cl_2 electrolyte. In addition, the TPP complex shows ligand-centred oxidation and reduction processes which are absent in the more saturated corrin ligand complex [Mo=O(L)].[31,32] The electrochemistry of $Cr^V{=}O$ analogues has also been investigated.[33]

Brief descriptions of cyclic voltammetric studies of complexes that contain the

$$\{Mo \overset{\overset{O}{\|} \diagup \overset{S}{\|}}{\underset{S}{\diagdown \diagup}} Mo\} \text{ and } \{Mo \overset{\overset{S}{\|} \diagup \overset{S}{\|}}{\underset{S}{\diagdown \diagup}} Mo\}$$

units have appeared.[34,35]

Molybdenum–Sulphur–Iron Complexes.—Nitrogenase, the enzyme which catalyses the reduction of N_2 to NH_3, has been the centre of intense investigation over the past fifteen years or so. Recent EXAFS spectroscopy on the Mo–Fe protein component of this enzyme, and on a low-molecular-weight Mo–Fe cofactor that was isolated from it, purport to show that the molybdenum is surrounded by three or four sulphur atoms, with an associated iron atom in the second co-ordination sphere. Determinations of metal and of sulphur in the Mo–Fe cofactor are not very accurate, but their sizes suggest that there are probably at least four iron atoms and four, or more, sulphur atoms associated with each molybdenum centre. The cofactor does not contain amino-acid residues or peptides, and other analyses suggest that there are no haem or flavin prosthetic groups associated with the Mo–Fe protein.[36] With this background, inorganic chemists have devoted considerable attention to the synthesis and characterization of Mo–S–Fe assemblies.

Initial studies were based upon attempts to substitute a molybdenum atom in the cubane-like Fe_4S_4 core of the synthetic 4 Fe,4 S ferredoxins, and this was achieved independently by Holm and Garner and their co-workers, who synthesized complexes of the types (1), (2), and (3). Complexes of type (2) have been studied by

(1)

---

[31] Y. Matsuda, S. Yamada, and Y. Murakami, *Inorg. Chem.*, 1981, **20**, 2239.
[32] Y. Matsuda and Y. Murakami, in *Chem. Uses Molybdenum, Proc. Int. Conf., 3rd*, ed. H. F. Barry and P. C. H. Mitchell, Climax Molybdenum Co., Ann Arbor, Michigan, 1979, p. 270.
[33] Y. Murakami, Y. Matsuda, and S. Yamada, *J. Chem. Soc., Dalton Trans.*, 1981, 853.
[34] K. F. Miller, A. E. Bruce, J. L. Corbin, S. Wherland, and E. I. Stiefel, *J. Am. Chem. Soc.*, 1980, **102**, 5102.
[35] A. W. Edelblut, F. Folting, J. C. Huffman, and R. D. Wentworth, *J. Am. Chem. Soc.*, 1981, **103**, 1927.
[36] 'Nitrogen Fixation', Vols. 1 and 2, ed. W. E. Newton and W. H. Orme-Johnson, University Park Press, Baltimore, 1980.

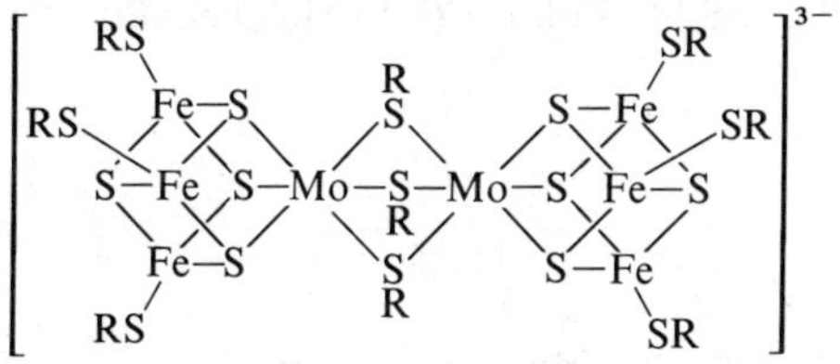

[Structure (2): cluster with formula charge $3-$]

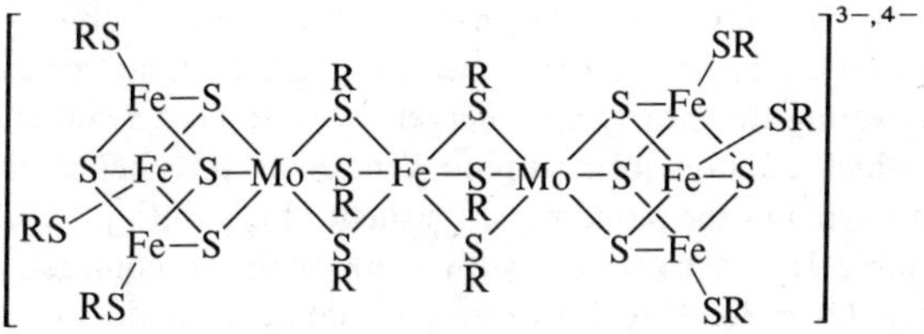

[Structure (3): cluster with formula charge $3-, 4-$]

cyclic and staircase voltammetry, generally in a DMF electrolyte, at a platinum electrode.[37-39] These, and their tungsten analogues, show two reversible one-electron-reduction waves, separated by 1—200 mV, and two reversible one-electron-oxidation waves, separated by about the same amount. At more negative potentials, two less well-defined reductions are also observed; this is summarized in Scheme 10. The small separation in the reversible potentials of the oxidation and

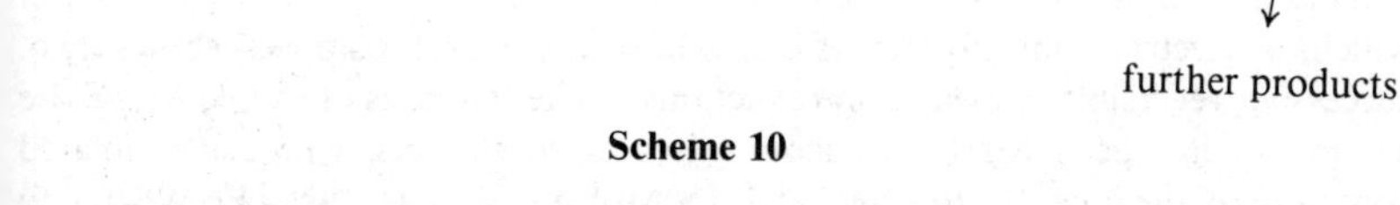

$$\{1-\} \underset{-e^-}{\overset{+e^-}{\rightleftharpoons}} \{2-\} \underset{-e^-}{\overset{+e^-}{\rightleftharpoons}} [Mo_2Fe_6S_8(EtS)_9]^{3-} \underset{-e^-}{\overset{+e^-}{\rightleftharpoons}} \{4-\} \underset{-e^-}{\overset{+e^-}{\rightleftharpoons}} \{5-\}$$

$\downarrow +e^-$ (ill-defined)

$\downarrow +e^-$

further products

Scheme 10

reduction pairs, e.g. the $3-/4-$ couple and the $4-/5-$ couple, together with the Mössbauer and isotropic n.m.r. shift data,[38] suggest that the cubes are only weakly coupled. This is reinforced by the similar electrochemical behaviour of complexes of the type (1).[38] The iron-bridged clusters (3) also show a reversible reduction, associated with the unique iron atom (Scheme 11).

These species are undoubtedly very interesting in their own right, and have X-ray absorption characteristics that are promisingly close to those observed for the

[37] G. Christou, C. D. Garner, R. M. Miller, C. E. Johnson, and J. D. Rush, *J. Chem. Soc., Dalton Trans.*, 1980, 2363.
[38] T. E. Wolff, P. P. Power, R. B. Frankel, and R. H. Holm, *J. Am. Chem. Soc.*, 1980, **102**, 4694.
[39] R. H. Holm, *Chem. Soc. Rev.*, 1981, **10**, 455.

$$[(EtS)_3Fe_3S_4Mo(EtS)_3Fe^{III}(EtS)_3MoS_4Fe_3(SEt)_3]^{3-}$$

$$+e^- \updownarrow -e^-$$

$$\{Fe^{II}\}$$

Conditions: Glassy carbon electrode, MeCN, [NBu$_4$]$^+$ [BF$_4$]$^-$, at ca 25 °C

Scheme 11

enzyme; however, they show little propensity for reaction with dinitrogen, or even with carbon monoxide. Their electrochemistry suggests that they would be ideal as reagents for delivering two nearly 'iso-energetic' electrons to a substrate. Various hydrogenases which contain iron–sulphur centres as prosthetic groups reduce protons to hydrogen, but the monomeric synthetic [Fe$_4$S$_4$(SR)$_4$]$^{3-}$ clusters do not produce detectable H$_2$ on treatment with weak acids. In contrast, the dimeric Fe$_3$S$_4$Mo clusters (2), reduced to their quinqueanionic form, do so,[37] as shown in reaction (2).

$$[Mo_2Fe_6S_8(SPh)_9]^{5-} + PhSH \longrightarrow \tfrac{1}{2}H_2 + [Mo_2Fe_6S_8(SPh)_9]^{4-} + PhS^- \quad (2)$$

The inactivity of the dimeric Mo–Fe clusters towards substrates (N$_2$, CO, HC≡CH, etc.) of nitrogenase is usually rationalized in terms of the lack of an exposed molybdenum site, and various attempts have been made to synthesize monomeric MoFe$_3$S$_4$ clusters with labile ligands on the molybdenum.[40]

Observations that tetrathiomolybdate, MoS$_4^{2-}$, can be isolated by degradation of the Mo–Fe protein have led to the suggestion that assemblies based upon this unit may be involved at the active site in nitrogenase. Studies of the co-ordination chemistry of MoS$_4^{2-}$ and WS$_4^{2-}$ as *ligands* have led to the synthesis of complexes of the type [Pd(MoS$_4$)$_2$]$^{2-}$, [Pd(WS$_4$)$_2$]$^{2-}$, [Pt(MoS$_4$)$_2$]$^{2-}$, and [Ni(MoS$_4$)$_2$]$^{2-}$ in which the metallothiolato-ligand is bidentate. In non-aqueous electrolytes, the platinum complex shows a single, irreversible two-electron reduction; the palladium complexes show two well-separated one-electron reductions, the first of which is electrochemically reversible, whilst the nickel complex shows two, successive, reversible one-electron reductions.[41] The synthesis of simple Mo–S–Fe assemblies has been led by Coucouvanis and co-workers, who have isolated complexes of the type [S$_2$MoS$_2$FeCl$_2$]$^{2-}$, [S$_2$MoS$_2$Fe(SPh)$_2$]$^{2-}$, and [Fe(WS$_4$)$_2$]$^{2-}$; however, the electrochemistry of these species has been only briefly characterized;[42–44] the two Mo–Fe assemblies are reduced irreversibly in an electrolyte of NN-dimethylacetamide at -1.34 and -1.33 V, respectively,[43] whereas the W–Fe–W species is reduced reversibly at -0.15 V *versus* the SCE[42] in a CH$_2$Cl$_2$ electrolyte.

[40] T. E. Wolff, J. M. Berg, and R. H. Holm, *Inorg. Chem.*, 1981, **20**, 175.
[41] K. P. Callaghan and P. A. Piliero, *Inorg. Chem.*, 1980, **19**, 2619.
[42] P. Stremple, N. C. Baenziger, and D. Coucouvanis, *J. Am. Chem. Soc.*, 1981, **103**, 4601.
[43] R. H. Tieckelmann, H. C. Silvis, T. A. Kent, B. M. Huynh, J. V. Waszczak, B-K. Teo, and B. A. Averill, *J. Am. Chem. Soc.*, 1980, **102**, 5550.
[44] D. Coucouvanis, *Acc. Chem. Res.*, 1981, **14**, 201.

Direct electrochemical studies on the two component proteins of nitrogenase itself have so far been impossible, although with the advent of surface mediators such as 4,4'-bipyridyl this may not always be the case. Indirect potentiostatic methods have, however, been employed, and these have been reviewed by Watt.[45] Coulometric studies, using dye mediators, suggest that the Mo–Fe protein in its resting state can be oxidized in two three-electron steps.[45] Reduction of the Mo–Fe protein (to any condition) has not been effected in the absence of the smaller Fe-protein component of nitrogenase as the electron-transfer mediator. Acetylene, a substrate of nitrogenase, has been shown to be reduced to ethylene in an electrochemical reaction that is catalysed by the Mo–Fe cofactor [reaction (3)].[46]

$$C_2H_2 \xrightarrow[\substack{\text{aqueous buffer} \\ \text{Mo–Fe cofactor} \\ \text{Hg electrode}}]{-1.7 \text{ V } versus \text{ the S.H.E.}} C_2H_4 \qquad (3)$$

Complexes of Dinitrogen and its Derivatives.—The chemistry of dinitrogen as a functional group ligating molybdenum has been elucidated almost exclusively by the study of its low-valent tertiary phosphine complexes, e.g. cis-$[Mo(N_2)_2(PMe_2Ph)_4]$. In addition to providing the primary basis for the discussion of mechanisms of action of nitrogenase at the molecular level, the reactions of these complexes and their derivatives with protic acids, alkyl halides, aldehydes, and ketones may afford a route to ammonia, amines, and hydrazines from gaseous N_2 under mild conditions.

The complex trans-$[Mo(N_2)_2(dppe)_2]$ (4) reacts with 1,5-dibromopentane in benzene to give trans-$[MoBr\{NNCH_2(CH_2)_3CH_2\}(dppe)_2]^+$ Br^- (5) in good yield. Electrochemical reduction to (5), in THF, under argon, affords a reactive Mo^{II} species $[Mo\{N\overline{NCH_2(CH_2)_3}CH_2\}(dppe)_2]$ (6) in solution. The reaction of (6) with anhydrous HBr gives the amine piperidine and the imido-complex trans-$[MoBr(NH)(dppe)_2]^+$ $[Br]^-$ (7), both in 60—70% yield. In contrast, electrochemical reduction of (5) under an atmosphere of carbon monoxide or dinitrogen yields the hydrazine N-aminopiperidine in ca 70% yield, together with the Mo^0 carbonyl or dinitrogen complexes $[Mo(CO)_2(dppe)_2]$ (8) and (4), respectively, as shown in Scheme 12. The bearing of these reactions on possible mechanisms of action of nitrogenase and on the conversion of dinitrogen into organo-nitrogen products in a cyclic sequence has been briefly discussed.[47,48]

The complexes trans-$[M(N_2)_2(dppe)_2]$ (M = Mo or W) react with aldehydes or ketones to give diazoalkane complexes, e.g. trans-$[WF(NNCHPh)(dppe)_2]^+[BF_4]^-$ and the cyclic voltammetry of fifteen such complexes with tertiary diphosphine and monophosphine co-ligands has been investigated. In general, each complex shows more or less reversible primary one-electron reduction and oxidation steps in CH_2Cl_2 and MeCN electrolytes, and further irreversible secondary reduction and oxidation processes are also observed at more extreme potentials. The radical and

---

[45] G. D. Watt, in Moly. Chem. Bio. Signif. [Proc. Int. Symp.], ed. W. E. Newton and S. Otsuka, Plenum, New York, 1980, p. 3.
[46] G. N. Petrova, L. M. Uzenskaya, L. A. Syrtsova, and O. N. Efimov, Elektrokhimiya, 1980, 16, 1079.
[47] C. J. Pickett and G. J. Leigh, J. Chem. Soc., Chem. Commun., 1981, 1033.
[48] W. Hussain, G. J. Leigh, and C. J. Pickett, J. Chem. Soc., Chem. Commun., 1982, 747.

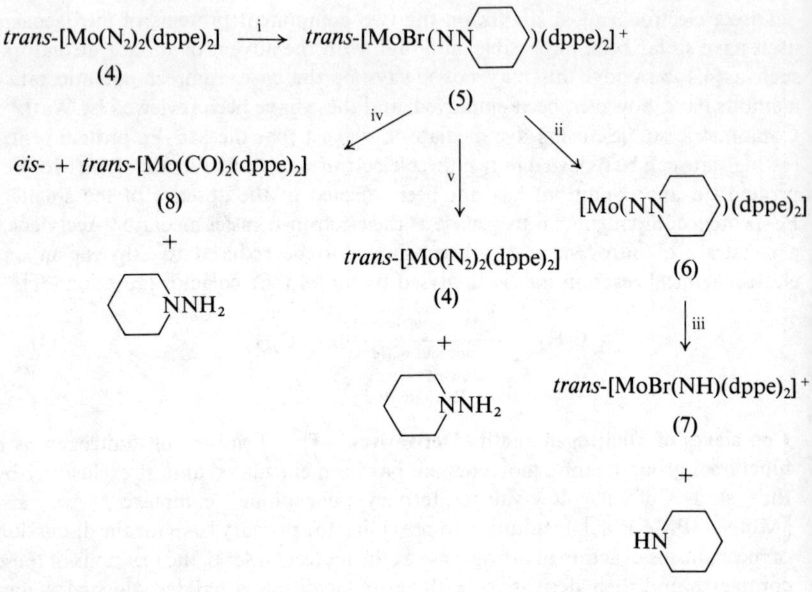

trans-[Mo(N$_2$)$_2$(dppe)$_2$] $\xrightarrow{i}$ trans-[MoBr(NN⟨⟩)(dppe)$_2$]$^+$
(4)   (5)

cis- + trans-[Mo(CO)$_2$(dppe)$_2$]
(8)
+
⟨⟩NNH$_2$

$\xrightarrow{iv}$ ↙   ↓v   ↘^{ii}

[Mo(NN⟨⟩)(dppe)$_2$]
(6)

trans-[Mo(N$_2$)$_2$(dppe)$_2$]
(4)
+
⟨⟩NNH$_2$

↓iii

trans-[MoBr(NH)(dppe)$_2$]$^+$
(7)
+
HN⟨⟩

Reagents: i, Br(CH$_2$)$_5$Br; ii, 2e$^-$, THF, -1.8 V vs the S.C.E., Pt or Hg electrodes, under argon, or LiBu^t; iii, anhydrous HBr; iv, 4e$^-$, THF, -1.8 V vs the S.C.E., Pt or Hg electrodes, under carbon monoxide; v, 4e$^-$, THF, -1.8 V vs the S.C.E., Pt or Hg electrodes, under N$_2$

Scheme 12

radical-dication intermediates were generated by controlled-potential electrolysis and their e.s.r. spectra were recorded. From these data it was concluded that the primary reduction is localized on the nitrogen atoms of the diazoalkane ligands whereas oxidation is metal-centred.[49]

The overall products of electro-reduction of diazoalkane complexes have been investigated by Pickett and George. The one-electron reduction of trans-[WF(N$_2$CH$_2$)(dppe)$_2$]$^+$ [BF$_4$]$^-$ at a mercury pool cathode gives [WF(dppe)$_2$N$_2$CH$_2$CH$_2$N$_2$WF(dppe)$_2$], isolated as a bright yellow solid in ca 55% yield (Scheme 13). In contrast, trans-[WF(N$_2$CMe$_2$)(dppe)$_2$]$^+$ gives both a dimeric species, and, via the transfer of a further electron, a monomeric diazenido-complex. trans-[WF{N$_2$CH(Fc)}(dppe)$_2$]$^+$ [BF$_4$]$^-$ (Fc = ferrocenyl group) gives only mono-

trans-[WF(N$_2$CH$_2$)(dppe)$_2$]$^+$ $\longrightarrow$ [WF(dppe)$_2$N$_2$CH$_2$CH$_2$N$_2$WF(dppe)$_2$]

Conditions: 1e$^-$, -1.8 V vs the S.C.E., Hg electrode, MeCN, [NBu$_4$]$^+$ [BF$_4$]$^-$

Scheme 13

[49] Y. Mizobe, R. Ono, Y. Uchida, M. Hidai, M. Tezuka, S. Moue, and A. Tsuchiya, *J. Organomet. Chem.*, 1981, **204**, 377.

meric diazenido-complexes in an overall two electron-reduction pathway[50] (Scheme 14).

$$[WF(N_2CR_2)(dppe)_2]^+ \xrightarrow{\text{1e pathway}} [WF(dppe)_2N_2CR_2CR_2N_2WF(dppe)_2]$$

$$\xrightarrow[\text{(or RX)}]{\text{2e pathway} \atop H^+} [WF(dppe)_2(N_2CR_2H)]$$

Conditions: THF or MeCN electrolyte, Pt electrode

Scheme 14

Dinitrogen complexes of Mo^0 and W^0 cannot be reduced electrochemically; they do, however, undergo facile oxidation in non-aqueous electrolytes. Thus the complexes trans-$[M(N_2)_2(R_2PCH_2CH_2PR_2)_2]$ and trans-$[M(N_2)(L)(dppe)_2]$ (M = Mo or W, R = alkyl or aryl, L = neutral or anionic two-electron-donor ligand) are oxidized in a THF electrolyte in two well-separated one-electron steps. The primary process is electrochemically reversible whilst the secondary process is more or less so, as shown in Scheme 15.

$$\text{trans-}[Mo(N_2)(L)(dppe)_2] \underset{-e^-}{\overset{+e^-}{\rightleftarrows}} \text{trans-}[Mo(N_2)(L)(dppe)_2]^+$$

$$\downarrow {-e^- \atop +e^-}$$

$$\text{trans-}[Mo(N_2)(L)(dppe)_2]^{2+}$$

(L = e.g., SCN^- or PhCN)

Conditions: Pt electrode, THF, 0.2M-$[NBu_4]^+[BF_4]^-$

Scheme 15

Oxidation potentials for these processes are sensitive to the nature of the chelating phosphine ligand and also to the nature of L, but are not particularly sensitive to the metal (Mo or W). The stability of the M^I–N_2 and M^{II}–N_2 complexes is determined by the rate of loss of N_2 from their co-ordination spheres. The variations in primary reversible oxidation potentials for a series of complexes trans-$[Mo(N_2)(L)(dppe)_2]$ have also been related to the net donicity of L, to the changes in $v(N_2)$, and to the reactivity of these complexes towards protic acids and alkyl halides. Tertiary monophosphine complexes, e.g. cis-$[Mo(N_2)_2(PMe_2Ph)_4]$, are oxidized irreversibly at a platinum electrode in THF electrolytes. Cyclic

---

[50] C. J. Pickett, J. E. Tolhurst, A. Copenhaver, T. A. George, and R. K. Lester, *J. Chem. Soc., Chem. Commun.*, 1982, 1071.

voltammetry has proved to be a convenient method for monitoring the substitution and alkylation reactions of co-ordinated N_2 in various molybdenum and tungsten complexes.[51—54]

Diazenido-complexes trans-$[M(N_2R)X(dppe)_2]$ (X = Br or I, M = Mo or W, R = alkyl) are synthesized from trans-$[M(N_2)_2(dppe)_2]$ and RX, a reaction which proceeds via a rate-determining dissociation of one of the N_2 ligands and homolytic cleavage of the alkyl halide to give a $M^I–N_2$ species and an alkyl radical, which can subsequently couple to form an N–C bond.[55] They are electro-inactive at reducing potentials but undergo reversible one-electron oxidations to give unstable green cation-radicals in THF electrolytes at a platinum electrode;[56] the products of electro-oxidation are not known.

Dinitrogen complexes of molybdenum in a predominantly sulphur-ligand environment are unknown, although dithiocarbamato-complexes that bear organo-diazenido(2 −)- and -hydrazido(2 −)-ligands have been synthesized. The electrochemistry of a series of such complexes, e.g. $[Mo(NNMePh)(S_2CNMe_2)_3]^+$ $[BF_4]^-$ (9), has been investigated by cyclic voltammetry and by controlled-potential electrolysis in non-aqueous electrolytes. Electroreduction of (9), followed by quenching with anhydrous HCl and base work-up, yields the free hydrazine $H_2NNMePh$ in an overall two-electron process. The mechanism of the electroreduction and the relationships between structure and redox potentials for a range of these complexes have been discussed.[57,58]

Russian workers have reported the formation of NH_2NH_2 from N_2 via electrolysis of buffered methanolic solutions of molybdenum(v).[59]

Nitrosyl complexes $[M(NO)(L)(dppe)_2]^y$ (M = Cr, Mo, or W; y = 0; L = F, Cl, Br, I, H, OH, or N_3; and y = 1+ ; L = MeCN or CO, but not all combinations) have been briefly investigated by cyclic voltammetry in non-aqueous electrolytes. All show reversible one-electron oxidations and the variation of $E_{\frac{1}{2}}^{ox}$ for this process with changes in M, L, and ν(NO) has been discussed.[60,61] An unusual redox chemistry of the trispyrazolylborate nitrosyl complex $[Mo\{HB(3,5-Me_2C_3HN_2)_3\}(NO)I_2]$ (10) has been described. Cyclic voltammetry, controlled-potential electrolysis, and e.s.r. studies have shown that it is reduced as shown in Scheme 16.

The authors claim that acetone, alcohols, and even H_2O and pyridine are capable of reducing (10) to the radical anion, although the products of oxidation of the substrate have not been characterized. If this interpretation is correct then it would appear that H_2O could be catalytically oxidized at potentials close to +0.1 V versus

---

[51] J. Chatt, C. T. Kan, G. J. Leigh, C. J. Pickett, and D. R. Stanley, *J. Chem. Soc., Dalton Trans.*, 1980, 2032.
[52] J. Chatt, G. J. Leigh, H. Neukomm, C. J. Pickett, and D. R. Stanley, *J. Chem. Soc., Dalton Trans.*, 1980, 121.
[53] A. J. L. Pombeiro, C. J. Pickett, R. L. Richards, and S. A. Sangokoya, *J. Organomet. Chem.*, 1980, **202**, C15.
[54] J. Chatt, W. Hussain, G. J. Leigh, H. Neukomm, C. J. Pickett, and D. A. Rankin, *J. Chem. Soc., Chem. Commun.*, 1980, 1024.
[55] J. Chatt, R. A. Head, G. J. Leigh, and C. J. Pickett, *J. Chem. Soc., Chem. Commun.*, 1977, 299.
[56] J. Chatt, R. A. Head, G. J. Leigh, and C. J. Pickett, *J. Chem. Soc., Dalton Trans.*, 1978, 1638.
[57] B. A. L. Crichton, J. R. Dilworth, C. J. Pickett, and J. Chatt, *J. Chem. Soc., Dalton Trans.*, 1981, 419.
[58] J. R. Dilworth, B. D. Neaves, and C. J. Pickett, *Inorg. Chem.*, 1980, **19**, 2859.
[59] S. I. Kulakovskaya and O. N. Efimov, *Elektrokhimiya*, 1980, **19**, 2859.
[60] J. S. Nixon, C. J. Pickett, and D. R. Stanley, unpublished results.
[61] S. Clamp, N. G. Connelly, G. E. Taylor, and T. S. Louttit, *J. Chem. Soc., Dalton Trans.*, 1980, 2162.

$$[\text{Mo}\{\text{HB}(3,5\text{-Me}_2\text{C}_3\text{HN}_2)_3\}(\text{NO})\text{I}_2] \underset{+e^-}{\overset{-e^-}{\rightleftharpoons}} [\text{Mo}\{\text{HB}(3,5\text{-Me}_2\text{C}_3\text{HN}_2)_2\}(\text{NO})\text{I}_2]^-$$
(10) $[E_{1/2} = +0.1 \text{ V}]$

Conditions: THF, $[\text{NBu}_4]^+[\text{PF}_6]^-$, Pt or glassy carbon electrodes; the value of $E_{1/2}$ is *versus* the S.C.E.

Scheme 16

the S.C.E.; however, other interpretations of the redox chemistry may be more probable.[62]

Organometallic Compounds.—The reversible one-electron oxidation potentials of closed-shell octahedral complexes $[\text{Cr}(\text{CO})_5(\text{L})]$, where L is either a neutral or an anionic two-electron-donor ligand, have been used to define a free-energy scale of single-valued ligand constants, P_L, according to equation (4).

$$P_L/\text{V} = \{E_{1/2}^{ox}[\text{Cr}(\text{CO})_5(\text{L})] - E_{1/2}^{ox}[\text{Cr}(\text{CO})_6]\}/\text{V} \quad (4)$$

Values of this ligand constant, measured directly *via* the equation or estimated indirectly from other series of complexes, are listed in Table 1. The constant P_L is considered to be a measure of the influence of the ligand L on the binding energy of the electrons in the highest occupied molecular orbitals of the chromium species: that

Table 1 *Values of the ligand constant, P_L, for ligands in the complexes* $[\text{Cr}(\text{CO})_5\text{L}]$

Ligand	P_L/V	Ligand	P_L/V
N^+	+1.46	PhNC	−0.38
NO^+	+1.40	PPh_3	−0.39
PF_3	+0.14	PhCN	−0.40
CO	0.00	PEt_3	−0.47
N_2	−0.07	MeCN	−0.58
PF_2NMe_2	−0.16	pyridine	−0.59
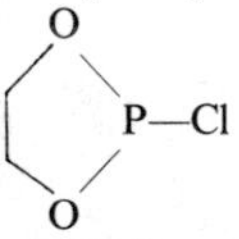 O—P—Cl (cyclic)	−0.16	NH_3	−0.77
		SCN^-	−0.88
		CN^-	−1.00
		$\text{I}^-, \text{Br}^-, \text{Cl}^-$	~ −1.17
		H^-	−1.22
P(OPh)_3	−0.18	OH^-	−1.55
P(OMe)_3	−0.35		

changes in solvation energy in the series do not seriously perturb this conclusion is supported by the good correlation of P_L with measurements of gas-phase vertical ionization potentials for volatile members of the series, as shown in Figure 1. The significance of the ligand constants has been interpreted in terms of the ability of L to donate or to remove charge from the metal centre: ligands which are traditionally considered to be good π-acceptors and poor σ-donors tend to have the more positive values of P_L, and *vice versa*. The constant P_L has been shown to correlate

---

[62] J. A. McCleverty and N. El Murr, *J. Chem. Soc., Chem. Commun.*, 1981, 960.

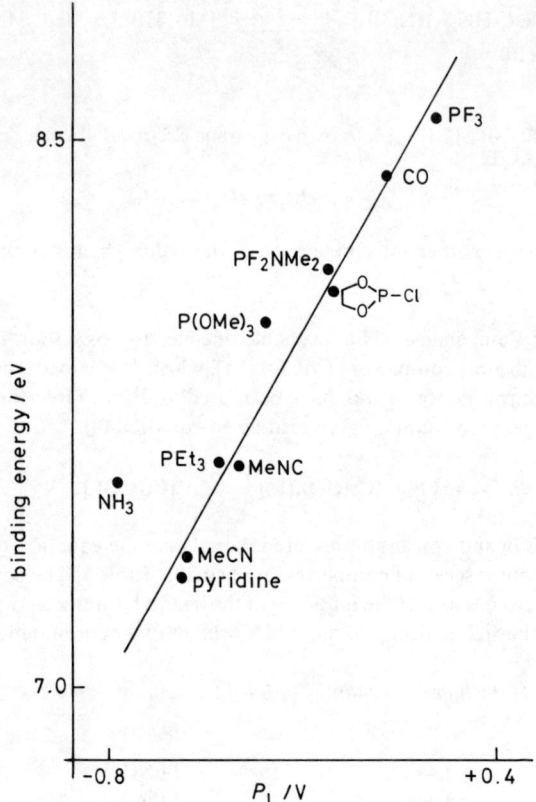

Figure 1 *The relationship between the photoelectron binding energy of 3d electrons of chromium and P_L for a range of ligands L in the complex* $[Cr(CO)_5L]$

reasonably well with $E_{\frac{1}{2}}^{ox}$ data for the primary one-electron oxidations of other series of closed-shell octahedral complexes $[M_s(L)]$ (M_s = a metal centre + auxiliary ligands), and these relationships have been analysed in terms of equation (5), where E_s and β are crudely taken as measures of the 'electron-richness' of the site $\{M_s\}$ and its 'polarizability', respectively.[51]

$$E_{1/2}^{ox}[M_s(L)] = E_s + \beta P_L \qquad (5)$$

Over the past few years, the variation in reversible one-electron oxidation potentials of geometric isomers of octahedral metal carbonyls has attracted some attention, both experimentally and theoretically. Bond has recently pointed out that *cis*- and *trans*-$[Mo(PBu^n_3)_2(CO)_4]$ show little difference in their oxidation potentials; thus structural isomerization following such electron-transfer reactions may be 'missed'.[63] Interestingly, *cis*-$[Mo(NO)(CO)(dppe)_2]^+$ undergoes well-resolved structural isomerization to give the *trans*-dication upon one-electron oxidation; this is analogous to the well-studied *cis*-$[Mo(CO)_2(dppe)_2]$ system.[51] The electrochemistry of [Mo(CO)$_2$(bipy)$_2$] has been briefly reported. Cyclic voltammetry shows that it undergoes reversible one-electron oxidation and a further irreversible one-electron oxidation at more positive potentials in a MeCN electrolyte. Furthermore, the complex shows a reversible single-electron reduction which is, presumably, ligand-based.[64] The mechanism of formation and the electrochemical properties of a seven-co-ordinate mixed-ligand Mo^{II}–CO complex, [Mo(CO)(S$_2$NMe$_2$)$_2$(dppe)$_2$], have been examined by cyclic-voltammetric, double-potential-step, coulometric, and low-temperature infrared spectroscopic methods.[65,66] Complexes containing the {M(CO)$_3$} unit (M = Cr, Mo, or W) that is facially co-ordinated to macrocyclic N-, S-, and P-donor ligands, e.g. (11), have been synthesized and their cyclic voltammetry has been briefly described.[67]

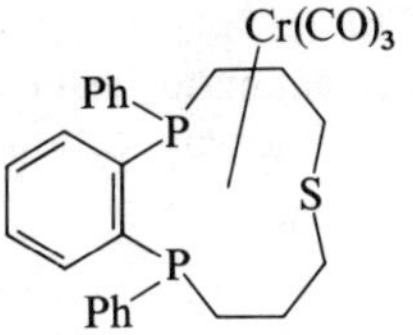

(11)

Arene(tricarbonyl)chromium derivatives (12)—(14) of [SnMe$_2$Ph$_2$] and [SnMePh$_3$] show interesting redox chemistry: each complex undergoes reversible oxidations in a propylene carbonate electrolyte, associated with the removal of an electron from each of the chromium centre(s); thus the oxidation of (14) is a three-electron process which gives a relatively stable trication whilst (12) and (13) give mono- and di-cations, respectively. The electrochemical data suggest that the arenetricarbonyl groups are essentially non-interacting and that substitution of the phenyl ring in the 2-, 4-, and 6-positions by donor groups can increase the stability of the cations.[68]

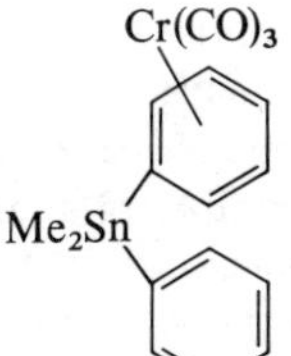

(12) 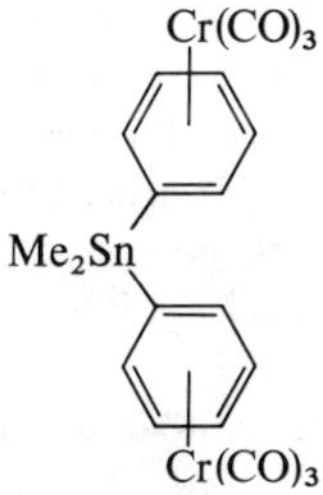   (13) 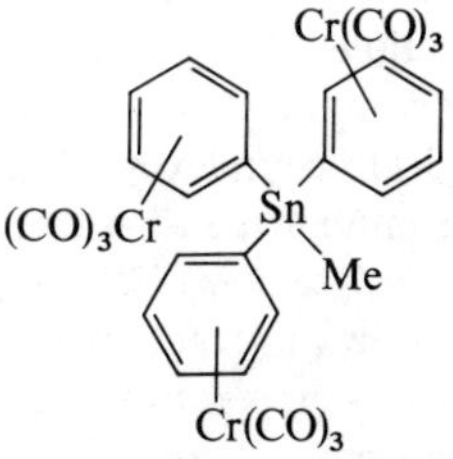   (14)

[63] A. M. Bond, D. J. Darensbourg, E. Mocellin, and B. J. Stewart, J. Am. Chem. Soc., 1981, **103**, 6827.
[64] T. T.-T. Li and C. H. Brubaker, Jr., J. Organomet. Chem., 1981, **216**, 223.
[65] B. A. L. Crichton, J. R. Dilworth, C. J. Pickett, and J. Chatt, J. Chem. Soc., Dalton Trans., 1981, 892.
[66] J. R. Dilworth, B. D. Neaves, J. Zubieta, and C. J. Pickett, Inorg. Chem., 1982, in the press.
[67] M. A. Fox, K. A. Campbell, and E. P. Kyba, Inorg. Chem., 1981, **20**, 4163.
[68] R. D. Reike, S. N. Milligan, I. Tucker, K. A. Dowler, and B. R. Willeford, J. Organomet. Chem., 1981, **218**, C25.

Bis-arene-chromium(0) and bis-arene-chromium(I) cations have been investigated in detail, using rotating-disc electrodes in a DMSO electrolyte. The $E_{\frac{1}{2}}$ data of the Cr^I/Cr^0 couple for some twenty of these compounds correlated with Hammett's *meta*-substituent constants. The authors conclude that the electronic effects of the arene substituents are transferred to the (metal) reaction centre by an inductive mechanism and that conjugation of substituents with the co-ordinated ring is weaker in these bis-arene complexes than in ferrocenes.[69] In another thorough study, a linear correlation between $E_{\frac{1}{2}}$ data for the Cr^I/Cr^0 couple of similar series of chromium complexes and the sum of Hammett's *meta*-substituent constants, $\Sigma \sigma_m$, was observed. The linear correlation of $E_{\frac{1}{2}}$ (and of $\Sigma \sigma_m$) with gas-phase ionization-potential data, kinetics, and activation parameters for the outer-sphere self-exchange process is discussed.[70]

Two-electron reduction of $[Cr(\eta^5\text{-}C_7H_7)(CO)_3]^+$ in a MeCN electrolyte affords an anion which attacks the parent cation to yield a carbon–carbon-bonded dimer (15); in contrast, the *free* cycloheptatrienyl cation dimerizes *via* radical–radical coupling.[71]

(15)

A fascinating chemistry of a dimolybdenum bis-η^6-arene complex (16) has been described by Silverthorn. Cyanide or methoxide ions attack one arene ring to give the η^5-ligated species, the electro-oxidation of which effects the overall substitution of H in the co-ordinated benzene ring of (16), as shown in Scheme 17.[72]

An excellent series of papers by Kochi and co-workers on the electrochemical reactions of organometallic compounds have been published over the past few years; amongst these are two publications on the electrochemistry of acyl, aryl, and hydrido-complexes of Group 6A.[73,74] Electrochemical oxidation of benzoyl-chromium(0) and -tungsten(0) carbonyls gives hydroxycarbene complexes in an electrolyte of MeCN that contains $[NEt_4]^+ [ClO_4]^-$, at a platinum electrode, *e.g.* as shown in reaction (6).

Low-temperature cyclic voltammetric studies and e.p.r. measurements, using a microporous Ag electrode flow-cell, show that these reactions proceed *via* $[M(COR)(CO)_5]$ (M = Cr or W), which are more or less thermally labile; these

---

[69] L. P. Yur'eva, S. M. Peregudova, L. N. Nekrasov, A. P. Korotkov, N. N. Zaitseva, N. V. Zakurin, and A. Yu. Vasil'kov, *J. Organomet. Chem.*, 1981, **219**, 43.
[70] T. T.-T. Li and C. H. Brubaker, Jr., *J. Organomet. Chem.*, 1981, **216**, 223.
[71] A. N. Romanin, A. Venzo, and A. Ceccon, *J. Electroanal. Chem. Interfacial Electrochem.*, 1980, **112**, 147.
[72] W. E. Silverthorn, *J. Am. Chem. Soc.*, 1980, **102**, 842.
[73] R. J. Klingler, J. C. Huffman, and J. K. Kochi, *Inorg. Chem.*, 1981, **20**, 34.
[74] J. K. Kochi, R. J. Klingler, and J. C. Huffman, *J. Am. Chem. Soc.*, 1980, **102**, 208.

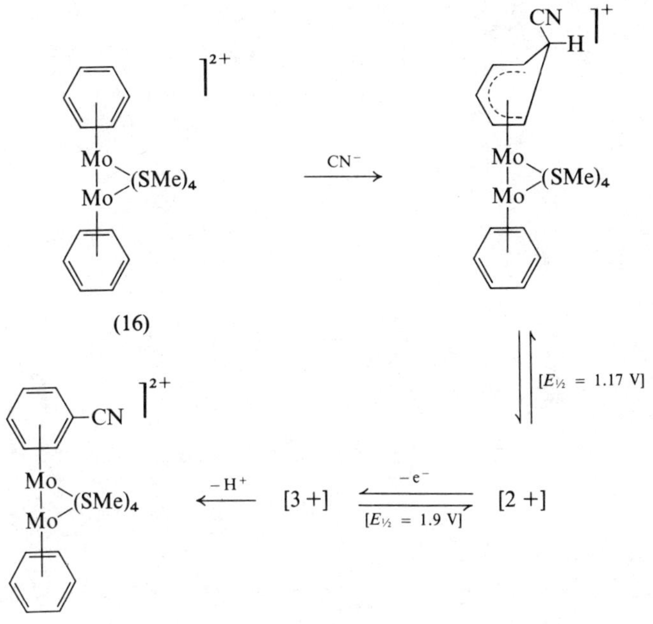

Scheme 17

Conditions: Pt electrode, propylene carbonate, $[NEt_4]^+ [PF_6]^-$; values of $E_{1/2}$ are relative to the S.C.E.

$$[M(-\overset{O}{\overset{\|}{C}}-R)(CO)_5]^- \xrightarrow{-e^-} [M(=C\underset{R}{\overset{OH}{\diagdown}})(CO)_5] \quad (6)$$

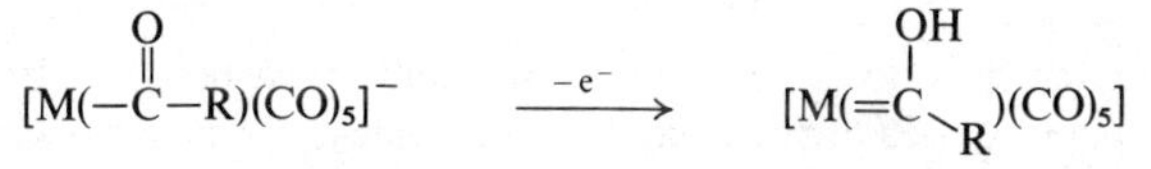

(R = alkyl or aryl; M = Cr or W)

neutral radicals decay *via* a first-order pathway which involves abstraction of H˙ from the supporting electrolyte, as shown in Scheme 18.[73]

Cleavage of metal–hydride and metal–carbon bonds in molybdenum and tungsten cyclopentadienyl complexes, *e.g.* $[W(\eta^5\text{-}C_5H_5)_2H_2]$ (17) and $[W(\eta^5\text{-}C_5H_5)_2(Ph)(H)]$ (18), has been investigated in MeCN that contains $[NBu_4]^+ [BF_4]^-$ at a platinum electrode. Following one-electron oxidation, the dihydrido-complex (17) rapidly loses a proton, whereas the more stable phenyl hydrido-cation that is obtained from (18) abstracts H˙ from the medium to give benzene. Under similar conditions, loss of a proton, rather than of CO, takes place upon one-electron oxidation of (19), as shown in Scheme 19.[74]

Linear trimetallic complexes that contain $\{M(\eta^5\text{-}C_5H_5)(CO)_3\}$ units (where M is Cr, Mo, or W), *e.g.* $[\{Cr(\eta^5\text{-}C_5H_5)(CO)_3\}_2Hg]$, have been reported to undergo cleavage of metal–metal bonds in most cases upon electroreduction.[75]

---

[75] P. Lemoine, A. Giraudeau, M. Gross, and P. Braunstein, *J. Chem. Soc., Chem. Commun.*, 1980, 77.

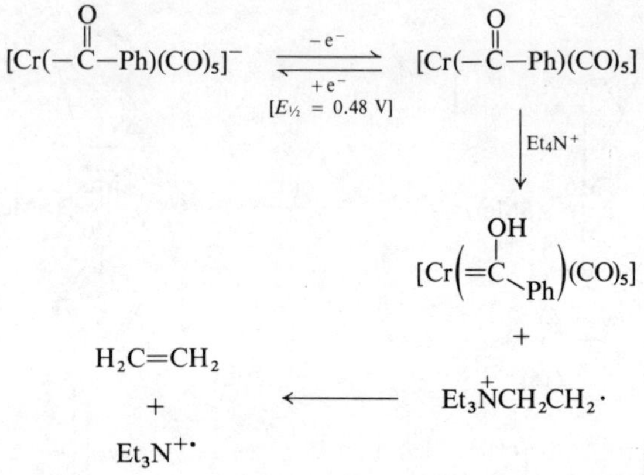

$$[Cr(-\overset{O}{\overset{\|}{C}}-Ph)(CO)_5]^- \underset{+e^-}{\overset{-e^-}{\rightleftarrows}} [Cr(-\overset{O}{\overset{\|}{C}}-Ph)(CO)_5]$$

$$[E_{1/2} = 0.48 \text{ V}]$$

$$\downarrow Et_4N^+$$

$$[Cr(=C\underset{Ph}{\overset{OH}{|}})(CO)_5]$$

$$H_2C=CH_2 \quad + $$
$$+ \quad \longleftarrow \quad Et_3\overset{+}{N}CH_2CH_2\cdot$$
$$Et_3N^{+\cdot}$$

Conditions: MeCN, [NEt$_4$]$^+$ [ClO$_4$]$^-$; value of $E_{1/2}$ is *versus* the S.C.E.

Scheme 18

$$[Mo(\eta^5\text{-}C_5H_5)H(CO)_3] \xrightarrow[-H^+]{-e^-} \tfrac{1}{2}[\{Mo(\eta^5\text{-}C_5H_5)(CO)_3\}_2]$$

(19) $\quad [E = -0.5 \text{ V}]$

Conditions: MeCN, [NEt$_4$]$^+$ [ClO$_4$]$^-$, Pt electrode; value of E is *versus* the S.C.E.

Scheme 19

Various voltammetric studies of penta-, tetra-, and tri-carbonyl carbene,[76—78] isocyanide,[79,80] and σ-bonded organometallic[81,82] complexes of metals of Group 6A have been briefly reported.

Other Studies of Complexes of Metals of Group 6A.—The effect of substituting D$_2$O for H$_2$O on the formal potentials of redox couples of some transition metals has been examined: the couples $[Cr(H_2O)_6]^{3+/2+}$ and $[Cr(D_2O)_6]^{3+/2+}$ show the greatest difference in $E^{\ominus\prime}$ (57 mV at 25 °C)[83]

The intensely coloured 'Pyrazine Green' complexes, *e.g.* $[Cr(H_2O)_5(pz)]^{3+}$ (pz = pyrazine), undergo pH-dependent reductions.[84]

---

[76] E. O. Fischer, F. J. Gammel, J. O. Besenhard, A. Frank, and D. Neugebauer, *J. Organomet. Chem.*, 1980, **191**, 261.
[77] R. Rieke, H. Kojima, and K. Oefele, *Angew. Chem.*, 1980, **93**, 550.
[78] G. Gritzner, P. Rechberger, and V. Gutman, *J. Electroanal. Chem. Interfacial Electrochem.*, 1980, **114**, 129.
[79] A. Bell, D. D. Klendworth, R. E. Wild, and R. A. Walton, *Inorg. Chem.*, 1981, **20**, 4457.
[80] T. E. Wood, J. C. Deaton, J. Corning, R. E. Wild, and R. A. Walton, *Inorg. Chem.*, 1980, **19**, 2614.
[81] A. Rusina, A. A. Vlcek, and K. Schmiedeknecht, *J. Organomet. Chem.*, 1980, **192**, 3(7.
[82] M. F. Lappert, C. L. Raston, B. W. Skelton, and A. H. White, *J. Chem. Soc., Chem. Commun.*, 1981, 485.
[83] M. J. Weaver and S. M. Nettles, *Inorg. Chem.*, 1980, **19**, 1641.
[84] F. Anson and J. Swartz, *Inorg. Chem.*, 1981, **20**, 2250.

Kadish has investigated the influence of substituted pyridine ligands on the redox chemistry of chromium tetraphenylporphinato-complexes.[85]

Reversible one-electron oxidations of various chromium, molybdenum, and tungsten species that contain quadruple M–M bonds, *e.g.* [Cr_2L_4] (L = anion of 2-amino-5-methylpyridine), in non-aqueous electrolytes have been described.[86,87] The reversible electrochemical oxidation of [Mo_6Cl_{14}]$^{2-}$ at +1.29 V (*vs* Ag/0.1M-AgNO$_3$, in MeCN) has been reported.[88] Chromium(III) has been immobilized on an (SN)$_x$ electrode[89] and [$Mo(CN)_8$]$^{4-/3-}$ has been electrostatically bound to a poly(vinylpyridine)-coated graphite electrode.[90]

4 Manganese, Technetium, and Rhenium

The evolution of oxygen during photosynthesis is dependent upon the presence of manganese (probably in oxidation states III or IV); the involvement of this element in photosystem II and in other manganese-dependent enzymes such as superoxide dismutase, which catalyses reaction (7), has stimulated the synthesis of various

$$2H^+ + 2O_2^{\bar{}} \longrightarrow H_2O_2 + O_2 \qquad (7)$$

manganese complexes, investigations of their redox chemistry, and studies of their interactions with O_2. The advent of ^{99}Tc diagnostic nuclear medicine has encouraged some recent new chemical and electrochemical studies of compounds of this rare element. Rhenium has a comparatively well-developed chemistry and there have been several recent studies of the electrochemical reactions of its compounds.

Manganese.—The syntheses of a vast range of quadri- and quinque-dentate Schiffs-base complexes of Mn^{II} and Mn^{III}, *e.g.* (20) and (21), have been described and redox potentials for the Mn^{II}/Mn^{III} couple reported. Chemical reactivity towards O_2 in terms of structure and redox properties is also discussed.[91–93]

(20)   (21)

---

[85] L. A. Bottomley and K. M. Kadish, *J. Chem. Soc., Chem. Commun.*, 1981, 1212.
[86] J. Watanabe, T. Saji, and S. Aoyagui, *J. Electroanal. Chem. Interfacial Electrochem.*, 1981, **129**, 369.
[87] T. C. Zietlow, D. D. Klendworth, T. Nimry, D. J. Salmon, and R. A. Walton, *Inorg. Chem.*, 1981, **20**, 947.
[88] A. W. Maverick and H. B. Gray, *J. Am. Chem. Soc.*, 1981, **103**, 1298.
[89] R. J. Nowak, W. Kutner, J. F. Rubinson, A. Voulgaropoulos, H. B. Mark, and A. G. MacDarmid, *J. Electrochem. Soc.*, 1981, **128**, 1927.
[90] N. Oyama, K. Sato, and H. Matsuda, *J. Electroanal. Chem. Interfacial Electrochem.*, 1980, **115**, 149.
[91] R. K. Boggess, J. W. Hughes, W. M. Coleman, and L. T. Taylor, *Inorg. Chim. Acta*, 1980, **38**, 183.
[92] W. M. Coleman, R. K. Boggess, J. W. Hughes, and L. T. Taylor, *Inorg. Chem.*, 1981, **20**, 700.
[93] W. M. Coleman, R. K. Boggess, J. W. Hughes, and L. T. Taylor, *Inorg. Chem.*, 1981, **20**, 1253.

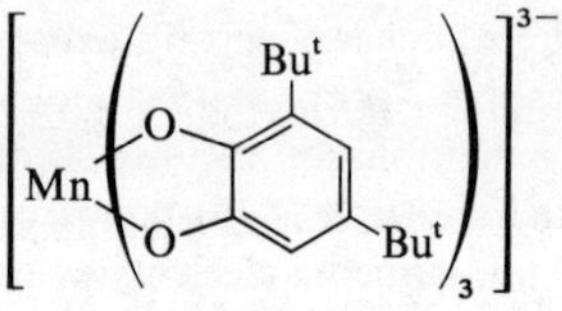

$$[Mn(\underset{O}{\overset{O}{<}}\underset{Bu^t}{\overset{}{\bigcirc}}Bu^t)_3]^{3-}$$

(22)

Catechol complexes (22) of Mn^{III} are electrochemically oxidized to Mn^{IV} species (Scheme 20), which are reported to bind O_2 reversibly as a superoxide ligand.[94,95]

The reversible binding of O_2 by complexes of manganese(II) with tertiary phosphines has been the subject of some controversy, and there have been two electrochemical studies pertinent to this.[96,97]

$$[Mn(DTBC)_3]^{3-} \underset{+e^-}{\overset{-e^-}{\rightleftarrows}} [Mn(DTBC)_3]^{2-}$$
(22) $[E_{1/2} = -0.45 \text{ V}]$

(DTBC = di-t-butylcatecholato)

Conditions: Me_2SO, $[NEt_4]^+$ $[ClO_4]^-$, Pt electrode; value of $E_{1/2}$ is *versus* the S.C.E.

Scheme 20

The electrochemistry of $[Mn(\eta^5\text{-}C_5H_5)(CO)_2(L)]$ (L = CO, N_2, N_2H_4, or NH_3) and of $[\{Mn(\eta^5\text{-}C_5H_5)(CO)_2\}_2NH_2NH_2]$ and $[\{Mn(\eta^5\text{-}C_5H_5)(CO)_2\}_2NHNH]$ in THF electrolytes, at a platinum electrode, has been explored. The ammino-complex is oxidized irreversibly to give the hydrazine species (23) in low yield, whilst the oxidation of the hydrazine complex (23) affords the dinitrogen compound (24), as shown in Scheme 21.

Interestingly, the reduction of (24) has been reported to be easier than that of its carbonyl analogue $[Mn(\eta^5\text{-}C_5H_5)(CO)_3]$.[98]

Technetium and Rhenium.—The electrochemistry of $[M=O(SR)_4]^-$ (M = Tc or Re, R = alkyl or aryl) in a MeCN electrolyte has been briefly described[99] and the electroreduction of perrhenate in weakly acidic citrate and oxalate media reported.[100]

The electrochemical oxidation of carbonyl complexes of technetium(I) in a MeCN electrolyte, at a platinum electrode, has been investigated; $[TcCl(CO)_2(PMe_2Ph)_3]$ and $[TcCl(CO)_3(PMe_2Ph)_2]$ both undergo an overall two-

---

[94] K. D. Magers, C. G. Smith, and D. T. Sawyer, *Inorg. Chem.*, 1980, **19**, 492.
[95] S. E. Jones, D-H. Chin, and D. T. Sawyer, *Inorg. Chem.*, 1981, **20**, 4257.
[96] A. Hosseiny, C. A. McAuliffe, K. Minten, M. J. Parrot, R. Pritchard, and J. James, *Inorg. Chim. Acta*, 1980, **39**, 227.
[97] R. M. Brown, R. E. Bull, M. L. H. Green, P. D. Grebenik, J. J. Martin-Polo, and D. M. P. Mingos, *J. Organomet. Chem.*, 1980, **201**, 437.
[98] T. Würminghausen and D. Sellmann, *J. Organomet. Chem.*, 1980, **199**, 77.
[99] A. Davison, C. Orvig, H. S. Trop, M. Sohn, B. V. De Pamphilis, and A. G. Jones, *Inorg. Chem.*, 1980, **19**, 1988.
[100] J. J. Vajo, D. A. Aikens, L. Ashley, D. E. Poeltl, R. A. Bailey, H. M. Clark, and S. C. Bunce, *Inorg. Chem.*, 1981, **20**, 3328.
[101] U. Mazzi, E. Roncari, R. Seeber, and G. A. Mazzocchin, *Inorg. Chim. Acta*, 1980, **41**, 95.

The Electrochemistry of Transition-metal Complexes

$[Mn(\eta^5-C_5H_5)(CO)_2(NH_3)] \xrightarrow[i]{-ne^-}$

$[Mn(\eta^5-C_5H_5)(CO)_2(N_2H_4)] + [Mn(\eta^5-C_5H_5)(CO)_2(N_2)]$
(23) (24)

$[Mn(\eta^5-C_5H_5)(CO)_2(N_2H_4)] \xrightarrow[ii]{-ne^-} [\{Mn(\eta^5-C_5H_5)(CO)_2\}_2NHNH]$
(23)

$\downarrow -ne^-$

$[Mn(\eta^5-C_5H_5)(CO)_2(N_2)]$
(24)

Conditions: i, THF, Pt electrode; ii, THF, $[NBu_4]^+ [PF_6]^-$, Pt electrode

Scheme 21

electron oxidation, with the formation of Tc^{III} species, as shown in Scheme 22.[101]

In a related study, complexes of technetium(III) with tertiary phosphines and arsines have been shown to be reduced reversibly to Tc^{II} complexes (Scheme 23), and it is reported that such species are some 200—300 mV *easier* to reduce than their Re^{III} analogues.[102]

$[TcCl(CO)_2(PMe_2Ph)_3] \underset{+e^-}{\overset{-e^-}{\rightleftarrows}} [TcCl(CO)_2(PMe_2Ph)_3]^+$

Conditions: MeCN, $[Na]^+ [ClO_4]^-$, Pt electrodes

Scheme 22

$[TcCl_2(dppe)_2]^+ \underset{-e^-}{\overset{+e^-}{\rightleftarrows}} [TcCl_2(dppe)_2]$
$[E_{1/2} = -0.040 \text{ V}]$

Conditions: DMF, $[NEt_4]^+ [ClO_4]^-$; the value of $E_{1/2}$ is *versus* the NaCl S.C.E.

Scheme 23

The complexes $[M(SnPh_3)(CO)_3(phen)]$ (M = Mn or Re) and their germanium analogues have been shown to undergo reversible one-electron reduction and ligand-localized irreversible oxidation.[103]

The electrochemistry of a wide range of carbonyl, tertiary phosphine, dinitrogen, tertiary phosphite, nitrile, and isocyanide mixed-ligand complexes of rhenium(I) has been described. Each complex usually undergoes a reversible one-electron oxidation in a THF electrolyte at a platinum anode to give more or less stable Re^{II} species. Further, more or less reversible, one-electron oxidations to Re^{III} are also

---

[102] R. W. Hurst, W. R. Heineman, and E. Deutsch, *Inorg. Chem.*, 1981, **20**, 3298.
[103] J. C. Luong, R. A. Faltynek, and M. S. Wrighton, *J. Am. Chem. Soc.*, 1980, **102**, 7892.

observed at more positive potentials. Relationships between structure, $\nu(N_2)$, and the redox potentials of these complexes are described, and the susceptibility of their carbonyl ligands to attack by MeLi is related to these properties.[104,105]

Polyhydride dirhenium complexes, e.g. $[Re_2H_8(PPh_3)_4]$, undergo interesting redox and substitution chemistry. Thus this complex is oxidized reversibly at a platinum electrode in a CH_2Cl_2 electrolyte, and e.s.r. studies of the cation at $-160\,°C$ are in accord with the highest occupied molecular orbital being delocalized and metal-based. Electrochemical oxidation of $[ReH_5(PPh_3)_3]$ and related complexes activates them towards attack by isocyanides,[106] as shown in Scheme 24.

$$[ReH_5(PPh_3)_3] \xrightarrow[{[E\ =\ +0.37\ V]}]{Bu^tNC} [Re(Bu^tNC)_3(PPh_3)_3]$$

Conditions: CH_2Cl_2, $[NBu_4]^+$ $[PF_6]^-$; the value of E is versus the S.C.E.

Scheme 24

The reduction of $[Re_2(O_2CR)_4X_2]^-$ complexes (R = alkyl or aryl; X = Cl, Br, or I) has been reported, and $E_\frac{1}{2}$ potentials are discussed in terms of the energy of the δ–δ* electronic transition for these complexes. Whereas the former measurements span a range of -0.42 to -0.24 V versus the S.C.E., $\lambda(\delta$–δ*) is rather insensitive to structural change.[107] Electrochemical measurements related to the luminescent excited state of $[Re_2Cl_8]$ have been reported, and it is concluded that photoexcitation of this species generates a powerful oxidant.[108]

5 Iron, Ruthenium, and Osmium

Iron.—Iron porphyrins and iron–sulphur centres are ubiquitous components of the redox proteins that are involved in electron, oxygen, hydrogen, and nitrogen metabolism in prokaryotic and eukaryotic cells. Electrochemical study of iron complexes which are 'models' of these centres therefore continues to be an active area of research. Recently, P-450 mono-oxygenases which contain haem iron and which are capable of activating alkanes have attracted much attention. Direct electrochemical methods for studying the redox chemistry of the simpler iron proteins are now emerging.

The organometallic electrochemistry of iron compounds has developed rapidly, and several new interesting electrochemical reactions involving the transformation of ligands have been discovered.

Iron Porphyrins and Related Compounds. Redox potentials for the Fe^{III}/Fe^{II} couples of $[Fe(TPP)X]$ and $[Fe(OEP)X]$ (X = halide or pseudohalide) have been shown to

---

[104] J. Chatt, G. J. Leigh, C. J. Pickett, and D. R. Stanley, *J. Organomet. Chem.*, 1979, **184**, C64; J. Chatt, G. J. Leigh, R. H. Morris, C. J. Pickett, and D. R. Stanley, *J. Chem. Soc., Dalton Trans.*, 1981, 800.
[105] A. Pombeiro, C. J. Pickett, and R. L. Richards, *J. Organomet. Chem.*, 1982, **224**, 285.
[106] J. D. Allison, C. J. Cameron, R. E. Wild, and R. A. Walton, *J. Organomet. Chem.*, 1981, **218**, C62.
[107] V. Srinivasan and R. A. Walton, *Inorg. Chem.*, 1980, **19**, 1635.
[108] D. G. Nocera and H. B. Gray, *J. Am. Chem. Soc.*, 1981, **103**, 7349.

correlate linearly with $Fe(^2p_{\frac{3}{2}})$ binding-energy data. This correlation is related to the change in charge density about the Fe^{III} centre, which is reflected in the value of $E_{\frac{1}{2}}$ for reduction.[109] The potentials for reversible one-electron oxidation for a range of tetraphenylporphyrin and octaethylporphyrin complexes of iron(III) have been measured and it has been concluded that this electron-transfer reaction is associated with the porphyrin ligand rather than with the metal centre.[110,111]

The first synthesis of an iron(III) complex (25) of the isobacteriochlorin family has been described and its redox properties have been investigated in CH_2Cl_2 and in

(25)

Pr^nCN electrolytes. Three redox processes have been identified, *i.e.* oxidation and reduction of the ring, to give a radical cation and a radical anion respectively, and metal-centred reduction of iron(III) to iron(II). It is found that hydroporphyrins are easier to oxidize but harder to reduce than their porphyrin analogues, that the gap between the highest occupied and lowest unoccupied orbitals is approximately constant for the porphyrin, chlorin, and isobacterin complexes of iron, and, that the saturation of the C=C bonds of pyrrole results in the macrocyclic ligand exerting its influence primarily through the σ-framework.[112,113]

Kadish and co-workers have attempted to correlate the temperature-dependent change in spin-state for d^5 Fe^{III} complexes ($^2T \rightleftharpoons {^6A}$) of the type [Fe(X-salen)$_2$(trien)]$^+$, where X-salen is a substituted salen ligand, with the values of $E_{\frac{1}{2}}$ for the Fe^{III}/Fe^{II} couples. They suggest that their study may have relevance to the biological transfer of electrons. Magnetic moments in solution, however, which reflect the population of the states ^2T and ^6A, correlate poorly with values of $E_{\frac{1}{2}}$, although they are sensitive to the solvent, to the substituents on salen, and to temperature. The correlation of values of $E_{\frac{1}{2}}$ with Donor Numbers is somewhat better, although the maximum variation, $\Delta E_{\frac{1}{2}}$, is only 90 mV, and plots of Hammett σ-constant *versus* $E_{\frac{1}{2}}$ are linear. The rate constants for the low-spin–high-spin interchange are *ca* 10^7 to 10^8 s^{-1}; such interchanges are probably sufficiently fast to mask a dependence of $E_{\frac{1}{2}}$ upon spin-state.[114]

The electrochemistry of an iron porphyrin complex (26) and its Fe–alkyl

---

[109] K. M. Kadish, L. A. Bottomley, J. G. Brace, and N. Winograd, *J. Am. Chem. Soc.*, 1980, **102**, 4341.
[110] L. A. Bottomley and K. M. Kadish, *Inorg. Chem.*, 1981, **20**, 1348.
[111] M. A. Phillippi, E. T. Shimomura, and H. M. Goff, *Inorg. Chem.*, 1981, **20**, 1322.
[112] C. K. Chang and J. Fajer, *J. Am. Chem. Soc.*, 1980, **102**, 848.
[113] A. M. Stolzenberg, L. O. Spreer, and R. H. Holm, *J. Am. Chem. Soc.*, 1980, **102**, 364.
[114] K. M. Kadish, K. Das, D. Schaeper, C. L. Merrill, B. R. Welch, and L. J. Wilson, *Inorg. Chem.*, 1980, **19**, 2816.

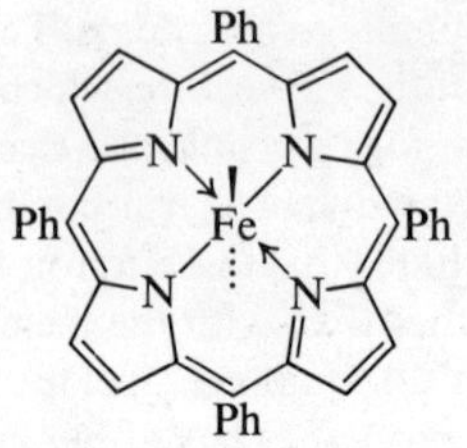

(26)

{Fe(TPP)}

derivatives in DMF that contains LiClO$_4$ or LiCl, at a platinum electrode, has been studied. Values of $E^{\ominus\prime}$ for Fe^{III}–alkyl/Fe^{II}–alkyl and Fe^{IV}–alkyl/Fe^{III}–alkyl couples have been reported: the Fe^{IV}–alkyl species are unstable and have lifetimes of < 1 ms. The mechanism of formation of the iron–alkyl bond has been investigated and it was concluded that S_N2 attack of the Fe^I species upon Bu^{n}I, Bu^{n}Br, or Bu^{n}Cl is involved [reaction (9)] rather than an initial transfer of an electron [reaction (8)].[115]

$$[Fe(TPP)] \underset{-e^-}{\overset{+e^-}{\rightleftarrows}} [Fe(TPP)]^- \qquad (8)$$

$$[Fe(TPP)]^- + RX \longrightarrow [FeR(TPP)] + X^- \qquad (9)$$

One-electron oxidation of a carbene iron porphyrin complex (27) is reported to lead to N-substitution of the porphyrin ring: this is thought to be related to the degradation of carbene complexes of cytochrome P-450 during the metabolism of CCl$_4$ in the liver.[116]

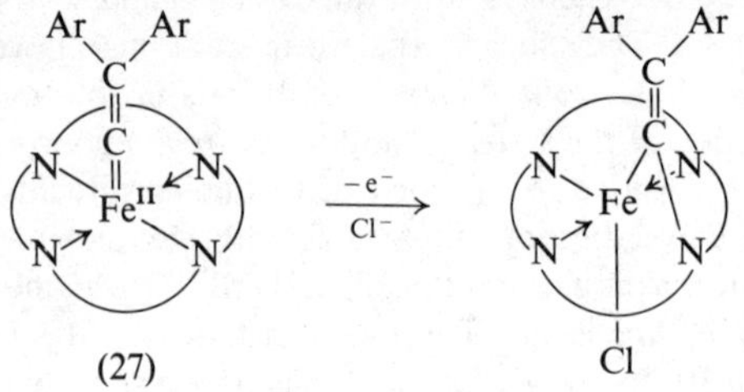

(27)

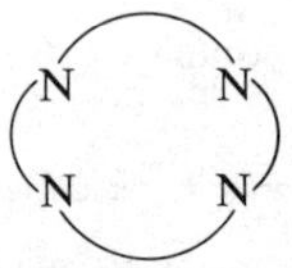

represents the porphyrin ligand

---

[115] D. Lexa, J. Mispelter, and J-M. Saveant, *J. Am. Chem. Soc.*, 1981, **103**, 6806.
[116] M. Lange and D. Mansuy, *Tetrahedron Lett.*, 1981, **22**, 2561.

The axial ligand 'speciation' of Fe^{III}, Fe^{II}, and Fe^{I} porphyrins by anions, pyridines, water, and various donor solvents has attracted considerable attention. The redox potentials of the Fe^{III}/Fe^{II} and Fe^{II}/Fe^{I} couples are particularly sensitive to the nature of the axial ligands. Kadish and co-workers have investigated the variation in these with pyridine concentration and Kuwana and co-workers have investigated the pH dependence of the redox behaviour of the water-soluble iron tetrakis-(N-methyl-4-pyridyl)porphyrin by electrochemical and spectroelectrochemical techniques.[117] The influence of counter-ions and solvent upon the $Fe^{III}(TPP)/Fe^{II}(TPP)$ couple has been systematically investigated and the axial ligand speciation in solvents with donor numbers >12 to <29 has been discussed.[118]

The electrochemistry of $\{Fe(TPP)\}$ dimers with bridging N, O, and SO_4^{2-} ligands has been described;[119–123] [$\{Fe(TPP)\}_2N$] shows two reversible one-electron reductions which are sensitive to the nature of co-ordinated axial ligands (e.g. amines), and base-off–base-on studies have been reported.[119,120] The ion [$\{Fe(TPP)\}_2O$]$^+$ is reduced at potentials 690 mV anodic of the μ-nitrido dimer.

The effects of medium on the electrochemical behaviour of horse heart cytochrome c have been investigated by differential pulse polarography; in all cases, an absorption pre-peak and a normal reduction peak were observed.[124] The cyclic voltammetry and differential pulse polarography of cytochrome c_3 from *Desulfovibrio vulgaris*, which contains four non-equivalent redox sites, were investigated and, using a digital simulation, values of $E^{\ominus\prime}$ for the four centres were estimated.[125,126]

Various studies concerned with the mechanism of reduction and oxidation of cytochrome c via gold electrodes that are modified by 4,4'-bipyridyl have been reported.[127,128] Chemically modified electrodes with a covalently attached viologen have also been constructed and used to mediate the transfer of electrons to horse heart cytochrome c.[129]

Microsomal and bacterial forms of cytochrome P-450 are rapidly reduced in the presence of 4,4'-bipyridyl at a nickel electrode to give the same products as are obtained by chemical reduction with dithionite. No reduction is observed in the absence of 4,4'-bipyridyl, but substantial overvoltages are necessary to reduce the two forms of P-450.[130]

---

[117] P. A. Forshey and T. Kuwana, *Inorg. Chem.*, 1981, **20**, 693.
[118] L. A. Bottomley and K. M. Kadish, *Inorg. Chem.*, 1981, **20**, 1348.
[119] K. M. Kadish, R. K. Rhodes, L. A. Bottomley, and H. M. Goff, *Inorg. Chem.*, 1981, **20**, 3195.
[120] L. A. Bottomley, *J. Electrochem. Soc.*, 1980, **127**, 307C.
[121] K. M. Kadish, L. A. Bottomley, J. G. Brace, and N. Winograd, *J. Am. Chem. Soc.*, 1980, **102**, 4341.
[122] I. A. Cohen, D. K. Lavallee, and A. B. Kopelove, *Inorg. Chem.*, 1980, **19**, 1098.
[123] M. A. Phillippi, N. Baenziger, and H. M. Goff, *Inorg. Chem.*, 1981, **20**, 3904.
[124] P. A. Serre, J. Haladfian, and P. Bianco, *J. Electroanal. Chem. Interfacial Electrochem.*, 1981, **122**, 327.
[125] W. F. Sokol, D. H. Evans, N. Katsumi, and Y. Tatsuhiko, *J. Electroanal. Chem. Interfacial Electrochem.*, 1980, **108**, 107.
[126] M. J. Eddowes, H. Elzanowska, and H. A. O. Hill, *Biochem. Soc. Trans.*, 1979, **7**, 735.
[127] W. J. Albery, M. J. Eddowes, H. A. O. Hill, and A. R. Hillman, *J. Am. Chem. Soc.*, 1981, **103**, 3904.
[128] M. J. Eddowes, H. A. O. Hill, and K. Ulosaki, *Bioelectrochem. Bioenerg.*, 1980, **7**, 527.
[129] N. S. Lewis and M. S. Wrighton, *Science*, 1981, **211**, 944.
[130] A. I. Archakov, B. V. Kuznetsov, M. V. Izotov, and I. I. Karuzina, *Dokl. Akad. Nauk SSSR*, 1981, **258**, 216.

Brief electrochemical studies of iron that is ligated by various bi-,[131,132] ter-,[133] and quadri-dentate[134] nitrogen-donor ligands have been reported.

Iron–Sulphur Compounds. Thiohydroxamato-complexes, *e.g.* (28), which contain high-spin Fe^{III} ($S=\frac{5}{2}$) can be obtained from cultures of *Pseudomonas fluorescens* and may be involved in the bacterial transport of iron. Electrochemical studies of this and of related complexes $[Fe\{ONR^1CR^2(S)\}_3]$ ($R^1 = Me$, $R^2 = H$, Ph, or Me) show that they undergo reversible one-electron reduction at a platinum electrode in acetone at potentials in the range that is accessible by reduction of NADH.[135] Mono-thiocarbamato-complexes of Fe^{III}, *e.g.* (29) show little difference in redox behaviour to their di-thiocarbamato-analogues.[136]

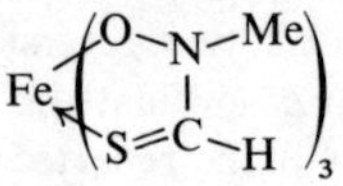

(28) 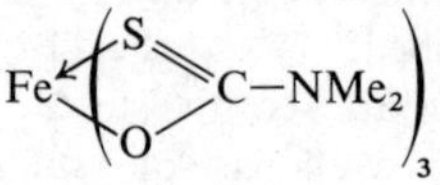   (29)

Electrochemical studies of complexes that contain iron, sulphur, and molybdenum have been discussed in an earlier section. Complexes that contain the Fe_2S_2 and Fe_2Se_2 cores have been briefly studied by cyclic voltammetry. Complex (30) and related $[Fe_2S_2(SR)_4]^{2-}$ or $[Fe_2Se_2(SR)_4]^{2-}$ species undergo reversible one-electron reductions to the trianions. Further reduction of these species is less well defined.[137–139]

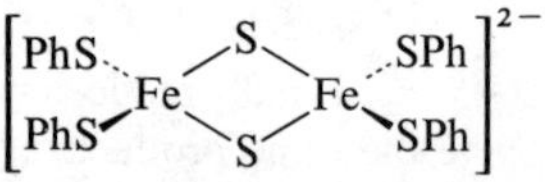

(30) 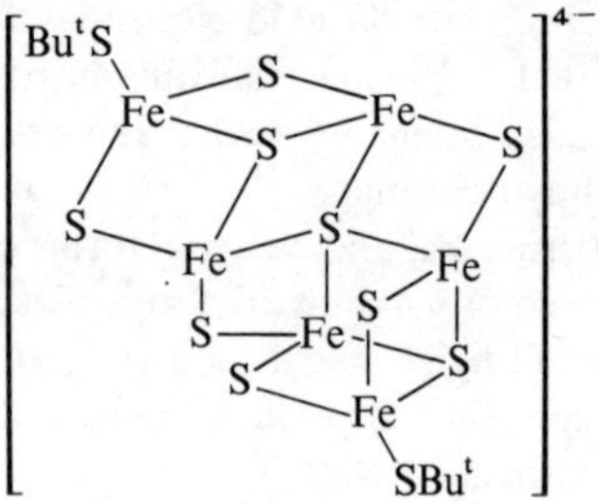   (31)

The reversible one-electron reduction of seleno-clusters, *e.g.* $[Fe_4Se_4(SPh)_4]^{2-}$, and of the water-soluble $[Fe_4S_4(SCH_2CH_2CO_2)_4]^{6-}$ species have been reported.

---

[131] P. Gouzerh, Y. Jeannin, C. Rocchiccioli-Deltcheff, and F. Valentini, *J. Coord. Chem.*, 1979, **9**, 221.
[132] J. M. Rao, D. J. Macero, and M. C. Hughes, *Inorg. Chim. Acta*, 1980, **41**, 221.
[133] R. R. Gagne, W. A. Marritt, D. N. Marks, and W. O. Siegl, *Inorg. Chem.*, 1981, **20**, 3260.
[134] K. M. Kadish, D. Schaeper, L. A. Bottomley, M. Tsutsui, and R. L. Bobsein, *J. Inorg. Nucl. Chem.*, 1980, **42**, 469.
[135] D. Brockway, K. S. Murray, and P. J. Newman, *J. Chem. Soc., Dalton Trans.*, 1980, 1112.
[136] D. L. Perry and S. R. Cooper, *J. Inorg. Nucl. Chem.*, 1980, **42**, 1356.
[137] P. K. Mascharak, G. C. Papaefthymiou, κ. B. Frankel, and R. H. Holm, *J. Am. Chem. Soc.*, 1981, **103**, 6110.
[138] R. H. Tieckelmann and B. A. Averill, *Inorg. Chim. Acta*, 1980, **46**, 135.
[139] J. G. Reynolds and R. H. Holm, *Inorg. Chem.*, 1980, **19**, 3257.

Surprisingly, solutions of the latter complex in a mercaptide-buffered aqueous electrolyte have been reported to show a further reversible one-electron reduction, corresponding to the 7 − /8 − couple; however, the cyclic voltammogram that is shown to support this observation is unconvincing.[140] The electrochemistry of a 'high potential' ferredoxin has been reported.[141]

An unusual cluster (31), containing an Fe_6S_9 core, has recently been synthesized and shown to undergo reversible oxidation, as shown in Scheme 25.[142]

$$\text{products} \underset{[E \,=\, -1.72\text{ V}]}{\overset{\text{irreversible}}{\longleftarrow}} [Fe_6S_9(SBu^t)_2]^{4-} \underset{+e^-}{\overset{-e^-}{\rightleftarrows}} (3-)$$
$$(31) \qquad [E_{1/2} = -0.62 \text{ V}]$$

Conditions: Me_2SO, vitreous carbon electrode; values of $E_{1/2}$ are *versus* the S.C.E.

Scheme 25

Organometallic Compounds. The adoption of ferrocene as an 'internal standard' in non-aqueous electrolytes has been recommended. This is already a widespread custom and, where applicable, its use should be encouraged.[143] A ferricinium dication, $[Fe(\eta^5\text{-}C_5Me_5)_2]^{2+}$ (32), has been prepared by controlled-potential oxidation in an $AlCl_3$-1-butylpyridinium chloride melt (Scheme 26).[144]

$$[Fe(\eta^5\text{-}C_5H_5)_2]^+ \underset{+e^-}{\overset{-e^-}{\rightleftarrows}} [Fe(\eta^5\text{-}C_5H_5)_2]^{2+}$$
$$[E_{1/2} = +1.39 \text{ V}] \qquad (32)$$

Conditions: $AlCl_3$ – 1-butylpyridinium chloride melt (1.5 : 1), vitreous carbon electrode; the value of $E_{1/2}$ is *versus* the Al electrode

Scheme 26

The reduction of ferrocene and of a substituted ferrocene has also been reported.[145,146] In the latter case, an unusual monocyclopentadienyl species (33) has been proposed as the product resulting from transfer of a single electron: further reduction of this species under CO gives the dicarbonyl anion (34), as shown in Scheme 27.[146]

A range of hydrocarbon-bridged di-ferrocenes and keto-ferrocenes have been studied by cyclic voltammetry and redox potentials have been discussed in terms of structure.[147] 1,2,3-Triferrocenylcyclopropenes have been synthesized and characterized by cyclic voltammetry and ^{13}C n.m.r. spectroscopy. Each complex shows reversible waves that are associated with the step-wise oxidation of the three

---

[140] R. A. Henderson and A. G. Sykes, *Inorg. Chem.*, 1980, **19**, 3104.
[141] B. Feinberg and V-K. Lan, *Bioelectrochem. Bioenerg.*, 1980, **7**, 187.
[142] G. Christou, R. H. Holm, M. Sabat, and J. Ibers, *J. Am. Chem. Soc.*, 1981, **103**, 6269.
[143] R. R. Gagne, C. A. Koval, and G. C. Lisensky, *Inorg. Chem.*, 1980, **19**, 2854.
[144] R. J. Gale, P. Singh, and R. Job, *J. Organomet. Chem.*, 1980, **199**, C44.
[145] Y. Mugnier, C. Moise, J. Tirouflet, and E. Laviron, *J. Organomet. Chem.*, 1980, **186**, C49.
[146] N. El Murr and A. Chaloyard, *J. Organomet. Chem.*, 1980, **193**, C60.
[147] E. Fujita, B. Gordon, M. Hillman, and A. G. Nagy, *J. Organomet. Chem.*, 1981, **218**, 105.

$$[\text{Fe}(\eta^5\text{-C}_5\text{H}_4\text{R})_2] \underset{-e^-}{\overset{+e^-}{\rightleftharpoons}} [\text{Fe}(\eta^5\text{-C}_5\text{H}_4\text{R})_2]^-$$

$$\downarrow$$

$$[\text{Fe}(\eta^5\text{-C}_5\text{H}_4\text{R})(\text{CO})_2]^- \overset{\text{CO}}{\longleftarrow} \{\text{Fe}(\eta^5\text{-C}_5\text{H}_4\text{R})\} + [\text{C}_5\text{H}_4\text{R}]^-$$

(34) (33)

(R = MeCO)

Scheme 27

iron sites. The separation in the $E_{\frac{1}{2}}$ potentials for these processes suggests that electronic interactions between iron centres are weak. Surface-confined monolayers and multilayers of ferrocene derivatives[148] have continued to attract considerable interest in the chemistry, electrochemistry, and photoelectrochemistry of modified electrodes.[149] The kinetics of transfer of charge and the rates of reduction of surface-bound ferricinium centres by reagents in solution have been studied,[150—152] plasma-polymerized vinylferrocene electrodes have been prepared,[153] and photo-assisted electrocatalytic reduction of $CHCl_3$ and of CCl_4 has been investigated.[154] A novel development has been the preparation of bilayer films that contain ferrocene groups as one redox partner; the theory and properties of such assemblies, and their behaviour as rectifying interfaces, have been studied in detail.[155—157]

Cyclopentadienyl-arene-iron(II) cations, e.g. $[\text{Fe}(\eta^5\text{-C}_5\text{H}_5)(\eta^6\text{-C}_6\text{H}_6)]^+$, show particularly interesting reduction chemistry. Several years ago, it was shown that such complexes undergo reversible one-electron reductions to the neutral species and, in strictly aprotic media, an unstable anion could be generated at more extreme potentials.[158] A study of such complexes by Astruc and co-workers has led them to propose that the neutral species are true nineteen-electron complexes, with planar and parallel cyclic ligands.[159] Nevertheless, these species behave as if there is radical-anion character on the arene ring. Thus, the reduction of $[\text{Fe}(\eta^5\text{-C}_5\text{H}_5)(\eta^6\text{-C}_6\text{H}_6)]$ in the presence of carbon dioxide or an alkyl halide is regio- and stereo-specific, and has been reported to proceed via the pathway shown in Scheme 28.[160]

Interestingly, the reaction of $[\text{Fe}(\eta^5\text{-C}_5\text{Me}_5)(\eta^6\text{-C}_6\text{Me}_6)]$ with air or O_2 leads to activation of the C—H, and this is suggested to occur via the generation of the

---

[148] K. W. Willman, R. D. Rocklin, R. Nowak, K-N. Kuo, F. A. Schultz, and R. W. Murray, J. Am. Chem. Soc., 1980, **102**, 7629.
[149] J-N. Chazalviel and T. B. Truong, J. Am. Chem. Soc., 1981, **103**, 7447.
[150] N. S. Scott, N. Oyama, and F. C. Anson, J. Electroanal. Chem. Interfacial Electrochem., 1980, **10**, 549.
[151] T. Ikeda, C. R. Leidner, and R. W. Murray, J. Am. Chem. Soc., 1981, **103**, 7422.
[152] M. Sharp and M. Petersson, J. Electroanal. Chem. Interfacial Electrochem., 1981, **122**, 409.
[153] G. J. Samuels and T. J. Meyer, J. Am. Chem. Soc., 1981, **103**, 307.
[154] M. F. Dautartas, K. R. Mann, and J. F. Evans, J. Electroanal. Chem. Interfacial Electrochem., 1980, **110**, 379.
[155] P. Denisevich, K. W. Willman, and R. W. Murray, J. Am. Chem. Soc., 1981, **103**, 4727.
[156] H. D. Abruna, P. Denisevich, M. Umana, T. J. Meyer, and R. W. Murray, J. Am. Chem. Soc., 1981, **103**, 1.
[157] C. D. Ellis, R. W. Murray, and T. J. Meyer, J. Am. Chem. Soc., 1981, **103**, 7480.
[158] A. N. Nesmeyanov, N. A. Vol'kenau, P. V. Petrovskii, L. S. Kotova, V. A. Petrakova, and L. T. Denisovich, J. Organomet. Chem., 1981, **210**, 103.
[159] J-R. Hamon, D. Astruc, and P. Michaud, J. Am. Chem. Soc., 1981, **103**, 758.
[160] N. El Murr, J. Chem. Soc., Chem. Commun., 1981, 251.

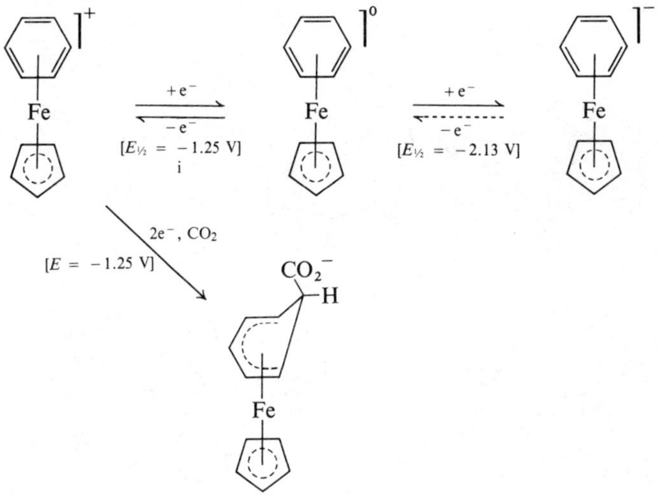

The Electrochemistry of Transition-metal Complexes

Conditions: i, THF, $[NBu_4]^+ [PF_6]^-$, Pt, Hg, or C electrode; values of E and $E_{1/2}$ are *versus* the S.C.E.

Scheme 28

superoxide radical anion,[161] as shown in reaction (10). Electro-reduction of related compounds in which functional organic groups are attached to the cyclic ligands is regiospecific, in contrast to chemical reductions, and localized on the functional group. Thus the reduction of $[Fe(\eta^5\text{-}C_5H_5)(\eta^6\text{-}C_6H_5COOH)]^+$ on mercury in protic

$$[Fe(\eta^6\text{-}C_6Me_6)(\eta^5\text{-}C_5H_5)] \xrightarrow[\text{THF}]{O_2} \text{product} \quad (10)$$

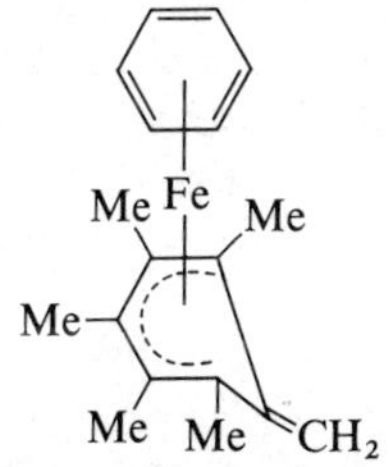

media gives the primary alcohol (Scheme 29) whilst the reduction of (35) gives the *endo*-alcohol (Scheme 30).[162,163]

Miholova and Vlcek have investigated in great detail the electrochemistry of cyclopentadienyl iron carbonyls of the type $[Fe(\eta^5\text{-}C_5H_5)(CO)_2X]$ (X = Cl, Br, I, H, CN, GeCl$_3$, GePh$_3$, SnCl$_3$, SnPh$_3$, or SiPh$_3$) and the dimers $[\{Fe(\eta^5\text{-}C_5H_5)(CO)_2\}_2]$ and $[Hg\{Fe(\eta^5\text{-}C_5H_5)(CO)_2\}_2]$, in THF that contains 0.1M-$[NBu_4]^+ [ClO_4]^-$, at a dropping mercury electrode. Whereas the *overall* reduction of the monomeric complexes and the dimeric species can lead to the formation of the stable anion

---

[161] D. Astruc, J-R. Hamon, E. Roman, and P. Michaud, *J. Am. Chem. Soc.*, 1981, **103**, 7502.
[162] E. Roman, D. Astruc, and A. Darchem, *J. Organomet. Chem.*, 1981, **219**, 221.
[163] C. Moinet, E. Roman, and D. Astruc, *J. Electroanal. Chem. Interfacial Electrochem.*, 1981, **121**, 241.

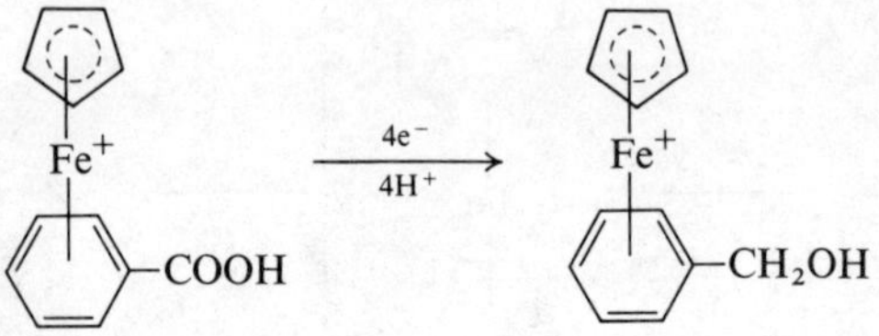

Scheme 29

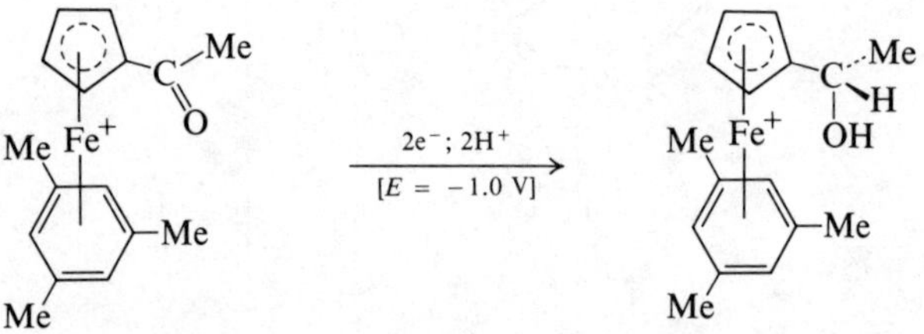

(35)

Conditions: Aqueous H_2SO_4, pH 0; the value of E is *versus* the S.C.E.

Scheme 30

$[Fe(\eta^5\text{-}C_5H_5)(CO)_2]^-$, this anion can also attack $[Hg\{Fe(\eta^5\text{-}C_5H_5)(CO)_2\}_2]$ to give an unusual trimeric species (36). The complex $[Hg\{Fe(\eta^5\text{-}C_5H_5)(CO)_2\}_2]$ is a product of the attack of the 'krypto-radical' (37) on the mercury electrode. These electrode reactions are summarized in Scheme 31.[164,165]

Alkyl and acyl complexes of the type $[Fe(\eta^5\text{-}C_5H_5)(CH_3)(CO)_2]$, $[Fe(\eta^5\text{-}C_5H_5)(COMe)(CO)_2]$, and $[Fe(\eta^5\text{-}C_5H_5)(COMe)(CO)(PPh_3)]$ have been shown to undergo single-electron oxidation at a platinum electrode in a MeCN electrolyte to give a variety of products, depending upon the site of nucleophilic attack (Scheme 32).[166,167]

An interesting electrochemistry is shown by thio- and seleno-carbamato-complexes of Fe^{II}, *e.g.* $[Fe(\eta^5\text{-}C_5H_5)(S_2CNMe_2)(CO)_2]$. Electrochemical oxidation and reduction reactions in MeCN involve the transformation of the carbamato-ligand from unidentate to bidentate, and *vice versa*, as solvent or CO are lost from, or co-ordinated to, the metal centre.[168] Complexes $[Fe(\eta^5\text{-}C_5H_5)(RN_3R)(L)]$ (R = aryl, L = tertiary phosphine or tri-alkyl or -aryl phosphite) undergo a reversible one-electron oxidation in a CH_2Cl_2 electrolyte.[169]

Bond and co-workers have examined the electro-oxidation of a range of triphenyl-phosphine, -arsine, and -stibine complexes $[Fe(CO)_4(L)]$ and

---

[164] D. Miholova and A. A. Vlcek, *Inorg. Chim. Acta*, 1980, **41**, 119.
[165] D. Miholova and A. A. Vlcek, *Inorg. Chim. Acta*, 1980, **43**, 43.
[166] R. J. Klingler and J. K. Kochi, *J. Organomet. Chem.*, 1980, **202**, 49.
[167] R. H. Magnuson, S. Zulu, W-M. T'sai, and W. P. Giering, *J. Am. Chem. Soc.*, 1980, **102**, 6887.
[168] G. Nagao, K. Tanaka, and T. Tanaka, *Inorg. Chim. Acta*, 1980, **42**, 43.
[169] J. G. M. van der Linden, A. H. Dix, and E. Pfeiffer, *Inorg. Chim. Acta*, 1980, **39**, 271.

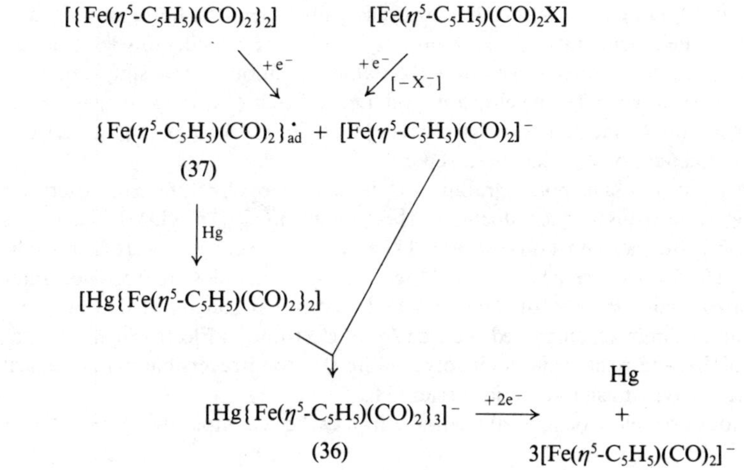

The Electrochemistry of Transition-metal Complexes

$[\{Fe(\eta^5\text{-}C_5H_5)(CO)_2\}_2]$ $[Fe(\eta^5\text{-}C_5H_5)(CO)_2X]$

$\searrow$ +e$^-$ $\swarrow$ +e$^-$ / [−X$^-$]

$\{Fe(\eta^5\text{-}C_5H_5)(CO)_2\}^{\bullet}_{ad}$ + $[Fe(\eta^5\text{-}C_5H_5)(CO)_2]^-$

(37)

$\downarrow$ Hg

$[Hg\{Fe(\eta^5\text{-}C_5H_5)(CO)_2\}_2]$

$\downarrow$

$[Hg\{Fe(\eta^5\text{-}C_5H_5)(CO)_2\}_3]^-$ $\xrightarrow{+2e^-}$ Hg + $3[Fe(\eta^5\text{-}C_5H_5)(CO)_2]^-$

(36)

Scheme 31

$[(\eta^5\text{-}C_5H_5)Fe(-\overset{O}{\underset{\|}{C}}\diagdown Me)(CO)_2]$ $\xrightarrow[\text{[+1.3 V }vs\text{ the S.C.E.]}]{-3e^-}$ $EtO\overset{O}{\underset{\|}{C}}Me$ + unidentified metal products

i

$[(\eta^5\text{-}C_5H_5)Fe(Me)(CO)_2]$ $\xrightarrow[\text{[0.88 V }vs\text{ the S.C.E.]}]{-e^-}$ $[(\eta^5\text{-}C_5H_5)Fe(-\overset{O}{\underset{\|}{C}}\diagdown Me)(CO)(MeCN)]^+$

ii

$[(\eta^5\text{-}C_5H_5)Fe(CN)(-\overset{O}{\underset{\|}{C}}\diagdown Me)(CO)]^-$ $\underset{+e^-\ [+0.40\text{ V }vs\text{ the S.C.E.}]}{\overset{-e^-}{\rightleftarrows}}$ $[(\eta^5\text{-}C_5H_5)Fe(CN)(-\overset{O}{\underset{\|}{C}}\diagdown Me)(CO)]$

iii

$\downarrow$ iv

$Me\overset{O}{\underset{\|}{C}}Me$ + MeCHO
(*ca* 40%)   (*ca* 6%)

+

unidentified metal products

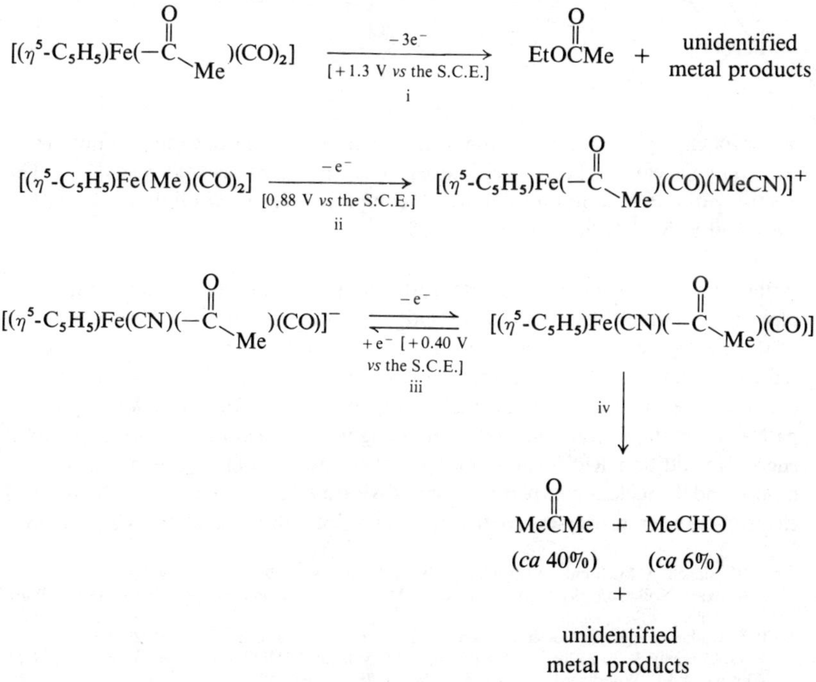

Conditions: i, MeCN + 10%EtOH, [NBu$_4$]$^+$ [ClO$_4$]$^-$, platinum electrode; ii, MeCN, [NBu$_4$]$^+$ [ClO$_4$]$^-$, platinum electrode; iii, MeCN, [NEt$_4$]$^+$ [ClO$_4$]$^-$, platinum electrode, at 0 °C; iv, warm to 20 °C under CO.

Scheme 32

[Fe(CO)$_3$(L)$_2$] in CH$_2$Cl$_2$ electrolyte by cyclic voltammetry on both platinum and mercury. The reversibility on platinum was found to be considerably less than on mercury, and the authors suggest that 'Hg-stabilised' cations are possibly formed.[170] More or less reversible one-electron oxidation of each of a range of phosphine, phosphite, and arsine derivatives of the type [Fe(CO)$_3$(L)$_2$] on a platinum electrode in non-aqueous solvents has been reported.[171]

Phase-sensitive a.c. polarography was used to measure rate parameters for transfer of electrons for the reductions of [Fe(η^5-COT)(CO)$_3$] in a DMF electrolyte. Heterogeneous rate constants for the 0/1− and 1−/2− processes were found to be 0.24 and 0.15 cm s^{-1} respectively, and the values of α were close to 0.5. These rates compared well with that for the free ligand, and it was concluded that no gross structural change accompanied the transfer of electrons.[172] Electrochemical oxidation of this and related species involves a one-electron irreversible step; chemical oxidation gave unusual dimeric species (38).[173]

Various brief electrochemical studies of iron carbene complexes, *e.g.* (39),[174] and

[structure (38): bicyclic diene system with Fe(CO)$_3$ groups, charge 2+]

$$\left[\text{Fe}(\overset{|}{\underset{\text{Me}}{\text{C}}}\text{NCH}_2\text{CH}_2\text{NMe})(\text{CO})_3(\text{PPh}_3)\right]$$

(38) (39)

of tetraphenylcyclobutadienyl complexes,[175] mixed-ligand isocyanide complexes,[176] and polynuclear carbonyl complexes of iron have been described.[177] The electrosyntheses of some unusual MeCN complexes of iron *via* anodic dissolution of the metal in MeCN have been reported.[178,179]

Ruthenium and Osmium.—Certain ruthenium(II) complexes, usually with 2,2′-bipyridyl or related ligands, undergo charge-transfer transitions in the visible region of the electronic spectrum to generate excited-state species that are capable of reducing certain substrates, *e.g.* methyl viologen, in solution. The existence of these exploitable excited states is of considerable interest as they provide a possible pathway for the useful conversion of sunlight into chemical or electrochemical energy. In addition, Ru^{II} forms numerous other water-soluble and, generally robust, mono- and di-nuclear complexes which provide model compounds for the study of electron-transfer processes, particularly intramolecular charge-transfer reactions

---

[170] S. W. Blanch, A. M. Bond, and R. Colton, *Inorg. Chem.*, 1981, **20**, 755.
[171] P. K. Baker, N. G. Connelly, B. M. R. Jones, J. P. Maher, and K. R. Somers, *J. Chem. Soc., Dalton Trans.*, 1980, 579.
[172] B. Tulyathan and W. E. Geiger, *J. Electroanal. Chem. Interfacial Electrochem.*, 1980, **109**, 325.
[173] N. G. Connelly, R. L. Kelly, M. D. Kitchen, R. M. Mills, R. F. D. Stansfield, M. W. Whiteley, S. M. Whiting, and P. Woodward, *J. Chem. Soc., Dalton Trans.*, 1981, 1317.
[174] M. F. Lappert, J. J. MacQuitty, and P. L. Dye, *J. Chem. Soc., Dalton Trans.*, 1981, 1583.
[175] N. G. Connelly, R. L. Kelly, and M. W. Whiteley, *J. Chem. Soc., Dalton Trans.*, 1981, 34.
[176] J. Hanzlik, G. Albertin, E. Bordignon, and A. A. Orio, *J. Organomet. Chem.*, 1982, **224**, 49.
[177] P. A. Dawson, D. M. Peake, B. H. Robinson, and J. Simpson, *Inorg. Chem.*, 1980, **19**, 465.
[178] A. Drummond, J. F. Kay, J. H. Morris, and D. Reed, *J. Chem. Soc., Dalton Trans.*, 1980, 284.
[179] J. F. Kay, J. H. Morris, and D. Reed, *J. Chem. Soc., Dalton Trans.*, 1980, 1917.

and redox reactions of complexes that are confined to an electrode surface. Finally, interesting ligand-centred redox reactions of various Ru^{II} complexes have been discovered which may throw light upon some aspects of the biological chemistry of iron.[180] Recent electrochemical studies of osmium complexes are few, and they generally tend to parallel those of ruthenium.

The complex $[Ru(bipy)_3]^{2+}$ is central to much of the work on photo-redox energy conversion. A recent study of the electronic spectrum of the three reduced states of this complex in an optically transparent thin-layer cell has led to the suggestion that successive one-electron reductions give *localized* bipyridyl radical-anion ligated species, *e.g.* $[Ru^{II}(bipy)(bipy\ radical\ anion)_2]$, rather than delocalized structures, as previously proposed on the basis of e.p.r. data.[181] Interestingly, a self-consistent-field molecular-orbital calculation on $[Ru(bipy)_3]^{2+}$, and on other tris-bipyridyl complexes of d^6 metals, treats the π-electron system of the three ligand molecules as a whole by taking account of $2p\pi$ atomic orbitals that belong to different ligands L. When changes in solvation energy are allowed for, values of $E_\frac{1}{2}$ correlate linearly with the calculated charges on the metal and on the ligand.[182]

Electrochemical data have been included in papers which describe the synthesis and other properties of a wide range of complexes of Ru^{II} with predominantly N-donor ligands.[183—191]

An interesting multi-electron oxidation of hydrocarbon substrates, mediated by a Ru^{IV} redox catalyst, has been reported. *cis*-$[Ru(bipy)_2(py)(H_2O)]^{2+}$ (40) is oxidized in an overall chemically reversible two-electron process to give a reactive $\{Ru^{IV}O\}$ complex (41), which can attack toluene and cyclohexene (Scheme 33). These reactions may possibly relate to the function of FeO groups in the mono-oxygenases.[192]

One of the best examples of the electrochemical activation of a co-ordinated molecule is provided by the multi-electron oxidation of ligating NH_3 through to co-ordinated nitrate at a single ruthenium site (Scheme 34). The redox processes involved in this conversion may bear upon the bacterial degradation of NH_3 by bacteria of the genera *Nitrosomonas* and *Nitrobacter*.[193,194]

The energetics of transfer of electrons to, from, and between metal centres that are joined by a bridging ligand ('bridge') has received much theoretical and experimental attention over the past few years, and the facile synthesis of numerous complexes that contain the $\{Ru(bridge)Ru\}$ moiety has provided the foundation for studies of mixed-valence species.

[180] T. J. Meyer, *Acc. Chem. Res.*, 1978, **11**, 94.
[181] G. A. Heath, L. J. Yellowlees, and P. S. Braterman, *J. Chem. Soc., Chem. Commun.*, 1981, 287.
[182] T. Saji and S. Aoyagui, *J. Electroanal. Chem. Interfacial Electrochem.*, 1980, **108**, 223.
[183] S. S. Isied, *Inorg. Chem.*, 1980, **19**, 911.
[184] S. Goswami, A. R. Chakravarty, and A. Chakravorty, *Inorg. Chem.*, 1981, **20**, 2246.
[185] P. Belser and A. Von Zelewsky, *Helv. Chim. Acta*, 1980, **63**, 1675.
[186] A. R. Chakravarty and A. Chakravorty, *Inorg. Nucl. Chem. Lett.*, 1981, **17**, 97.
[187] R. A. Krause and K. Krause, *Inorg. Chem.*, 1980, **19**, 2600.
[188] B. P. Sullivan, J. M. Calvert, and T. J. Meyer, *Inorg. Chem.*, 1980, **19**, 1404.
[189] P. Bernhard, H. Lehmann, and A. Ludi, *J. Chem. Soc., Chem. Commun.*, 1981, 1216.
[190] A. R. Chakravarty and A. Chakravorty, *Inorg. Chem.*, 1981, **20**, 275.
[191] C-K. Poon, C-M. Che, and Y-P. Kan, *J. Chem. Soc., Dalton Trans.*, 1980, 128.
[192] B. A. Moyer, M. S. Thompson, and T. J. Meyer, *J. Am. Chem. Soc.*, 1980, **102**, 2310.
[193] F. R. Keene, D. J. Salmon, J. L. Walsh, H. D. Abruna, and T. J. Meyer, *Inorg. Chem.*, 1980, **19**, 1896.
[194] M. S. Thompson and T. J. Meyer, *J. Am. Chem. Soc.*, 1981, **103**, 5577.

$[(bipy)_2(py)Ru(OH_2)]^{2+}$ $\xrightarrow[{[E\ =\ +0.47\ V]}]{-2e^-}$ $[(bipy)_2(py)Ru=O]^{2+}$
(40) (41)

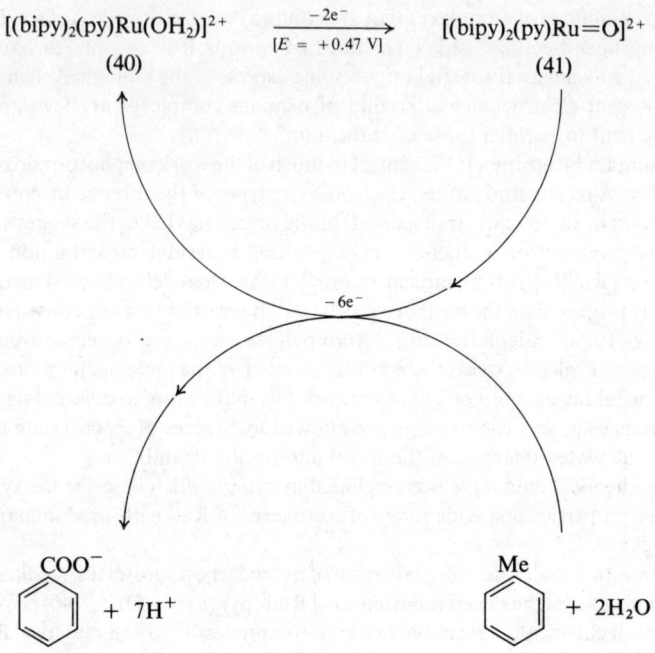

COO⁻ + 7H⁺

Me + 2H₂O

Conditions: [Li]⁺ [ClO₄]⁻, at pH 7, platinum anode; the value of E is *versus* the S.C.E.

Scheme 33

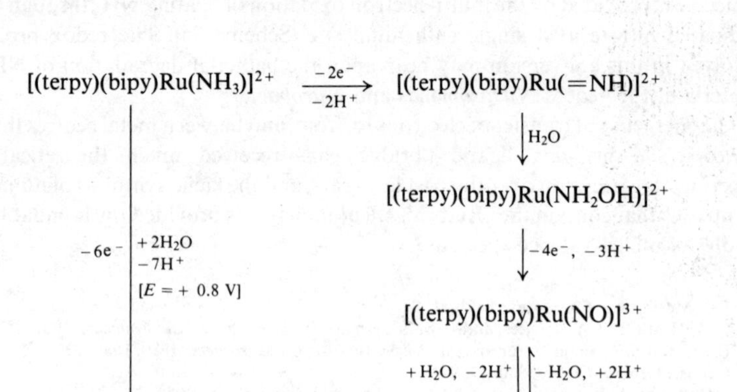

$[(terpy)(bipy)Ru(NH_3)]^{2+}$ $\xrightarrow[{-2H^+}]{-2e^-}$ $[(terpy)(bipy)Ru(=NH)]^{2+}$

↓ H₂O

$[(terpy)(bipy)Ru(NH_2OH)]^{2+}$

$-6e^-$ | $+2H_2O$, $-7H^+$ | $[E = +0.8\ V]$

↓ $-4e^-$, $-3H^+$

$[(terpy)(bipy)Ru(NO)]^{3+}$

+H₂O, −2H⁺ ↕ −H₂O, +2H⁺

$[(terpy)(bipy)Ru(ONO_2)]^{2+}$ $\xleftarrow[{+H_2O,\ -2H^+}]{-3e^-}$ $[(terpy)(bipy)Ru(NO_2)]^+$

Conditions: $H_2PO_4^-/HPO_4^{2-}$ buffer, at pH 6.8; value of E is *versus* the NaCl S.C.E.

Scheme 34

A method for determining the values of $E^{\ominus\prime}$ for the $Ru^{III}(bridge)Ru^{III}/Ru^{III}(bridge)Ru^{II}/Ru^{II}(bridge)Ru^{II}$ couples (and hence the comproportionation constants) by cyclic-voltammetric and differential-pulse-polarographic measurements has been reported.[195] Redox potentials for mono-, di-, and tri-nuclear complexes of ruthenium,[196] weakly interacting ruthenium centres,[197] delocalized systems,[198] species in which 'outer-sphere' intramolecular transfer of electrons obtains, e.g. $[\{bipy)_2ClRu\}_2(Ph_2PCH_2PPh_2)]^{3+}$,[199] and series of pyrazine-bridged phenanthroline complexes have been discussed in detail.[200]

Redox data for $[Os(OEP)(1-Me-im)]$ (1-Me-im = 1-methylimidazole) in CH_2Cl_2 electrolyte have been reported.[201] Redox and excited-state-lifetime data on carbonyl ruthenium porphyrin complexes have been discussed.[202] *Internal* transfer of electrons from Ru^{II} to OEP is induced by a change in the axial ligand: replacement of CO by a better *net* donor stabilizes the Ru^{III} centre (Scheme 35).[203]

$[Ru^{II}(OEP)(CO)(py)] \xrightarrow{-e^-} [Ru^{II}(OEP^{\div})(CO)(py)]^+$

$\downarrow PPh_3$

$[Ru^{II}(OEP)(PPh_3)_2] \underset{+e^-}{\overset{-e^-}{\rightleftarrows}} [Ru^{III}(OEP)(PPh_3)_2]^+ \underset{+e^-}{\overset{-e^-}{\rightleftarrows}} [Ru^{III}(OEP^{\div})(PPh_3)_2]^{2+}$

Conditions: CH_2Cl_2, $[NBu_4]^+$ $[ClO_4]^-$, platinum electrode

Scheme 35

Low-temperature electrochemical data on dithiocarbamato-complexes of osmium for the couples $Os^V/Os^{IV}/Os^{III}/Os^{II}$ have been reported.[204]

Numerous studies of ruthenium complexes that are covalently anchored,[205-207] electrostatically bound,[208, 209] and confined in a polymeric film[210-214] at electrode

---

[195] D. E. Richardson and H. Taube, *Inorg. Chem.*, 1981, **20**, 1278.
[196] W. F. Sokol, D. H. Evans, K. Niki, and T. Yagi, *J. Electroanal. Chem. Interfacial Electrochem.*, 1980, **108**, 107.
[197] J. E. Sutton and H. Taube, *Inorg. Chem.*, 1981, **20**, 3125.
[198] J. A. Baumann and T. J. Meyer, *Inorg. Chem.*, 1980, **19**, 345.
[199] B. P. Sullivan and T. J. Meyer, *Inorg. Chem.*, 1980, **19**, 752.
[200] K. Cloninger and R. W. Callaghan, *Inorg. Chem.*, 1981, **20**, 1611.
[201] J. Billecke, W. Kokisch, and J. W. Buchler, *J. Am. Chem. Soc.*, 1980, **102**, 3624.
[202] D. P. Rilleman, J. K. Neagle, L. F. Burringer, and T. J. Meyer, *J. Am. Chem. Soc.*, 1981, **103**, 56.
[203] M. Barley, J. Y. Becker, G. Domazetis, D. Dolphin, and B. R. James, *J. Chem. Soc., Chem. Commun.*, 1981, 982.
[204] S. H. Wheeler and L. H. Pignolet, *Inorg. Chem.*, 1980, **19**, 972.
[205] H. D. Abruna, J. L. Walsh, T. J. Meyer, and R. W. Murray, *Inorg. Chem.*, 1981, **20**, 1481.
[206] H. D. Abruna, J. L. Walsh, T. J. Meyer, and R. W. Murray, *J. Am. Chem. Soc.*, 1980, **102**, 3272.
[207] P. K. Ghosh and T. G. Spiro, *J. Am. Chem. Soc.*, 1980, **102**, 5543.
[208] I. Rubinstein and A. J. Bard, *J. Am. Chem. Soc.*, 1980, **102**, 6641.
[209] N. Oyama, T. Shimomura, and K. Shigehara, *J. Electroanal. Chem. Interfacial Electrochem.*, 1980, **112**, 271.
[210] J. M. Calvert and T. J. Meyer, *Inorg. Chem.*, 1981, **19**, 27.
[211] O. Haas and J. G. Vos, *J. Electroanal. Chem. Interfacial Electrochem.*, 1980, **113**, 139.
[212] O. Haas, M. Kriens, and J. G. Vos, *J. Am. Chem. Soc.*, 1981, **103**, 1318.
[213] N. S. Scott, N. Oyama, and F. C. Anson, *J. Electroanal. Chem. Interfacial Electrochem.*, 1980, **110**, 303.
[214] G. J. Samuels and T. J. Meyer, *J. Am. Chem. Soc.*, 1981, **103**, 307.

surfaces have recently been reported. Electro-initiated polymerization of vinylpyridine complexes of ruthenium and of vinylferrocene has allowed the construction of bilayer assemblies with a 'rectifying' interface.[215] Covalently bound nitrosyl, nitrito- and nitrato-complexes of ruthenium undergo similar redox reactions to their analogues in free solution,[206] and $[Ru(bipy)_3]^{4+}$, when generated in a polymer film, can effect the oxidation of alcohols to ketones.[214] Electrostatically bound $[Ru(bipy)_3]^{2+}$ in a Nafion polymer functions as an electrochemical luminescence device in aqueous solutions that contain oxalate.[208]

6 Cobalt, Rhodium, and Iridium

Much of the recently reported electrochemistry of cobalt complexes is associated with two research themes. First, studies of the cobalamins; secondly, studies of the interaction of dioxygen with cobalt centres. Apart from investigations relating to the alkyl-cobalamins, electrochemical studies of organometallic cobalt compounds are limited, and this area remains largely undeveloped. Electrochemical studies of complexes of rhodium and iridium are few.

Cobalt.—*Compounds that are Related to the Cobalamins.* Cobalt(III) in cyanocobalamin (Vitamin B_{12}, cobinamide) is chelated by four nitrogen atoms in a macrocyclic ring of the corrin type. The two axial co-ordination positions are occupied by a cyanide anion and by a benzimidazole base, which is itself anchored to the corrin ring. Two cobalamin cofactors, important in enzymatic methyl-transfer reactions and 1,2-shift reactions of hydrogen, are formed by replacement of the cyanide ligand by either a methyl group or by 5-deoxyadenosyl respectively, in the B_{12} form of the vitamin, (42).

The electrochemical behaviour of two inorganic 'models' of the cobalamins, a cobaloxime (43) and Costa's complex (44), has been compared to that of coenzyme B_{12} compounds. Effects of the concentration of base upon the stability of the Co–C bond were explored by cyclic voltammetry and it was concluded that Costa's complex (44) was a somewhat better mimic of cobalamin chemistry than was the dimethylglyoxime derivative.[216,217] Costa and co-workers have shown that the electrochemical generation of $[Co^I(salen)]^-$ in the presence of Bu^tCl (or Bu^tBr) yields $[Co^{II}(salen)]$, 2-methylpropene, and hydrogen as the main products (Scheme 36).[218]

Indirect electrochemical reduction of organic bromides, *e.g.* (45), is mediated by aquocobalamin (Vitamin B_{12a}) or a synthetic analogue (46); the process is proposed to occur *via* Co^I- and Co^{III}-alkyl intermediates (Scheme 37).[219]

One-electron oxidations of $[Co^{III}Et(salen)]$ and a range of related Schiffs-base and glyoxime complexes were shown by cyclic voltammetry to be electrochemically reversible in a MeCN electrolyte at a platinum electrode. It was found that pyridine reacts rapidly with the Co^V–alkyl complexes to displace the alkyl group *via*

---

[215] H. D. Abruna, P. Denisevich, M. Umana, T. J. Meyer, and R. W. Murray, *J. Am. Chem. Soc.*, 1981, **103**, 1.
[216] R. G. Finke, B. L. Smith, R. Droege, C. M. Elliott, and E. Hershenhart, *J. Organomet. Chem.*, 1980, **202**, C25.
[217] C. M. Elliott, E. Hershenhart, R. G. Finke, and B. L. Smith, *J. Am. Chem. Soc.*, 1981, **103**, 5558.
[218] A. Puxeddu, G. Costa, and N. Marsich, *J. Chem. Soc., Dalton Trans.*, 1980, 1489.
[219] R. Scheffold, M. Dike, S. Dike, T. Herold, and L. Walder, *J. Am. Chem. Soc.*, 1980, **102**, 3642.

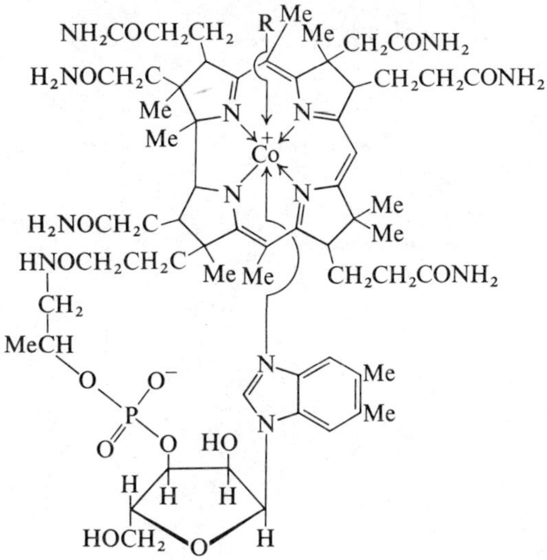

(42) R = [adenosyl group] or Me

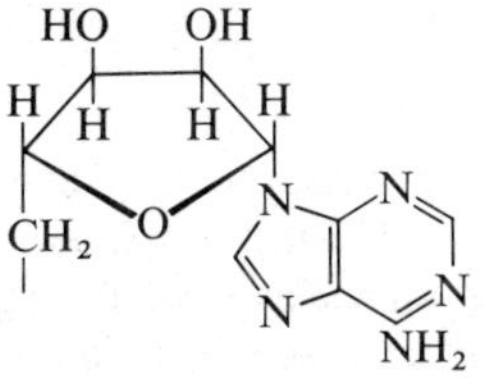

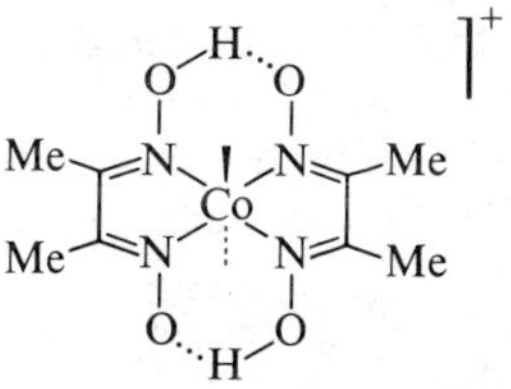

(43) 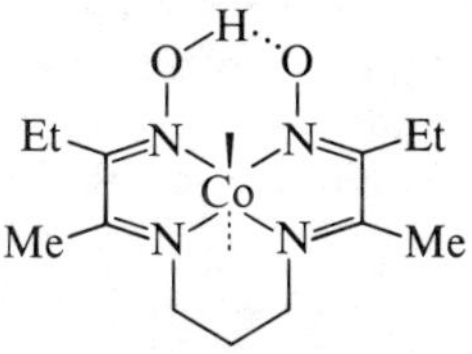 (44)

$$[Co^{II}(salen)] + e^- \longrightarrow [Co^I(salen)]^-$$

$$\downarrow Bu^t Br$$

$$C_4H_8 + \tfrac{1}{2}H_2 + Br^- \longleftarrow [CoBr(Bu^t)(salen)]^-$$

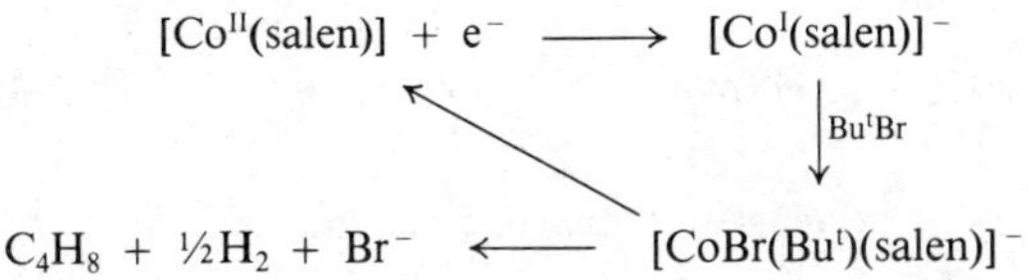

Scheme 36

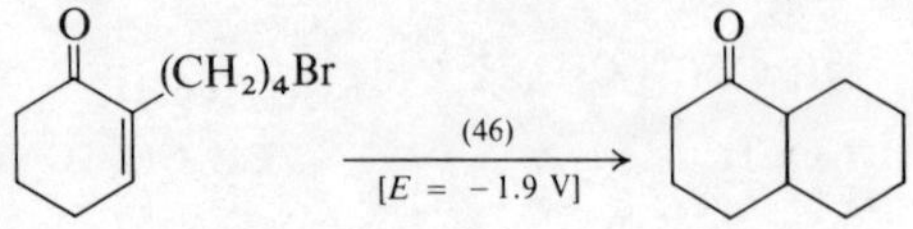

Conditions: DMF, 0.1M-[Li]$^+$ [ClO$_4$]$^-$ +0.05M-[NH$_4$]$^+$ [Br]$^-$, Hg pool electrode; the value of E is *versus* the S.C.E.

Scheme 37

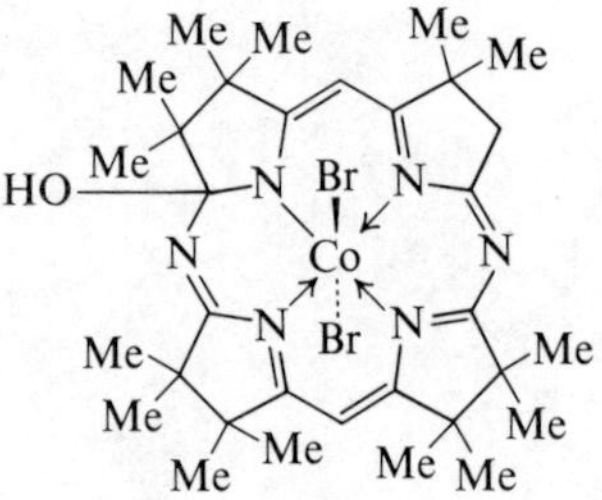

nucleophilic substitution at the α-carbon atom. Kinetic results suggested that an S_N2 mechanism was operative for the glyoxime complexes whereas the salen complexes reacted *via* an internal nucleophilic displacement, S_Ni. In some cases the Co^{IV}–C bond undergoes rapid unimolecular heterolytic or homolytic dissociation. These pathways are summarized by Scheme 38.[220]

In a very detailed electrochemical and kinetic study, Kochi and co-workers have shown that dimethyl derivatives of complexes of Co^{III} with macrocyclic N-donor ligands undergo irreversible one-electron oxidation in a MeCN electrolyte with the concomitant homolytic cleavage of one Co–C bond. In the absence of donors of hydrogen or chlorine atoms, the predominant organic product is ethane, formed by

$$[RCo^{III}(chel)] \underset{+e^-}{\overset{-e^-}{\rightleftarrows}} [RCo^{IV}(chel)]^+$$

$$[RCo^{IV}(chel)]^+ + py \rightleftarrows [RCo^{IV}(chel)(py)]^+$$

$$[RCo^{IV}(chel)(py)]^+ \longrightarrow [R(py)]^+ + [Co^{II}(chel)]$$

(chel = a chelating macrocycle)

Scheme 38

---

[220] M. E. Vol'pin, I. Ya. Levitin, A. L. Sigan, J. Halpern, and G. M. Tom, *Inorg. Chim. Acta*, 1980, **41**, 271.

radical–radical coupling. The comparison of electrochemical with chemical kinetic data allowed the elucidation of the intimate mechanism.[221]

The validity of various cobalt complexes as models for the cobalamins has been questioned by Heineman, Mark, and co-workers on the grounds that the reduction of Co^{III} to Co^{II} in vitamin B_{12} is unusually slow, except for the aquocobalamin Co^{III} species. Substitution of the cobalt in the B_{12} corrin by other metals, such as copper, nickel, or rhodium, results in complexes that are electro-inactive in the range -0.96 to $+1.25$ V *versus* the N.H.E., even in the presence of mediators. The authors conclude that there is a high activation energy for transfer of electrons, and their explanation for this is that the periphery of the cobalamin molecule is completely saturated, and the only pathway for transfer of charge is through the axial ligands.[222] Electrochemical studies of 1,18-dimethyldehydrocorrin complexes of cobalt have been reported.[223]

Recently, detailed studies of aquo- and cyano-cobalamins, rather than of models of them, have been described. These spectroelectrochemical studies have been principally concerned with the speciation of the axial ligand of the oxidized and reduced forms of cobalamin at various values of pH and pCN^- in buffered aqueous solution. The diaquocobalamin Co^{III} species is reduced at a potential some 700 mV negative of the Co^{III}/Co^{II} couple for the aquocobalamin. At low pH, the Co^{III}/Co^{II} waves are clearly separated from that for Co^{II}/Co^{I}, whilst at high pH the $Co^{III}/Co^{II}/Co^{I}$ couples merge. Spontaneous reduction of Co^{III} to Co^{II} occurs at pH 10. The speciation is discussed in terms of the benzimidazole base-on/base-off forms. The electrochemical two-electron reduction of dicyanocobalamin is suggested to occur *via* an *e.c.e.* mechanism in all cases where electrochemical methods allow its distinction from a disproportionation pathway.[224—227]

Cobalt Complexes and Dioxygen. The utilization of transition-metal sites to relax constraints of spin and of symmetry on the reactivity of molecular oxygen and the promotion of multi-electron reduction processes has a direct bearing upon fuel-cell technology, where a catalyst for the reduction of O_2 to H_2O by four electrons, at or near the reversible thermodynamic potential of $+1.23$ V *versus* the N.H.E., is sought (see Chapter 4). Present technology to effect the fast reduction of O_2 to water demands the use of expensive platinum electrodes at quite high overpotentials. Cobalt complexes appear to be particularly promising as catalysts for the multi-electron reduction of O_2; many stable dinuclear species that contain bridging peroxo-ligands are now known, and factors relating to their formation, to the mode of bonding of O_2, and to their redox behaviour are being actively studied. The design and synthesis of complexes to effect the rapid reduction of O_2 to H_2O have been briefly reviewed.[228]

[221] W. H. Tamblyn, R. J. Klingler, W. S. Hwang, and J. K. Kochi, *J. Am. Chem. Soc.*, 1981, **103**, 3161.
[222] K. A. Rubinson, J. Caja, R. W. Hurst, E. Itabashi, T. M. Kenyhercz, W. R. Heineman, and H. B. Mark, Jr., *J. Chem. Soc., Chem. Commun.*, 1980, 47.
[223] Y. Murakami, Y. Aoyama, and K. Tokunaga, *J. Am. Chem. Soc.*, 1980, **102**, 6736.
[224] D. Lexa, J. M. Saveant, and J. Zickler, *J. Am. Chem. Soc.*, 1980, **102**, 2654.
[225] D. Lexa, J. M. Saveant, and J. Zickler, *J. Am. Chem. Soc.*, 1980, **102**, 4851.
[226] R. L. Birke and S. Vankatesan, *J. Electrochem. Soc.*, 1981, **128**, 984.
[227] C. Amatore, D. Lexa, and J. M. Saveant, *J. Electroanal. Chem. Interfacial Electrochem.*, 1980, **111**, 81.
[228] J. P. Collman, F. C. Anson, S. Bencosme, A. Chong, T. Collins, P. Denisevich, E. Evit, T. Geiger, and J. A. Ibers, in *Org. Synth.: Today Tomorrow* [*Proc. IUPAC Symp. Org. Synth., 3rd, 1980*], ed. B. M. Trost and G. R. Hutchinson, Pergamon Press, Oxford, 1981, p. 29.

Octahedral cobalt(II) complexes, generally with predominantly ammine or polyamine ligands, react with molecular oxygen in aqueous solution to give stable peroxo-bridged dinuclear complexes that contain the structural unit (47). The redox potentials of a wide range of such species have been determined and related to the nature of the bonding of $Co(\eta^2\text{-}O_2)Co$ and to the concept of transfer of charge from Co^{II} to O_2 upon oxygenation of the mononuclear precursor complex.[229—231] The thermodynamics and kinetics of uptake of O_2 by Co^{II} complexes, e.g. (48), have been

studied amperometrically and formal oxidation potentials measured in an aqueous electrolyte, at 25 °C.[232] However, such stable complexes tend not to give H_2O upon reduction. Electrocatalytic reduction of O_2 has been studied, using homogeneous and surface-confined complexes of cobalt.[233—235] Anson, Collman, and their co-workers have prepared a series of dimeric metalloporphyrin molecules which were deposited on graphite and studied electrochemically at a ring-disc electrode, in an aqueous electrolyte, in the presence of O_2. Most effected the less interesting two-electron reduction of O_2 to H_2O_2; however, the dimeric Co^{II} complex (49) was found

(49) (Each ellipse represents a porphyrin ring)

to be capable of catalysing the four-electron reduction of O_2 to H_2O.[236] Relationships between structure and catalytic activity were discussed.[236] Cobalt tetra(aminophenyl)porphyrins have been covalently bonded to carbon electrodes, but the transfer of electrons in the Co^{III}/Co^{II} step is extraordinarily slow, possibly because of the co-ordination of surface carboxylate groups to the metal centre.[237]

[229] W. R. Harris, G. L. McLendon, A. E. Martell, R. C. Bess, and M. Mason, *Inorg. Chem.*, 1980, **19**, 21.
[230] G. McLendon and W. F. Mooney, *Inorg. Chem.*, 1980, **19**, 12.
[231] S. R. Pickens and A. E. Martell, *Inorg. Chem.*, 1980, **19**, 15.
[232] C-L. Wong, J. A. Switzer, K. P. Balakrishnan, and J. F. Endicott, *J. Am. Chem. Soc.*, 1980, **102**, 5511.
[233] T. Geiger and F. C. Anson, *J. Am. Chem. Soc.*, 1981, **103**, 7489.
[234] J. H. Zagal, R. Bindra, and E. B. Yeager, *J. Electrochem. Soc.*, 1980, **127**, 1506.
[235] A. Bettelheim, R. J. H. Chan, and T. Kuwana, *J. Electrochem. Soc.*, 1980, **110**, 93.
[236] J. P. Collman, P. Denisevich, Y. Konai, M. Marrow, C. Koval, and F. C. Anson, *J. Am. Chem. Soc.*, 1980, **102**, 6027.
[237] K. W. Willman, R. D. Rocklin, R. Nowak, K-N. Kuo, F. A. Schultz, and R. W. Murray, *J. Am. Chem. Soc.*, 1980, **102**, 7629.

Other Studies of Cobalt Compounds. The polarography of the $[Co(en)_3]^{3+/2+}$ couple has been studied in a range of solvents[238] and the electrode kinetics of EDMA, iminodiacetate, and diethylenetriamine complexes of the $Co^{III/II}$ couple have been investigated.[239,240]

Electrosyntheses of cobalt complexes *via* anodic dissolution of the metal in the presence of various ligands have been described.[241,242]

Electrochemical reduction of $[Co\{S_2C_2(CN)_2\}_2]^{2-}$ affords the trianion, which reacts rapidly with weak acids to give the unstable hydrido-species.[243]

Cobalt nitrosyl complexes of the type $[CoI\{P(OEt)_3\}_3(NO)]^+$ have been reported to undergo irreversible oxidation in a non-aqueous electrolyte to give a range of Co^{III} and Co^{II} products.[244] The e.s.r. spectra of radical anions generated *via* the electrolytic one-electron reduction of various dinuclear acetylene-bridged cobalt carbonyl complexes in a THF electrolyte at $-60\,°C$ have been discussed.[245] The fluorophosphine-bridged complex $[(CO)_2Co\{MeN(PF_2)_2\}Co(CO)_2]$ undergoes successive electrochemically reversible reductions to give the green radical anion and the pale-yellow dianion: interactions of these anions with O_2, MeI, Li^+, and H^+ are briefly reported.[246] Electrochemically generated radical anions of cobalt carbonyls undergo rapid substitution of CO by phosphines or phosphites to give products which are electro-oxidized to the closed-shell species at the potential that is necessary for the generation of the radical anion: this affords an electrode-induced substitution pathway.[247]

Finally, the electrocatalytic reduction of CO_2 to CO and H_2 with up to 90% current efficiency *via* cobalt (and nickel) complexes, *e.g.* (50), has been described (Scheme 39).[248]

Rhodium and Iridium.—The interaction of Vaska's complex, $[IrCl(CO)(PPh_3)_2]$, with Hg^{2+}, Cu^{2+}, or Fe^{3+} in CH_2Cl_2 and PhH/EtOH media has been studied

(50)

---

[238] A. Kotocova and U. Mayer, *Collect. Czech. Chem. Commun.*, 1980, **45**, 335.
[239] T. Ohsaka, N. Oyama, S. Yamaguchi, and H. Matsuda, *Bull. Chem. Soc. Jpn.*, 1981, **54**, 2475.
[240] A. Yamada, T. Yoshikuni, and N. Tanaka, *Inorg. Chem.*, 1981, **20**, 2090.
[241] A. Cinquanit, R. Seeber, R. Cini, and P. Zanello, *Inorg. Chim. Acta*, 1981, **53**, L201.
[242] J. J. Habeeb, F. F. Said, and D. G. Tuck, *J. Chem. Soc., Dalton Trans.*, 1981, 118.
[243] A. Vlcek, Jr., and A. A. Vlcek, *Inorg. Chim. Acta*, 1980, **41**, 123.
[244] R. Seeber, G. A. Mazzocchin, G. Albertin, and E. Bordignon, *J. Chem. Soc., Dalton Trans.*, 1980, 979.
[245] B. M. Peake, P. H. Rieger, B. H. Robinson, and J. Simpson, *J. Am. Chem. Soc.*, 1980, **102**, 156.
[246] A. Chaloyard and N. El Murr, *J. Organomet. Chem.*, 1980, **188**, C13.
[247] G. J. Bezems, P. H. Rieger, and S. Visco, *J. Chem. Soc., Chem. Commun.*, 1981, 265.
[248] B. Fisher and R. Eisenberg, *J. Am. Chem. Soc.*, 1980, **102**, 7361.

$$CO_2 \xrightarrow[{[E\ =\ -1.6\ V]}]{(50)} CO + H_2$$

[93% current efficiency]

Conditions: H_2O – MeCN (2 : 1), 0.1M-$[K]^+$ $[NO_3]^-$, Hg pool electrode; the value of E is versus the S.C.E.

Scheme 39

polarographically;[249] trans-$[IrCl_4(PMe_2Ph)_2]$ has been shown to be a convenient and reasonably strong oxidant in non-aqueous solvents ($E^{\ominus\prime} = +0.90$ V versus Ag/AgCl).[250] Schiffs-base complexes of Rh^I and Ir^I have been studied by polarography and by cyclic voltammetry in CH_2Cl_2 and in MeCN.[251] The complex $[\{Rh_2(CNCH_2CH_2CH_2NC)_4\}_2Cl]$ produces H_2 from acidic aqueous solution upon irradiation with visible light. This species was generated at a rotating-disc electrode, and its reactions with H_3O^+ were studied.[252] The ions $[Rh(phen)_3]^+$, $[Rh(bipy)_3]^{3+}$, and $[Rh(bipy)_2(OH)_2]^+$ have been studied by cyclic voltammetry in aqueous solution and redox properties have been discussed in terms of the thermodynamics of the photo-generation of H_2.[253] The electronic structures of the multiply bonded complex $[Rh_2(O_2CEt)_4(PPh_3)_2]^+$ and related cation radicals were studied by generating them electrolytically in the cavity of an e.s.r. spectrometer.[254]

7 Nickel, Palladium, and Platinum

The electrochemistry of nickel complexes has received comparatively little attention recently. Studies have centred around the redox reactions of quadridentate square-planar complexes of nickel(II) and tetrahedral complexes of nickel(II) with tertiary phosphines, e.g. $[NiCl_2(PBu^t_3)_2]$. The reactions of electrogenerated reduced derivatives of these species with alkyl and aryl halides have continued to be studied, unabashed by the general observation that the organic products are almost invariably derived from alkyl and aryl radicals.

Studies of complexes of platinum and palladium have been few.

Nickel.—Nickelocene, $[Ni(\eta^5-C_5H_5)_2]$, is oxidized in two, successive, reversible one-electron steps in a melt of $AlCl_3$ and 1-butylpyridinium chloride.[255] Reduction of $[Ni(\eta^5-C_5H_5)(PPh_3)_2]^+$ affords the intermediates $[Ni(\eta^5-C_5H_5)(PPh_3)_2]$ and $[Ni(\eta^5-C_5H_5)(PPh_3)]$, which undergo ring- and phosphine-transfer reactions to give stable $[Ni(PPh_3)_4]$ and $[Ni(\eta^5-C_5H_5)_2]$.[256] The trinuclear complexes $[Ni_2(\eta^5-C_5H_5)_2(\mu^3-$

---

[249] J. Vecernik, J. Masek, and A. A. Vlcek, *Collect. Czech. Chem. Commun.*, 1980, **45**, 1468.
[250] C. E. Briant, K. A. Rowland, C. T. Webber, and D. M. P. Mingos, *J. Chem. Soc., Dalton Trans.*, 1981, 1515.
[251] J. Hanzlik, A. Camus, G. Mestroni, and G. Zassinovich, *J. Organomet. Chem.*, 1981, **204**, 115.
[252] K. R. Mann and B. A. Parkinson, *Inorg. Chem.*, 1981, **20**, 1921.
[253] R. Ballardini, G. Varani, and V. Balzani, *J. Am. Chem. Soc.*, 1980, **102**, 1719.
[254] T. Kawamura, K. Fukamachi, T. Sowa, S. Hayashida, and T. Yonezawa, *J. Am. Chem. Soc.*, 1981, **103**, 364.
[255] R. J. Gale and R. Job, *Inorg. Chem.*, 1981, **20**, 45.
[256] E. K. Barefield, D. A. Krost, D. S. Edwards, D. G. Van Derveer, R. L. Trytko, S. P. O'Rear, and A. N. Williamson, *J. Am. Chem. Soc.*, 1981, **103**, 6219.

CO)$_2$Co(η^5-C$_5$H$_{5-n}$Me$_n$)]$^+$ (n = 0, 1, or 5) are reduced reversibly at a platinum electrode in a dimethoxyethane electrolyte.[257]

The electrochemical reduction of the complexes [NiX$_2$(PR$_3$)$_2$] (X = halide, R = alkyl or aryl) in non-aqueous solvents, in the presence of PR$_3$, affords a variety of Ni^I and Ni^0 species. The reduction of such Ni^{II} species in the presence of alkyl or aryl halides can be electrocatalytic, and can afford the usual organic products that arise from the formation of alkyl or aryl radicals.[258—261]

A range of square-planar complexes of nickel(II) with quadridentate Schiffs-base ligands have been examined and shown to undergo metal- or ligand-based redox reactions.[262] Ligand-induced intermolecular transfer of electrons has been demonstrated for the macrocyclic complex (51), as shown in Scheme 40, and related

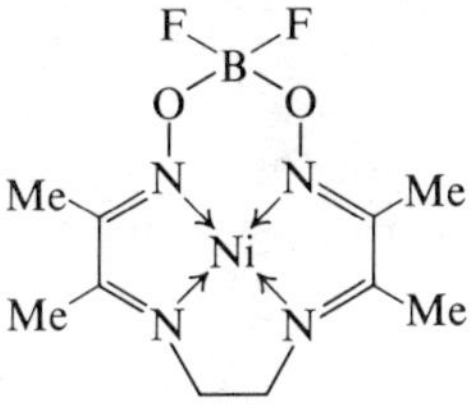

(51)

[Ni^{II}(mac)] $\underset{-e^-}{\overset{+e^-}{\rightleftarrows}}$ [Ni^{II}(mac$^-$)] $\xrightarrow{CO}$ [Ni^I(CO)(mac)]
(51)
  (mac = macrocycle)

Conditions: DMF, [NBu$_4$]$^+$ [ClO$_4$]$^-$, platinum electrode

Scheme 40

species,[263, 264] and the first example of a dynamic equilibrium between metal-oxidized–ligand-reduced and metal-reduced–ligand-oxidized partners has been identified.[263] The electrochemistry of two symmetrical 'insulated' nickel(II) centres, contained in bi-macrocyclic Schiffs-base ligands, has been reported.[265] The anodic generation of a water-soluble complex of nickel(III) and a macrocyclic ligand, and of its Cu^{III} analogue, has been shown.[266]

[257] L. R. Byers, V. A. Uchtman, and L. F. Dahl, *J. Am. Chem. Soc.*, 1981, **103**, 1942.
[258] G. Bontempelli, F. Magno, M. De Nobili, and G. Schiavon, *J. Chem. Soc., Dalton Trans.*, 1980, 2288.
[259] S. Sibille, M. Troupel, J-F. Fauvarque, and J. Perichon, *J. Chem. Res. (S)*, 1980, 147.
[260] M. Troupel, Y. Rollin, S. Sibille, J. Perichon, and J-F. Fauvarque, *J. Organomet. Chem.*, 1980, **202**, 435.
[261] S. Sibille, J-C. Folest, J. Coulombeix, M. Troupel, J-F. Fauvarque, and J. Perichon, *J. Chem. Res. (S)*, 1980, 268.
[262] L. Fabbrizzi and A. Poggi, *Inorg. Chim. Acta*, 1980, **39**, 207.
[263] R. R. Gagne and D. M. Ingle, *J. Am. Chem. Soc.*, 1980, **102**, 1443.
[264] R. R. Gagne and D. M. Ingle, *Inorg. Chem.*, 1981, **20**, 420.
[265] D. H. Busch, G. G. Christoph, L. L. Zimmer, S. C. Jackels, J. J. Grzybowski, R. C. Callahan, M. Kojima, K. A. Holter, J. Mocak, N. Herron, M. Chavan, and W. P. Schammel, *J. Am. Chem. Soc.*, 1981, **103**, 5107.
[266] L. Fabbrizzi and A. Poggi, *J. Chem. Soc., Chem. Commun.*, 1980, 646.

Indirect electrochemical reduction of alkyl bromides to give good yields of dimeric products, mediated by complexes of nickel(II) with Schiffs bases, occurs via a radical pathway;[267] in the presence of an alkene, a mixture of organic products is obtained.[268] The [Ni(acac)$_2$]-catalysed conjugate addition of alkenyl-zirconium reagents to $\alpha\beta$-unsaturated ketones has been studied electrochemically.[269] Studies of amino-acid– and thiolato–nickel(II) complexes have been described.[270,271]

Palladium and Platinum.—The reductions of various bis-halido-complexes of Pt^{II} and Pd^{II} that contain isocyanide[272,273] and 2-aminopyridine[274] and of the bis(arylazo-oximato) ligands[275] of Pt^0, Pd^I, and Pd^0 species have been reported. The porphyrin complex [Pt(TPP)Cl$_2$] undergoes an irreversible two-electron reduction to [Pt(TPP)].[276] A co-ordinatively unsaturated platinacyclobutane [Pt(C$_3$H$_6$)(bipy)] has been electrosynthesized from [PtCl$_2$(C$_3$H$_6$)(bipy)]. The Pt^{II} complex can be further electroreduced to an e.s.r.-active anion in a reversible one-electron process.[277] Trimetallic complexes, e.g. [Pt(Bu^{t}NC)$_2${Mn(CO)$_5$}$_2$], usually undergo irreversible one-electron reductions in non-aqueous media at platinum or gold electrodes with rupture of one Pt–M bond and the expulsion of an anion, e.g. [Mn(CO)$_5$]$^-$.[278—280]

8 Conclusions

This Report has categorized the electrochemical behaviour of transition-metal complexes according to the metal centre; here it is appropriate to focus on some general themes which interrelate the electrochemistry.

Organometallic Complexes.—Electrochemical reactions of metal carbonyl complexes are fairly well represented in each Group of the transition metals, but their electrochemistry offers few surprises. Electrochemical activation of the CO ligand towards nucleophilic or electrophilic attack has yet to be demonstrated, although a novel insertion reaction has been reported.[167] There are now several interesting electrochemical reactions which involve transformations of co-ordinated cyclopentadienyl or arene ligands, and undoubtedly this area will develop rapidly.[72,160] There have been several elegant electrochemical studies of the mechanism of formation and cleavage of σ-bonds between a transition metal and

---

[267] C. Gosden and D. Pletcher, *J. Organomet. Chem.*, 1980, **186**, 401.
[268] C. Gosden, J. B. Kerr, D. Pletcher, and R. Rosas, *J. Electroanal. Chem. Interfacial Electrochem.*, 1981, **117**, 101.
[269] F. M. Dayrit, D. E. Gladkowski, and J. Schwartz, *J. Am. Chem. Soc.*, 1980, **102**, 3976.
[270] D. M. Roundhill, *Inorg. Chem.*, 1980, **19**, 557.
[271] D. Negoin, J. Camus, and N. Muresan, *Rev. Roum. Chim.*, 1980, **25**, 839.
[272] J. E. Dubois, P. C. Lacaze, and M. Pham, *J. Electroanal. Chem. Interfacial Electrochem.*, 1981, **117**, 233.
[273] P. Lemoine, A. Giraudeau, M. Gross, and P. Braunstein, *J. Organomet. Chem.*, 1980, **202**, 447.
[274] G. Horn and H. P. Schroer, *Stud. Biophys.*, 1980, **81**, 83.
[275] P. Bandyopadhyay, P. K. Mascharak, and A. Chakravorty, *J. Chem. Soc., Dalton Trans.*, 1981, 623.
[276] J. W. Buchler, K-L. Lay, and H. Stoppa, *Z. Naturforsch., Teil B*, 1980, **35**, 433.
[277] R. J. Klingler, J. C. Huffman, and J. K. Kochi, *J. Organomet. Chem.*, 1981, **206**, C7.
[278] A. Giraudeau, P. Lemoine, M. Gross, and P. Braunstein, *J. Organomet. Chem.*, 1980, **202**, 455.
[279] P. Lemoine, A. Giraudeau, M. Gross, R. Bender, and P. Braunstein, *J. Chem. Soc., Dalton Trans.*, 1981, 2059.
[280] P. Lemoine, A. Giraudeau, M. Gross, and P. Braunstein, *J. Chem. Soc., Chem. Commun.*, 1980, 77.

carbon in a wide range of complexes.[115,267] Unfortunately, homolytic scission seems to dominate the cleavage reactions, and consequently organic products are usually derived from free radicals, which couple, disproportionate, or abstract hydrogen atoms from another substrate to give hydrocarbon products. The conversion of acyl or aroyl ligands into hydroxycarbene groups via electrochemical oxidation has been described;[74] however, reactions which generate carbonium ion or carbanionic character in a M–C bond are few, and useful syntheses that are based on these remain elusive.

Complexes with Co-ordinated Nitrogen and Oxygen Groups.—Electrochemical activation of co-ordinated nitrogen-containing ligands is beginning to appear relatively facile. Several new electrode reactions have been identified, e.g. the conversion of co-ordinated NH_3 into co-ordinated NO_3^-,[193] which involve multi-electron oxidation or reduction processes. A wide range of oxo-ligand species have been examined electrochemically, and this has been stimulated by metalloenzyme chemistry; for example, the oxidation of hydrocarbons by the P-450 mono-oxygenases. An electrochemically generated $Ru^{IV}=O$ species[192] has been shown to oxidize unsaturated hydrocarbons and, with little doubt, electrochemical studies of =O, —OH, and OH_2 complexes with this type of activation in mind will continue. The multi-electron reduction of co-ordinated dioxygen to water, mediated by transition-metal complexes, provides another important basis for electrochemical studies.[236]

The Influence of Ligands upon the Redox Properties of Transition-metal Complexes.—The influence of ligands on the redox potentials of transition-metal centres has long been recognized, and there have been several recent systematic studies. The speciation of axial ligands in porphyrin, corrin, and other macrocyclic ligand complexes has been extensively explored and an important observation in this area is that a single ligand change can induce intramolecular transfer of electrons between the macrocycle and the metal.[203,262] Electrode-induced substitution of metal complexes in which there is little net current flow is another novel development.[247] Finally, the electrochemistry of super-complexes in which a complex, e.g. $[Fe(CN)_6]^{4-}$, is held within a macrocyclic cavity promises to provide a fascinating area of research.

Photoelectrochemistry and the Electrochemistry of Modified Electrodes.—The importance of electrochemical studies of transition-metal complexes in the area of energy conversion hardly needs emphasis, and there have been numerous recent papers concerned with this.[228] Metal complexes that are anchored to electrodes or contained in polymeric films on an electrode surface continue to attract considerable attention. A novel development has been the construction of bilayer assemblies which have rectifying properties.[215] There are still relatively few examples of the use of electrodes that are modified with metal complexes to electrocatalyse (or mediate) the oxidation or reduction of a substrate,[214] other than those associated with the reduction of O_2 to peroxide or water.

Acknowledgement. The author thanks Mr Ian Latham for help in the preparation of this manuscript.

4
The Electrochemistry of Oxygen

BY D. J. SCHIFFRIN

1 Introduction

The electrochemistry of oxygen has lately received a great deal of attention. There are sound economic reasons for this, in particular in relation to the fuel-cell programme currently in progress in the U.S.A.[1] and the possibility of using oxygen cathodes in the chlor-alkali industry to replace the hydrogen-evolution reaction,[2] as discussed by Beck and Ruggeri.[3] The replacement of the hydrogen-evolution reaction by the reduction of oxygen would result, in principle, in a saving of 0.8 V of cell voltage, but an inexpensive, stable, and efficient air electrode is required for this application. The use of macrocyclic organic metal complexes as electrocatalysts is an interesting alternative to electrodes that are based on precious metals, and such complexes have been the subject of two recent reviews by van den Brink, Barendrecht, and Visscher[4] and by Tarasevich and Radyushkina;[5] also, the relationship between the electrochemistry of macrocycles and the biochemical aspects of the utilization of oxygen is becoming increasingly clear. Spiro has recently edited a book on these problems[6] in which the main current ideas are reviewed. These include transport of oxygen, oxygen-insertion reactions, the role of oxygen in the respiratory chain, and the reactivity of radicals that are derived from oxygen. These problems have also been the subject of the Second BOC Priestley Conference, in 1980.[7]

The corrosion of metals in aqueous environments has far-reaching practical consequences, but the study of the basic electrochemistry of the usual cathodic process, *i.e.* reduction of oxygen, on engineering materials has been very limited. In particular, the understanding of the effects of the usual contaminants, oxide layers, *etc.*, on the mechanism of reduction of O_2 has received very little attention.

This Report covers approximately the period from the end of 1979 to October 1981, and reference to previous work has been kept to a minimum. A general description of the mechanism of reduction is first given, then the use of different electrode materials and different reaction media is described, and finally, some areas of applications are given. Within this context, corrosion due to dissolved oxygen has

---

[1] Editorial, *Consult. Eng.*, 1981, **45**, No. 6, p.38.
[2] E. Yeager and P. Bindra, *Chem.-Ing.-Tech.*, 1980, **52**, 384.
[3] T. R. Beck and R. T. Ruggeri, in 'Advances in Electrochemistry and Electrochemical Engineering', ed. H. Gerischer and C. W. Tobias, Wiley, New York, 1981, p. 263.
[4] F. van den Brink, E. Barendrecht, and W. Visscher, *Recl.: J. Roy. Neth. Chem. Soc.*, 1980, **99**, 253.
[5] M. R. Tarasevich and K. A. Radyushkina, *Russ. Chem. Rev. (Engl. Transl.)*, 1980, **49**, 718.
[6] 'Metal Ion Activation of Dioxygen', ed. T. G. Spiro, Wiley, New York, 1980.
[7] 'Oxygen and Life', (Special Publication No. 39), Royal Society of Chemistry, London, 1981.

2 A General Mechanism for the Reduction of Oxygen

Reference will be made to the generalized mechanism for the reduction of O_2; this is shown in Scheme 1 for alkaline solutions.[8] Here k_1 represents the direct reduction to

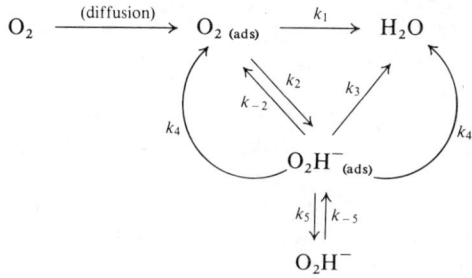

Scheme 1

water, *i.e.*, without the formation of an intermediate that can be desorbed and detected on the ring of a rotating-ring–disc electrode (RRDE) experiment; k_1 will describe equally well any reaction in which the reductive splitting of the O–O bond occurs entirely in the adsorbed state, and the RRDE diagnostic criteria will not allow a distinction to be made between an electrochemical reductive cleavage and dissociative adsorption of oxygen. The routes shown in Scheme 1 are therefore related to the possibilities offered by the RRDE method, and although this may appear to be a limitation, its usefulness has been repeatedly demonstrated. The factor k_2 is an overall rate constant for the formation of adsorbed peroxide, and may involve other rate constants that are related to both the intermediate formation of adsorbed superoxide and its disproportionation reactions; k_3 is the rate constant for reduction of peroxide, k_4 refers to the dismutation of adsorbed peroxide, and k_5 and k_{-5} represent rate constants for the processes of desorption and adsorption of peroxide.

The equations relating the disc and ring currents to the rate constants have been described[8] and used for many systems.[9]

3 Studies on Different Electrode Materials

Carbon.—Besides its application as a support for catalysts, the characteristics of a variety of carbons for the reduction of O_2 have been studied. Efforts have been made to find properties that can be related to their behaviour as electrode materials; for instance, Balej *et al.*[10] correlated the concentration of paramagnetic centres of

---

[8] H. S. Wroblowa, Y-C. Pan, and G. Razumney, *J. Electroanal. Chem. Interfacial Electrochem.*, 1976, **69**, 195.
[9] H. Behret, W. Clauberg, and G. Sandstede, *Ber. Bunsenges. Phys. Chem.*, 1979, **83**, 139.
[10] J. Balej, K. Balogh, P. Stopka, and O. Spalek, *Collect. Czech. Chem. Commun.*, 1980, **45**, 3249.

carbons used in the construction of porous electrodes with the yield of H_2O_2. A clear correlation was demonstrated; a high concentration of paramagnetic centres decreases the production of peroxide. Pyrolysis at 1200 °C in the presence of K_2S resulted in both a permanent suppression of the paramagnetic centres and a dramatic 250-fold increase in the yield of H_2O_2 as compared with some of the untreated samples.

The influence of BET area on electrochemical activity has been studied by Appleby et al.,[11] who tested 47 different types of carbons and graphites. The materials employed in this comprehensive study were graphites, thermal blacks, acetylene blacks, bone-black, furnace blacks, channel blacks, and activated carbons from several sources. The BET areas of most of the materials studied did not decrease much on teflonation. Nickel-supported porous electrodes were studied, in 6M-KOH, and the observed Tafel slopes were ~ -40 mV per decade for the materials of high specific surface area whereas the values were greater than -60 mV per decade for the carbons of low specific surface area. The reaction order with respect to oxygen was 1 and the Tafel slopes were pH-dependent. An important result that was obtained in this work was the demonstration of a direct proportionality between surface area and electrocatalytic activity for channel and furnace blacks, and for ground graphite, over two orders of magnitude of BET area. This indicates a very high accessibility to all the available surface of these materials. From such measurements, however, the authors concluded that, in the case of the activated carbons, only a fraction of the surface is accessible for sustaining an electrochemical process. However, it was proposed that the unused pores could serve as a reservoir of active material which could become exposed as the active areas were progressively destroyed with use, thus providing an enhanced life-span for practical devices.

The question of flooding of pores has been analysed by Gamburtsev et al.,[12] using a.c. impedance techniques for the estimation of the actual area of contact between the carbon surfaces and the electrolyte. Flooding and limitations on the transport of gas were more noticeable when air was used instead of pure oxygen, and the process of flooding was a result of irreversible changes in the surface on the active areas of the carbon.

In contrast with the results of Appleby et al.,[11] Kukushkina et al.[13] found two Tafel slopes in the reduction of oxygen on activated carbons of high specific area (1000 $m^2 g^{-1}$). These authors had previously developed a new technique for performing electrochemical studies on powders.[14] Usually, the study of the properties of disperse powders presents difficulties in defining a convenient and reproducible electrode surface. PTFE-bonded electrodes with a known loading have been used,[11] but the comparison of materials of different porosity is difficult and the actual area of the electrode in contact with the solution is uncertain. To overcome these problems, paraffin paste, solid pyrolytic graphite, or vitreous

---

[11] A. J. Appleby and J. Marie, *Electrochim. Acta*, 1979, **24**, 195.
[12] S. Gamburtsev, I. Iliev, A. Kaisheva, G. V. Shteinberg, and L. N. Mokrousov, *Sov. Electrochem. (Engl. Transl.)*, 1980, **16**, 912.
[13] I. A. Kukushkina, G. V. Shteinberg, M. R. Tarasevich, and V. S. Bagotskii, *Sov. Electrochem. (Engl. Transl.)*, 1981, **17**, 193.
[14] G. V. Shteinberg, I. A. Kukushkina, V. S. Bagotskii, and M. R. Tarasevich, *Sov. Electrochem. (Engl. Transl.)*, 1979, **15**, 443.

carbon electrodes have been employed in the past. The use of paste electrodes suffers from the obvious disadvantage of the possible alteration of the electrocatalytic properties of the dispersed materials by the paraffin that is employed in their preparation. Shteinberg et al.[14] overcame these difficulties by using a floating gas-diffusion electrode. This consisted of a very thin layer of carbon, usually a monolayer of particles of diameter 20 μm, placed on a porous, hydrophobic, conducting substrate. No binder was used, and the whole assembly floated on the solution under study, with the carbon particles placed between the solution and the floating current-collector. The actual area of contact was determined by measurements of double-layer capacitance. Using this technique, Tafel slopes of −55 to −60 mV per decade and of −110 to −150 mV per decade were observed for potentials below and above 0.75—0.80 V (vs a hydrogen electrode, at the same pH as the solution under study), respectively.[13] The behaviour of the activated carbons that were studied was very similar to that observed for platinum,[15] and the change in the Tafel slope was again ascribed to changes with potential in the nature of the surface functional groups on carbon.

It is interesting to compare these observations with electrochemical and spectroscopic evidence on the nature of oxidized carbon surfaces. Lowde et al.[16] observed the onset of a broad surface-reduction process for oxidized carbon, at a potential of ∼0.8 V, and proposed the formation of phenolic hydroxy-groups. The nature of the 'surface oxides' will depend strongly both on the method of preparation of the carbon and on the oxidation conditions employed;[17] the electrochemical surface activities of carbons have also been attributed to the presence of quinone/hydroquinone surface redox groups[18,19] (see the work cited in ref. 16).

At present, there is not a clear understanding of the relationship between surface groups that are formed on oxidized carbons and their electrocatalytic activity. Oxidation of the surface of carbon fibres has been studied, though from the point of view of their adhesion properties in polymer composites,[18] but further work is required to establish the relationship between the nature of surface groups and surface chemical properties of the carbons.

The mechanism of reduction of oxygen on different carbons is far from clear. As discussed by Appleby et al.,[11] a Tafel slope approaching −30 mV per decade can simply result from the establishment of steady-state diffusion conditions across the diffusion layer, when the surface equilibrium is the two-electron process:

$$O_2 + H_2O + 2e^- \rightleftharpoons O_2H^- + OH^- \qquad (1)$$

The argument presented is that, for diffusion of O_2H^- across a Nernst layer of thickness δ, the current is given by:

$$i = 2FD_{O_2H^-} C^\sigma_{O_2H^-}/\delta \qquad (2)$$

where F is the Faraday constant and $D_{O_2H^-}$ is the diffusion coefficient and $C^\sigma_{O_2H^-}$ the

---

[15] D. B. Sepa, M. V. Vojnovic, and A. Damjanovic, *Electrochim. Acta*, 1981, **26**, 781.
[16] D. R. Lowde, J. O. Williams, P. A. Attwood, R. J. Bird, B. D. McNicol, and R. T. Short, *J. Chem. Soc., Faraday Trans. 1*, 1979, **75**, 2312.
[17] H. Selig and L. B. Ebert, *Adv. Inorg. Chem. Radiochem.*, 1980, **23**, 281.
[18] E. Fitzer, K. H. Geigl, W. Huettner, and R. Weiss, *Carbon*, 1980, **18**, 389.
[19] J. P. Randin and E. Yeager, *J. Electroanal. Chem. Interfacial Electrochem.*, 1975, **58**, 313.

surface concentration of O_2H^-. Since equilibrium (1) prevails at the surface, the potential is given by:

$$E = E_{[1]}^{\ominus} - \frac{RT}{2F} \ln C_{O_2H^-}^{\sigma} C_{OH^-} \qquad (3)$$

where $E_{[1]}^{\ominus}$ is the standard potential of reaction (1) and C_{OH^-} the concentration of OH^- ions, which is considered to be constant. The combination of equations (2) and (3) gives the desired current–potential relationship, i.e.,

$$i = \text{const} \times (C_{OH^-})^{-1} \exp[-2(E - E_{[1]}^{\ominus})F/RT] \qquad (4)$$

These authors rejected this explanation since the observed value of the coefficient $(\partial E/\partial \text{pH})_i$ was RT/F instead of $RT/2F$. It is perhaps possible that the origin of this discrepancy is in the neglect of the acid–base properties of H_2O_2. Considering the equilibrium:

$$O_2H^- + H_2O \rightleftharpoons H_2O_2 + OH^- \qquad (5)$$

the equilibrium potential for reaction (1) is now:

$$E = E_{[1]}^{\ominus} - \frac{RT}{2F} \ln\left[C_T^{\sigma} \left\{ \frac{(C_{OH^-})^2}{K + C_{OH^-}} \right\} \right] \qquad (6)$$

where C_T^{σ} is the total concentration of peroxide at the electrode surface ($C_{H_2O_2}^{\sigma} + C_{O_2H^-}^{\sigma}$) and K is the equilibrium constant of reaction (5). From equations (2) and (6), the current is now given by:

$$i = \frac{2FD_{O_2H^-}}{\delta}(C_{OH^-})^{-2}(K + C_{OH^-}) \exp[-2(E - E_{[1]}^{\ominus})F/RT] \qquad (7)$$

For $C_{OH^-} \gg K$, the coefficient $(\partial E/\partial \text{pH})_i$ is $RT/2F$, whereas for $C_{OH^-} \ll K$, $(\partial E/\partial \text{pH})_i = RT/F$, and the different dependence on pH may be due to the acid–base properties of hydrogen peroxide. Similar arguments apply in the case of macrocycle catalysts that are supported on active carbon.

The mechanism proposed by Appleby et al.[11] involves the reduction of superoxide ion in the rate-determining step (r.d.s.), but the proposed reaction involving the formation of a hydroxyl radical:

$$O_2^- + H_2O + e^- \rightarrow O_2H^- + OH\cdot \quad \text{(r.d.s.)} \qquad (8)$$

is obviously incorrect. The reaction sequence that was proposed by Kukushkina et al.[13] was:

$$O_2 \rightleftharpoons O_{2(\text{ads})} \qquad (9)$$

$$O_{2(\text{ads})} + e^- \rightarrow O_2^-{}_{(\text{ads})} \quad \text{(r.d.s.)} \qquad (10)$$

$$O_2^-{}_{(\text{ads})} + H_2O \rightarrow HO_2 + OH^- \qquad (11)$$

$$HO_2 + H_2O + e^- \rightarrow H_2O_2 + OH^- \qquad (12)$$

The observed change in slopes occurs with a surface that is partially blocked by carbon oxides, the coverage being linearly dependent on potential.

The Electrochemistry of Oxygen

Cobalt.—The reduction of oxygen on passive cobalt in 1M-KOH has been studied by Fabjan et al.[20] The Tafel slope that was found was of -75 to -80 mV per decade; the reaction order with respect to oxygen was $\frac{1}{2}$, and cobalt was found to be a better electrocatalytic material than palladium, silver, gold, nickel, or iron. The same reaction order was found for nickel and iron, and this was taken to be indicative of a disproportionation step in the reduction sequence (either Scheme 2 or Scheme 3). Scheme 2 involves the disproportionation of superoxide to give hydroxyl radicals [step (3)], whereas in Scheme 3, decomposition of peroxide is suggested.

Scheme 2	Scheme 3
(1) $O_2 + e^- \rightarrow O_2^-$	(1) $O_2 + e^- \rightarrow O_2^-$
(2) $O_2^- + H_2O \rightarrow HO_2 + OH^-$	(2) $O_2^- + H_2O \rightarrow HO_2 + OH^-$
(3) $HO_2 \rightarrow OH\cdot + \frac{1}{2}O_2$	(3) $HO_2 + e^- \rightarrow O_2H^-$
(4) $OH\cdot + e^- \rightarrow OH^-$	(4) $O_2H^- \rightarrow OH^- + \frac{1}{2}O_2$

These reaction sequences are of great practical importance and will therefore be discussed in some detail. Although the disproportionation step (3) in Scheme 2 is an attractive proposition, there are several difficulties with this proposal.

(a) The pK of the acid–base equilibrium:

$$HO_2 \rightleftharpoons H^+ + O_2^- \tag{13}$$

is 4.69,[21] and therefore, in 1M-KOH, step (3) of Scheme 2 should read:

$$(3')\ O_2^- + O_2^- + 2H_2O \rightarrow 2OH\cdot + 2OH^- + O_2 \tag{14}$$

(b) The free energy for the above disproportionation reaction can be calculated, using the value of the standard potential for the O_2/O_2^- couple in aqueous solutions measured by Meisel and Czapski[22] ($E^{\ominus}_{O_2/O_2^-} = -0.325$ V), and taking other thermodynamic data from tables.[23] The value obtained for reaction (14) is $+52.6$ kJ mol^{-1}, indicating that, at least, the *homogeneous* disproportionation reaction [(3) of Scheme 2] is thermodynamically unfavourable. This is in contrast with the well-established decay pathways for the superoxide radical that have been observed from pulse radiolysis, i.e.[21]

$$O_2^- + O_2^- + H_2O \rightleftharpoons O_2 + H_2O_2 + 2OH^- \tag{15}$$

---

[20] Ch. Fabjan, M. R. Kazemi, and A. Neckel, *Ber. Bunsenges. Phys. Chem.*, 1980, **84**, 1026.
[21] B. H. J. Bielski, *Photochem. Photobiol.*, 1978, **28**, 645.
[22] D. Meisel and G. Czapski, *J. Phys. Chem.*, 1975, **79**, 1503.
[23] C.R.C. Handbook of Chemistry and Physics, 61st Edition, 1980, CRC Press Inc., Florida, USA.

Although these considerations indicate that the generation of OH˙ radicals by reaction (3) of Scheme 2 is unlikely, the possibility of this reaction occurring between adsorbed species must not be excluded. As an example of the difficulties arising from the previous discussion, it is interesting to mention previous work by Zurilla, Sen, and Yeager[24] on the reduction of oxygen on gold in alkaline solutions. These authors concluded that the rate at which the dismutation reaction (15) occurred with adsorbed species was at least two orders of magnitude faster than the reported homogeneous rate. As a consequence, they proposed that the second step in the reduction sequence could be the chemical dismutation process (15), and the mechanism corresponded thus to Scheme 3 without the protonation steps. In order to distinguish between the different reaction pathways it is probably necessary to use other techniques; for instance, spectroscopic methods for the analysis of intermediates *in situ*.

The analysis of the reaction order presents another difficulty; the value of $\frac{1}{2}$ that was found means that each oxygen molecule yields two species that are reduced in the rate-determining step (r.d.s.). The Tafel slope that was measured for cobalt can be taken as indicative of the r.d.s. corresponding to the first electron-transfer step [(1) in Scheme 2 or (2) in Scheme 3]. This is also the case for iron and nickel (see later) and therefore, in the sequential mechanisms proposed by Fabjan *et al.*,[20] the dismutation step should not affect the r.d.s., unless some of the reduction steps occur in quasi-equilibrium.

Copper and Its Alloys.—In spite of the great practical importance of this metal in its use in corrosion-resistant alloys for applications in seawater, only one kinetic study of the reduction of oxygen on this metal has been published. Balakrishnan and Venkatesan[25] investigated this process on copper and brass, using the rotating-ring-disc electrode technique. From this work, it is clear that the kinetics of reduction of oxygen are strongly influenced by the presence of Cu_2O on the surface. In neutral solutions, the anodic linear-sweep voltammetry of oxide-free electrodes presents a broad peak in the potential range -0.54 to -0.6 V *vs* the N.H.E., followed by rapid inhibition at more positive potentials. These features are absent if the electrode has been pre-passivated at the rest potential, and this probably indicates the involvement of soluble cuprous species in this potential range. It has been suggested that the anodic formation of passive films on copper in alkaline solutions occurs through a dissolution–precipitation mechanism,[26] involving the reactions:[27]

$$Cu + OH^- \rightarrow Cu(OH) + e^- \quad (16)$$

followed by:

$$2\,Cu(OH) \rightleftharpoons Cu_2O + H_2O \quad (17)$$

The complications that arise in the study of the reduction of oxygen on a metal that can dissolve anodically, yielding ions that can catalyse reactions of intermediates (such as H_2O_2), are staggering. Skinner *et al.*[28] showed that Cu^+ ions react

---

[24] R. W. Zurilla, R. K. Sen, and E. Yeager, *J. Electrochem. Soc.*, 1978, **125**, 1103.
[25] K. Balakrishnan and V. K. Venkatesan, *Electrochim. Acta*, 1979, **24**, 131.
[26] M. Yamashita, K. Omura, and D. Hirayama, *Surf. Sci.*, 1980, **96**, 443.
[27] S. M. Abd El Haleem and B. G. Ateya, *J. Electroanal. Chem. Interfacial Electrochem.*, 1981, **117**, 309.
[28] J. F. Skinner, A. Glasel, L-C. Hsu, and B. L. Funt, *J. Electrochem. Soc.*, 1980, **127**, 315.

with H_2O_2 according to:

$$Cu^+ + H_2O_2 \xrightarrow{k_1} Cu^{2+} + OH^- + OH\cdot \qquad (18)$$

followed by:

$$Cu^+ + OH\cdot \xrightarrow{k_2} Cu^{2+} + OH^- \qquad (19)$$

with $k_1 = 260$ dm^3 mol^{-1} s^{-1} at 25 °C. These reactions must occur in the region where the maximum in the current is observed, and the features of reduction of oxygen on copper must reflect the combined effects of the dissolution–precipitation reactions, the Cu^I–H_2O_2 reaction, and the normal kinetics of the reduction of O_2 on metals (whatever can be considered here as normal!). This may explain the unusual negative slopes of the diagnostic plots[8] of I_{disc}/I_{ring} *versus* $\omega^{-\frac{1}{2}}$ which Balakrishnan *et al.*[25] ascribed to the reaction of hydrogen atoms with oxygen, according to:

$$O_2 + 2H\cdot \rightarrow H_2O_2 \qquad (20)$$

Although this reaction could occur on electrodes where a significant surface concentration of adsorbed hydrogen atoms can exist, this is an unlikely explanation for copper electrodes in neutral solutions.

To add to these complications, the behaviour of copper and its alloys in seawater[29] differs significantly from that observed in saline solutions[25] in that no current maximum is observed, and a single Tafel slope of -0.14 V per decade is observed.

Gold.— The kinetic constants for the reduction of oxygen on gold in contact with high-purity solutions of 0.1M-KOH and 0.05M-H_2SO_4 have been studied by Fischer *et al.*[30,31] The RRDE technique was employed and the ring-disc results were analysed by the diagnostic method described by Wroblova *et al.*[8] The values for k_1 and k_2 (see Scheme 1) for the 'direct' and sequential pathways were 0.02 and 0.4 cm s^{-1}, respectively, in 0.1M-KOH whereas in the 0.05M-H_2SO_4 solution these values were 3×10^{-3} and 0.2 cm s^{-1}, respectively. It is surprising that these authors found experimental evidence for the 'direct' reduction mechanism on gold, since previous extensive work by Zurilla *et al.*[24] in 0.1M-NaOH failed to reveal this effect. A similar analysis of ring–disc data in NaOH solutions by Adzić *et al.*[32,33] did not show evidence for a direct reduction pathway either, and the origin of this discrepancy is not clear.

The modification of the electrocatalytic properties of metal surfaces by adsorption of foreign metal adatoms has received considerable attention in the past few years,[34] since it was discovered that adsorbed monolayers of metal ions had electrocatalytic properties which differed widely from those of the bulk metal.

A clear example of these interesting properties is the change in the mechanism for

---

[29] S. R. de Sanchez and D. J. Schiffrin, *Corrosion Sci.*, 1982, **22**, 585; also unpublished results.
[30] P. Fischer and J. Heitbaum, *J. Electroanal. Chem. Interfacial Electrochem.*, 1980, **112**, 231.
[31] P. Fischer, J. Heitbaum, and W. Vielstich, *Chem.-Ing.-Tech.*, 1980, **52**, 423.
[32] R. R. Adzić, A. V. Tripković, and N. M. Marković, *J. Electroanal. Chem. Interfacial Electrochem.*, 1980, **114**, 37.
[33] R. R. Adzić, N. M. Marković, and A. V. Tripković, *Bull. Soc. Chim. Beograd*, 1980, **45**, 399.
[34] R. R. Adzić, *Isr. J. Chem.*, 1979, **18**, 166.

reduction of oxygen on gold that is caused by adsorbed bismuth[33] and lead ions,[32] in alkaline solutions. Electrocatalysis by the deposition of these two ions at an underpotential is rather similar, and results in an almost two-fold increase of the reduction current. This was due to the change in the mechanism for reduction of O_2H^-: on gold surfaces, the rate of reduction of peroxide is limited by a slow chemical step, whereas in the modified surfaces it corresponds to an activated process with Tafel slope of -120 mV per decade. Adzić et al.[32,33] used the rotating-ring–disc technique to obtain diagnostic information based on the reactions shown in Scheme 1 and found that the results for the modified gold surfaces could be interpreted either as a direct + sequential mechanism, operating in parallel, or as a completely sequential reaction, with the peroxide strongly adsorbed. With this last assumption, the very low levels of peroxide that could be picked up at the ring (less than 2% of the disc current) resulted from both a slow desorption step of the intermediate and from the large increase in the rate constant for electrochemical reduction of O_2H^- that is caused by the adsorbed metal ions.

The mechanism of reduction on these modified surfaces is still uncertain; at sufficiently negative potentials, corresponding to the formation of a full monolayer, the two-electron reduction that is characteristic of bismuth- and lead-modified surfaces is observed, indicating that the catalytic effect originates in the interaction of the metal ions with a Au(OH) layer. It was suggested that the charge on the adsorbed ions is thus decreased, favouring the adsorption of O_2, O_2^-, and O_2H^- on the modified surface. Also, even at potentials where Au(OH) is not stable, the presence of adsorbed metal ions would have a similar effect of increasing the adsorption of the intermediates for reduction of O_2.

The catalytic effects of the ions of some heavy metals on the reduction of oxygen have been known for a long time;[35] in particular, the catalytic decomposition of hydrogen peroxide by alkaline solutions of lead(II) ions is well documented (see ref. 36 and the work cited in ref. 37). It is tempting, therefore, to consider the possibility of a mechanism involving the reaction between lead(II) and O_2H^-:

$$O_2H^- + Pb^{II} + H_2O \rightarrow Pb^{IV} + 3\,OH^- \qquad (21)$$

followed by the electrochemical step:

$$Pb^{IV} + 2\,e^- \rightarrow Pb^{II} \qquad (22)$$

in the catalysis caused by Au/Pb surfaces in the reduction of oxygen.

The difference between this sequence and the proposed mechanisms of homogeneous catalytic decomposition is that, in this case, the source of electrons is the metal surface instead of O_2H^- ion. Unfortunately, the mechanism of catalysis appears to be more complicated than this; in the case of reduction of oxygen on a mercury surface, Pieterse et al.[37] concluded that the catalytic effect could not be due to the oxidation–reduction cycle of Pb^{II}–Pb^{IV}. Further work is required to elucidate these important effects.

---

[35] F. Strnad, Collect. Czech. Chem. Commun., 1939, **11**, 391.
[36] 'Gmelins Handbuch der Anorganischen Chemie; Sauerstoff, Lieferung 7', Verlag Chemie, Weinheim/Bergstr., 1966, p. 2320.
[37] M. M. J. Pieterse, M. Sluyters-Rehbach, and J. H. Sluyters, J. Electroanal. Chem. Interfacial Electrochem., 1980, **109**, 41.

Iron and Steel.—The reduction of oxygen on iron and steel is one of the most important electrochemical reactions, since it determines the rate of deterioration of structures, motor vehicles, ships, *etc.* It is very surprising, therefore, that the amount of work published in recent years on this subject is minimal. Studies of the polarization of passive iron in alkaline solution showed a Tafel slope of -0.14 to -0.16 V per decade in alkaline solutions and reaction orders of $\frac{1}{2}$ with respect to O_2 and -1 with respect to the OH^- ions.[20] The negative reaction order with respect to OH^- was considered to result from specific adsorption of hydroxide ions on active centres and to a decrease in the solubility of O_2 with an increase in the concentration of alkali. This last consideration is not very clear; if the pressure of oxygen is kept constant, and the system is at equilibrium, the thermodynamic activity of dissolved O_2 should be almost constant as the concentration of alkali is varied. The change in solubility of O_2 should therefore have no effect, since the thermodynamic activity is the variable to be considered in the kinetic equation. This observation is, of course, applicable to many other reports where the decrease in solubility due to salting-out effects is considered to influence the reaction order, even though conditions of saturation by O_2 prevail. Probably, a high concentration of alkali does affect the kinetics of reduction of O_2 through a change in the activity of water, which can be very large when comparing dilute with very concentrated KOH solutions. This will be so if a water molecule is involved in the rate-determining step. Besides competition for active sites, the negative value that is found for the reaction order may reflect the changes in the oxide–solution inner potential difference that are due to the establishment of the O^{2-}–H_2O equilibrium at the interface.

Lead.—The interest in the reduction of oxygen on lead is in the design of sealed, maintenance-free, lead–acid batteries, where the oxygen that is evolved during overcharging can be reduced at the negative plate. The only recent study was that of Khomskaya *et al.*,[38] who used a rotating lead disc electrode in contact with 4.75M-H_2SO_4. A reaction order of 1 with respect to oxygen, and diffusion control, were observed in the potential range that was studied (-0.96 to -1.30 V *vs* the mercury/mercurous sulphate electrode in the same acid solution). The potential limitations are due both to the onset of the evolution of hydrogen and to the anodic dissolution process. For instance, in the potential scale used by these authors, the redox potential for the $Pb/PbSO_4$ couple is -0.972 V;[23] therefore it is clear that, in any practical application in a lead–acid battery, the reduction of oxygen is likely to be diffusionally controlled. The only evidence presented in this work for a four-electron reduction ($n=4$) is that the value of the diffusion coefficient that is calculated from the dependence of the limiting current on rate of rotation (8×10^{-6} cm^2 s^{-1}) is in agreement with data from the literature. The four-electron reduction of O_2 on lead in a $HClO_4$ solution has recently been confirmed by Zwetanova and Jüttner[39] by comparing the limiting current of silver and lead rotating-disc electrodes.

In alkaline solutions, however, a value of $n=2$ has been observed,[32] and the reduction mechanism is different. A similar behaviour has been observed also for the

---

[38] E. A. Khomskaya, N. F. Gorbacheva, and N. B. Tolochkov, *Sov. Electrochem. (Engl. Transl.)*, 1980, **16**, 48.
[39] A. Zwetanova and K. Jüttner, *J. Electroanal. Chem. Interfacial Electrochem.*, 1981, **119**, 149.

reduction of oxygen on precious metals.[40] Hayes and Kuhn[41] have briefly reviewed some of the previous work on lead.

Mercury.—The kinetics of reduction of oxygen on mercury at a pH of 12.35 have been studied by Pieterse et al.,[42] using an a.c. impedance technique. No evidence could be found for any intermediate chemical limiting step, and the value of the standard rate constant that was found was $k_{sh}^{\ominus} = 0.035$ cm s^{-1}. Care should be taken when comparing values of rate constants; values of $k_{sh}^{\ominus}$ that are given in this work refer to the standard potential for the couple:

$$O_2 + H_2O + 2e^- \rightleftharpoons O_2H^- + OH^- \quad (23)$$

whereas very often, in the literature, the exchange current densities are referred to the redox potential of O_2/OH^-. The overall transfer coefficient (αn), where n is the number of electrons in the total reaction and α the transfer coefficient, was found to be 0.42, which leads to a Tafel slope of -0.134 mV per decade, indicating that the first electron-transfer reaction is rate-determining. An interesting comparison is made in this work with previous estimates for the standard heterogeneous rate constant for transfer of the first electron:

$$O_2 + e^- \rightarrow O_2^{\cdot -} \quad (24)$$

The standard rate constant for this reaction was estimated at 1.34 cm s^{-1}, in good agreement with previous calculations in the literature.

The catalytic effect of adsorbed Pb(OH)$_2$ on mercury on the kinetics of reduction of oxygen has been studied by Pieterse et al.,[37] using a d.c. polarographic technique. The currents were analysed, using an approximate procedure for the coupling of diffusional and kinetic control processes at a dropping mercury electrode. Neither chemical decomposition of H$_2$O$_2$ by Pb(OH)$_2$, leading to the regeneration of O$_2$, nor chemical reaction of H$_2$O$_2$ with Pb(OH)$_2$, followed by the electrochemical reduction of the higher oxidation state that is generated (i.e. Pb^{IV}) could account for the experimental results. The main conclusion of this work was that the effect of adsorbed Pb^{II} on the mercury surface is to make possible the further reduction of H$_2$O$_2$ in the potential region of the oxygen wave. The dependence of the heterogeneous rate constant (k_f) on the concentration of Pb^{II} was found to be of the form $1/\log k_f \propto 1/[\text{Pb}^{II}]$, which was indicative of a first-order reaction with respect to adsorbed Pb^{II} on the mercury surface, if the adsorption of Pb^{II} ion obeys a Langmuir isotherm. Some quantitative information on the adsorption of Pb^{II} (using, for instance, the techniques developed by Barclay and Anson[43]) would be of great interest for the understanding of these catalytic effects. As the authors point out, the double-layer capacitance of the mercury/solution interface in the presence of adsorbed Pb^{II} is very complex.

The presence of trace concentrations of metal ions that can catalyse the decomposition of hydrogen peroxide can have a profound effect on its voltammetric

---

[40] D. B. Sepa, M. V. Vojnovic, and A. Damjanovic, *Electrochim. Acta*, 1981, **26**, 781.

[41] M. Hayes and A. T. Kuhn, in 'Electrochemistry of Lead', ed. A. T. Kuhn, Academic Press, London, 1979, p. 199.

[42] M. M. J. Pieterse, M. Sluyters-Rehbach, and J. H. Sluyters, *J. Electroanal. Chem. Interfacial Electrochem.*, 1980, **107**, 247.

[43] D. J. Barclay and F. C. Anson, *J. Electroanal. Chem. Interfacial Electrochem.*, 1970, **28**, 71.

behaviour. Bednarkiewicz and Kublik[44] studied the influence of Fe^{III} on the reduction of H_2O_2 on a hanging-mercury-drop electrode, in alkaline solutions. The first reduction peak occurs at the reversible potential of the Fe^{III}/Fe^{II} couple, and the kinetic complications that are often observed in studies of the reduction of H_2O_2 were attributed entirely to trace levels of Fe^{III} as impurity. The rate constant for the reaction of Fe^{II} with hydrogen peroxide was estimated at 5.6×10^9 dm^3 mol^{-1} s^{-1}. This reaction rate is close to that of diffusionally controlled bimolecular reactions, and is an important reminder of the experimental care that has to be taken in studies where the homogeneous formation of H_2O_2 is possible. The authors[44] disregarded the possibility of explaining the splitting of the reduction wave for H_2O_2 that was previously observed by considering the reduction of HO_2^- and of H_2O_2 at different potentials. This is a reasonable assumption, considering that the acid–base equilibrium of H_2O_2 is probably very rapidly achieved.

The influence of adsorbed organic molecules on a mercury electrode on the rate of electrode reactions, including the reduction of oxygen, has been studied by Kucherenko et al.[45] Here, the reduction of oxygen was used rather as a probe for the effects of adsorption of α-naphthol on electrochemical reactions. These authors found the interesting result that the adsorption of α-naphthol in the flat position altered the rate of reduction in the same way as the adsorption of aliphatic compounds. Therefore, no direct transfer of electrons from the metal to the O_2 molecule across the π-electron system was observed. This result is important in giving further evidence that the site of transfer of electrons in the reduction of oxygen on macrocyclic complexes is only the transition-metal centre.

Nickel.—In alkaline solutions, the reaction order for reduction of O_2 with respect to oxygen was $\frac{1}{2}$,[20] and no dependence of the rate on the concentration of hydroxide ions was found. The proposed mechanism for reduction is similar to that of cobalt. The rate of catalytic decomposition of intermediate peroxide has been reported to be very small, which is a result in agreement with the work of Littauer and Tsai.[46] Fischer et al.[47] have also reported that $Ni(OH)_2$ or hydrated nickel oxides, deposited on platinum, are poor electrocatalytic materials.

Palladium.—Fabjan et al.[20] have reported a Tafel slope of -40 mV per decade, and reaction orders of 1 and -1 to -1.5 with respect to O_2 and OH^- ion, respectively, in the reduction of oxygen in alkaline solutions. It was proposed that the negative reaction order was due to the specific adsorption of the OH^- ions on the catalytic surface. In acid solutions (0.8 to $3M$-H_2SO_4), the Tafel slope was -90 to -110 mV per decade.

Palladium has a large catalytic activity for the decomposition of peroxide,[30,46] and Vitvitskaya[48] has studied the relative contribution of the electrochemical and chemical routes for decomposition. The electrochemical rate (v_e) was determined by extrapolation of the cathodic and anodic polarization curves to the steady-state

---

[44] E. Bednarkiewicz and Z. Kublik, *Electrochim. Acta*, 1979, **24**, 121.
[45] L. D. Kucherenko, A. V. Lizogub, and B. N. Afanas'ev, *Sov. Electrochem. (Engl. Transl.)*, 1980, **16**, 252.
[46] E. L. Littauer and K. C. Tsai, *J. Electrochem. Soc.*, 1979, **126**, 1924.
[47] W. Fischer, W. Siedlarek, H. Hinüber, and R. Lünenschloss, *Werkst. Korros.*, 1980, **31**, 774.
[48] G. V. Vitvitskaya, *Sov. Electrochem. (Engl. Transl.)*, 1980, **16**, 876.

potential, whereas the chemical rate (v_c) was determined by subtracting the electrochemical rate from the overall rate of decomposition that was observed. The ratio v_e/v_c was ~0.1 at a pH of 8.1. The authors proposed that the 'chemical' reaction:

$$2 H_2O_2 \rightarrow 2 H_2O + O_2 \qquad (25)$$

occurs with weakly adsorbed peroxide, whereas the electrochemical process occurs only with strong adsorption of peroxide.

The reaction of oxygen and hydrogen peroxide with adsorbed hydrogen on palladium, platinum, and rhodium has been studied by Nikitin et al.[49] These authors followed the potential as a function of time, at open circuit, when electrodes that were covered with adsorbed hydrogen were put in contact with O_2- and H_2O_2-containing solutions. The potential–time transients that were observed were very similar to the usual charging curves observed under galvanostatic conditions, and therefore, from this work, it can only be concluded that the reaction of O_2 and of H_2O_2 with adsorbed hydrogen is diffusion-controlled.

Platinum.—Platinum is an efficient catalyst for the reduction of oxygen and is being used at present in the phosphoric acid fuel-cell programme currently under development in the U.S.A.[1]

The reduction of oxygen on this metal has been extensively studied in the past, and very disparate results have been reported regarding Tafel slopes, reaction orders, and rate constants. The extensive kinetic study of reduction of oxygen on platinum that has been carried out by Sepa, Vojnovic, and Damjanovic[50–52] is therefore very welcome. The kinetic information that was obtained is summarized in Table 1. Two well-defined polarization regions were found, having different Tafel

Table 1 *Kinetic parameters for the reduction of oxygen on platinum at different values of pH and current density*[52]

	pH Range	
	Acid	Alkaline
(a) *Tafel slope*/mV per decade		
high i	−120	−120
low i	−60	−60
(b) *Reaction order for* O_2		
high i	1	1
low i	1	1
(c) *Reaction order for* H^+		
high i	1	—
low i	1.5	—
(d) *Reaction order for* OH^-		
high i	—	−0.5
low i	—	0

[49] V. G. Nikitin, G. I. Elfimova, and G. A. Bogdanovskii, *Russ. J. Phys. Chem. (Engl. Transl.)*, 1979, **53**, 1787.
[50] A. Damjanovic, D. B. Sepa, and M. V. Vojnovic, *Electrochim. Acta*, 1979, **24**, 887.
[51] D. B. Sepa, M. V. Vojnovic, and A. Damjanovic, *Electrochim. Acta*, 1980, **25**, 1491.
[52] D. B. Sepa, M. V. Vojnovic, and A. Damjanovic, *Electrochim. Acta*, 1981, **26**, 781.

slopes, and the transition from one form of behaviour to the other was pH-dependent. In order to account for the measured kinetic parameters, it was considered that the first transfer of an electron was the rate-determining step at all potentials, but, in the potential region corresponding to low current densities (*i*), the reaction occurred under Temkin conditions of adsorption of the reaction intermediates.

The region of low current density is observed at potentials where the surface of the platinum is covered with oxide species, and Sepa et al.[52] studied the dependence of the surface coverage on pH and potential. A linear relationship was found, of the form:

$$\theta(E, \text{pH}) = 0.85E + [2.3 \times (0.85RT/F) \times \text{pH}] + 0.65 \qquad (26)$$

with the potential, E, referred to the S.H.E. The coverage represents the sum of all oxygen-containing species, to which the reduction of oxygen makes a negligible contribution. Thus the linear dependence of θ on the potential results in a Tafel slope of -60 mV per decade for Temkin conditions, and the changeover to the behaviour of high current density simply reflects the complete reduction of the oxide layer. In this respect, it is important to note[52] that the potential at which the change in mechanism occurs corresponds to a value of $\theta \simeq 0.05$ for the whole pH range studied. The unusual reaction orders with respect to the H^+ or OH^- ions are a direct consequence of equation (26). In acid solution, the 'true' reaction order with respect to the proton is 1, and H^+ ions are involved in the rate-determining step. For acid solutions, the mechanism that has been proposed is:

$$S + O_2 \rightleftharpoons S\text{-----}O_2 \qquad (27)$$

(S is a surface site), followed by the electrochemical step:

$$S\text{-----}O_2 + H^+ + e^- \rightarrow \text{products} \quad (\text{r.d.s.}) \qquad (28)$$

In alkaline solutions, the mechanism is:

$$S + O_2 \rightleftharpoons S\text{-----}O_2 \qquad (29)$$

$$S\text{-----}O_2 + e^- \rightarrow S\text{-----}O_2^- \rightarrow \text{products} \qquad (30)$$

or

$$S\text{-----}O_2 + H_2O + e^- \rightarrow S\text{-----}O_2H + OH^- \rightarrow \text{products} \qquad (31)$$

but, in both cases, the reaction rate is determined by the transfer of the first electron.

Fischer and Heitbaum[30] used the rotating-ring–disc electrode to study the reduction of O_2 on platinum from acid and alkaline solutions. The Tafel slopes that were found were -150 and -145 mV per decade in supra-pure 0.1M-KOH and 0.05M-H_2SO_4 solutions respectively. If the potential scale is expressed against a S.H.E., all the measurements that these authors appear to have made correspond to the region of high current density of the polarization curves studied by Sepa et al.,[52] and therefore these two sets of results are in fair agreement. In acid solutions, less than 1% of the disc current results in generation of H_2O_2, but, in alkaline solutions, reduction to peroxide was observed in the limited potential range where Pt—O is not reduced. Using the diagnostic criteria for the ring–disc electrode, the values of the rate constants k_1 and k_2 for the direct and sequential paths were evaluated. In the acid solution, $k_1 = 0.4$ and $k_2 = 0.03$ cm s^{-1}, whereas for the case of 0.1M-KOH, $k_1 = 0.4$ and $k_2 = 0$ cm s^{-1}. The reduction mechanism was the same as proposed by Sepa et al.[52] It is significant to note, as suggested by the latter authors, that the E–log i relationships for either of the two reaction paths are similar. This might indicate that the reaction steps up to the rate-determining step are the same for both the direct and the sequential reaction pathway. A possible explanation for this will be discussed further on.

The problems associated with the reduction of oxygen in weakly acidic unbuffered solutions were analysed by Fischer and Siedlarek,[53] using a rotating platinum disc electrode. The overall reduction process:

$$O_2 + 4H^+ + 4e^- \rightarrow H_2O \tag{32}$$

consumes protons; in the absence of a buffering system, this results in a change in pH across the diffusion layer. As a consequence, if the concentration of protons is less than that of oxygen, a splitting of the reduction wave should be observed, corresponding to the process in acid and alkaline solutions, respectively, as the potential is made more negative. This effect was observed in the present work,[53] and similar complications were noted by Sepa et al.[52] These results are of practical importance in corrosion studies; for instance, in natural waters, where the buffering capacity is usually rather low, the simple extension of electrokinetic studies for buffered systems may be quite erroneous. The local changes in pH may bring about the precipitation of metal hydroxides which can act as inhibitors to the reduction of oxygen, as was shown by Fischer et al.[47] In their work, the precipitation of $Ni(OH)_2$ from a weakly acid solution of Ni^{2+} when oxygen was reduced at a rotating platinum electrode resulted in a significant inhibition of the oxygen-reduction reaction, since $Ni(OH)_2$ (or the corresponding oxides) are worse electrocatalytic materials than platinum.

The formation of poor electrocatalytic layers for corrosion protection has been in use for many years, in the treatment of condenser tubes using seawater as coolant. It has been known that, in practice, the addition of $FeSO_4$ to the seawater results in significant improvements in the resistance to corrosion. The mechanism involves the formation of ferric hydroxide, which is a poor electrocatalytic material for the reduction of oxygen, on the metal surfaces.

The understanding of the kinetics of decomposition of H_2O_2 on electrocatalytic surfaces is of practical importance since the presence of this intermediate is undesirable, for instance, in fuel cells and in air-cathode applications. van den Brink, Barendrecht, and Visscher[54] developed a RRDE method to determine the rate of catalytic decomposition at the disc by measuring ring currents when no current is flowing through the disc. This is based essentially in measuring the limiting anodic and cathodic ring currents as a function of rotation rate and taking into account the shielding at the disc due to the decomposition reactions of H_2O_2. The value that was found for the heterogeneous rate constant for the reaction:

$$H_2O_2 \rightarrow \tfrac{1}{2}O_2 + H_2O \tag{33}$$

in 1M-NaOH at 25 °C was 0.016 cm s^{-1}. According to Vitvitskaya,[48] this

[53] W. Fischer and W. Siedlarek, *Werkst. Korros.*, 1979, **30**, 695.
[54] F. van den Brink, E. Barendrecht, and W. Visscher, *J. Electrochem. Soc.*, 1980, **127**, 2003.

decomposition mechanism represents ~90% of the rate of decomposition of hydrogen peroxide on oxidized platinum.

The rates of reduction and oxidation of hydrogen peroxide in 0.1M-KNO_3 solutions at various pH values have been studied by Prabhu et al.,[55] using potential-step, galvanostatic-transient, and RDE techniques. The rate of electrochemical decomposition, measured by extrapolation of the cathodic and anodic kinetic polarization curves to the rest potential, was 1 to 2×10^{-3} cm s^{-1} in the pH range 2—12.3. It is interesting to compare this result with that of the previous authors. The rate constant of electrochemical decomposition (see Scheme 1) is:

$$k_{\text{electrochem}} = (k_3 + k_{-2}) - k_2 C_{O_2}/C_{H_2O_2} \tag{34}$$

and, in the work of Prabhu et al.,[55] the last term was very small, since $(k_3 + k_{-2})$ is approximately ten times smaller than k_4;[48] the rate constant for the chemical decomposition derived from this work should be 0.01—0.02 cm s^{-1}, in very good agreement with the results of van den Brink et al.[54] The standard rate constant for the oxidation of H_2O_2 was 2 to 3×10^{-4} cm s^{-1} for values of pH between 2 and 10.1. From the similarity in the reduction characteristics of both O_2 and H_2O_2, it was concluded that the reduction of H_2O_2 was preceded by its decomposition, and that the measured currents corresponded to the oxygen that was formed. At pH < 6, however, there was evidence for the direct reduction of H_2O_2 according to:

$$H_2O_2 + 2H^+ + 2e^- \rightarrow 2H_2O \tag{35}$$

The diffusion coefficient of H_2O_2 was calculated from the potential step and from galvanostatic transients, and the average value that can be calculated in the pH range 2—8.2 is $(1.57 \pm 0.03) \times 10^{-5}$ cm^2 s^{-1} at 25 °C. At higher pH values, the diffusion coefficient decreases, probably due to the formation of the O_2H^- anion.

Platinum is a preferred catalytic material in many practical applications, and some of the problems associated with its use have received attention. Matveeva et al.[56] studied the poisoning of platinum surfaces by sulphurous acid, in relation to the reduction of oxygen. This is of importance for the understanding of the reactions occurring during the purification of industrial gases and of exhaust gases from motor vehicles, using platinum catalysts. In this case, the cathodic reaction is the reduction of O_2 and the anodic process is the oxidation of SO_2 to sulphate. These authors found that adsorption of H_2SO_3 shifted the wave for reduction of O_2 some 0.5 V negative, resulting in a large decrease in the rate of catalytic oxidation.

Perhaps one of the potentially most important uses of platinum catalysts will be in the fuel cells that are currently under active development. In the case of the phosphoric acid system, the high operating temperatures require particular stability of the catalyst as well as high exchange-current densities, and Ross[57] has studied the properties of intermetallic compounds between platinum and the elements of Groups 4B and 5B; VPt_3, $ZnPt_4$, and $TaPt_5$ were chosen, after a preliminary screening programme, as the most promising intermetallics. All of the materials investigated had a face-centred cubic structure, in order to introduce as little structural changes as possible in comparison with pure platinum. The rationale

[55] V. G. Prabhu, L. R. Zarapkar, and R. G. Dhaneshwar, *Electrochim. Acta*, 1981, **26**, 725.
[56] E. S. Matveeva, V. A. Shepelin, and E. V. Kasatkin, *Sov. Electrochem. (Engl. Transl.)*, 1981, **17**, 506.
[57] N. P. Ross, Jr., Report 1980, EPRI-EM-1553, AVAIL. NTIS.

behind this work was to attempt to alter the energy of formation of the catalyst–oxygen bond, since it was proposed that changes in the heat of adsorption of the rate-determining intermediate (assumed to be O_2^- or O_2H) would be related to changes in the free energy of formation of the oxygenated species that constitute the anodic oxide film. The intermetallic VPt_3 was found to give current densities an order of magnitude higher than that of pure platinum in the potential region where the platinum surface becomes covered with oxide. A comparison of some of the observed properties is summarized in Table 2, showing the increased performance

Table 2 *Comparative behaviour of platinum alloys, as carbon-supported catalysts, in the reduction of oxygen*[57]

Catalyst[a]	Specific activity/ mA mg^{-1}	Specific activity/ μA cm^{-2}	Tafel slope/ mV decade^{-1}
Colloidal Pt	21	20	−105
Heat-treated Pt	36	52	−110
HfPt$_4$, ZrPt$_4$	55	110	−120
VPt$_3$	50	75	−115
TaPt$_5$, PbPt$_5$	50	100	−120

(a) All measurements were made in 97% H_3PO_4, at 450 K, and at 0.9 V *vs* the reversible hydrogen electrode

that can be achieved. The results that were found were consistent with the 'volcano'-type plots that have been extensively used in the analysis of electrocatalytic processes, and it was concluded that the effect of alloying was to lower the heat of adsorption of intermediates while keeping the same reduction pathways as for platinum.

The stability of different carbon supports for platinum catalysts in fuel-cell applications has been studied by McBreen *et al.*,[58] using electrochemical techniques. The parameters that were determined were the surface area of the platinum (by measuring desorption curves for hydrogen coulometrically), the kinetics of reduction of oxygen, and the stability of the carbon (using cyclic voltammetry, with different anodic limits). These techniques are very advantageous in a programme of screening materials since they give rapid semi-quantitative comparative data. It was found that the rate of sintering of platinum was dependent on the carbon support; for instance, Vulcan XC-72R carbon gives a high initial degree of dispersion of platinum, owing to its high porosity, but results in electrodes that show a high rate of sintering in the first 1600 hours of operation. Other carbons, with lower surface areas, that were studied were more stable, owing to stronger carbon–platinum interactions. A cathodic peak was observed in the cyclic voltammograms of most materials that were studied when the anodic limit was greater than 1.2 V *vs* a reversible hydrogen electrode. This is probably due to the formation of intercalation compounds[59] in graphitic regions of the supporting material, which will certainly have a deleterious effect on the stability of the carbon. It is interesting to note that the material which showed better performance (Regal 660R) did not appear to form detectable intercalates on anodic polarization, and it can be suggested that one

[58] J. McBreen, H. Olender, S. Srinivasan, and K. V. Kordesch, *J. Appl. Electrochem.*, 1981, **11**, 787.
[59] F. Beck, H. Junge, and H. Krohn, *Electrochim. Acta*, 1981, **26**, 799.

parameter that should be considered when screening suitable materials should be the resistance of each material towards intercalation. The incorporation of anions in the lattice can result in rapid deterioration of the material.

Silver.—Silver was one of the first catalytic materials to be used, on a carbon support, in the air–zinc battery. There is disagreement in recent work regarding the kinetic parameters for the reduction of O_2 on this metal in alkaline solutions. Fabjan et al.[20] have reported Tafel slopes of -45 to -50 mV per decade whereas Fischer and Heitbaum[30] found -150 mV per decade. The exchange-current densities that can be evaluated from reference 20 are also many orders of magnitude different from the values reported by Fischer et al.,[30] and the origin of the discrepancy is not clear. Both the direct and sequential mechanisms occur on silver,[30] and the values of k_1 and k_2 (see Scheme 1) are 0.8 and 0.04 cm s^{-1}, respectively.

A maximum in the ring current was observed at $\sim +0.3$ V *versus* the reversible hydrogen electrode (R.H.E.) when the supra-pure solutions that were under study were left for some time in contact with the atmosphere. It is interesting to compare these results with the work of Sheblovinskii and Kicheev.[60] These authors studied the reduction and oxidation of H_2O_2 on silver in alkaline solutions and found that the current–potential curve for the reduction process was time-dependent. When the silver electrode was left for a short time (1 minute) at the rest potential, diffusional reduction currents were observed, but a clear inhibition of the H_2O_2 process could be seen in the potential range ~ 0.1—0.6 V *vs* the R.H.E. for aged electrodes. Although Fischer et al.[30] did not report ageing effects, the formation of surface silver oxides will alter the electrode kinetics of the intermediates in the reduction of O_2 on silver.[60]

Zwetanova and Jüttner[39] studied the reduction of oxygen on different single-crystal surfaces of silver in acid solutions and the effect of the adsorption of Pb^{2+} and Tl^+. The Tafel slopes that were found for the different planes were $b(111) = -0.100$, $b(100) = -0.095$, and $b(110) = -0.110$ V per decade, respectively. The deposition of these metals at an underpotential resulted in a change of reduction mechanism and a change in overall stoicheiometry, from four to two electrons. Furthermore, the almost complete inhibition of the reduction of hydrogen peroxide on the modified surfaces was demonstrated by direct polarization studies of the reduction of H_2O_2. The values of the Tafel slopes that were found for reduction of O_2 were indicative of a reduction sequence similar to that observed for platinum, with the first transfer of an electron as the rate-determining step, but the detailed mechanism of the reaction remains unclear. As the authors pointed out, the mechanism for the rapid decomposition of H_2O_2 could be either chemical or electrochemical. At a pH of 8.1, it is known that only 2% of the rate of decomposition can be regarded as following the electrochemical route;[48] in alkaline solutions, very high rates of heterogeneous decomposition are observed.[46] However, Zwetanova and Jüttner argued that, irrespective of the pathway for reduction of H_2O_2, a strong chemical interaction with the surface should be required to break the O–O bond, and modification of the surface will have a pronounced effect on this step. A four-electron reduction of O_2 occurs on bulk lead and on silver, but it is clear

---

[60] V. M. Sheblovinskii and A. G. Kicheev, *Sov. Electrochem. (Engl. Transl.)*, 1980, **16**, 1354.

that, in acid solutions, the catalytic properties of the lead adatom layer will differ from those of the bulk phase. The adsorption of Pb^{2+} was modified by the presence of chloride ions in solution, and therefore it indirectly affected the rate of reduction. It is important to compare the inhibition of the four-electron process that was observed in this work, leading to a two-electron process, with the reverse effect that is observed with gold in alkaline solutions, when Pb^{2+} ions are adsorbed to give monolayer coverage.[32] A rationalization of this difference in behaviour is required, but it is likely that some specific pH-dependent chemical interactions are operative.

4 Co-ordination Compounds of Transition Metals

The low values of the rate constants that are usually found for the electrochemical reduction of oxygen are in line with the relatively low chemical reactivity of this molecule. This can be regarded as a consequence of both the high stability of the O–O bond (dissociation energy of 988.4 kJ mol^{-1}) and spin restrictions in chemical transformations.

A simplified molecular-orbital description of the O_2 molecule is shown in Figure 1.[61] The ground state has two unpaired electrons, as required by Hund's rule, thus giving rise to a triplet ($^3\Sigma_g^-$) ground state. The singlet configurations $^1\Delta_g$ and $^1\Sigma_g^+$ are excited states, having energies of 94.7 and 157.8 kJ mol^{-1} above the ground state. Therefore, the chemistry and electrochemistry of oxygen is related to the properties of a triplet state, leading to fundamental restrictions on its reactivity. The reaction of a triplet molecule with a singlet to give singlet products is a spin-forbidden process, unless a change in spin state can occur in the course of the reaction. For oxygen, this would imply not only a substantial activation energy, but also a time constant for spin inversion that is not compatible with the lifetime of

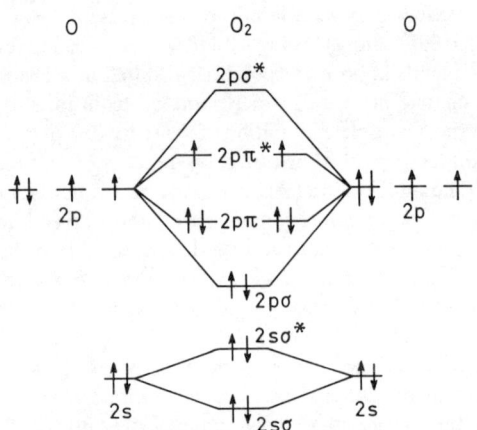

Figure 1 *A molecular-orbital diagram of the oxygen molecule* (Reproduced by permission from *Chem. Rev.*, 1973, **73**, 235)

[61] J. S. Valentine, *Chem. Rev.*, 1973, **73**, 235.

The Electrochemistry of Oxygen

many collision complexes; that is to say, 10^{-8} to 10^{-13} s for the latter case.[62] The poor chemical reactivity of oxygen results from the conservation of angular momentum in a chemical reaction; since spin angular momentum is weakly coupled with the environment, the initial state of the reaction of O_2 with, for example, a singlet organic molecule will result in an allowed triplet product, and the stability of this intermediate would have to be such that it survived for much longer than the time for a vibration in order to obtain a reaction product in the singlet state.

This spin restriction can be removed by co-ordination with a transition-metal centre that has unpaired electrons, and this is why transition-metal compounds have been extensively used in chemical oxidation reactions.[63] In the case of the electrochemical reduction of oxygen, the use of a co-ordinating transition-metal centre in the course of the reduction brings the possibility of some degree of control on the energetics of the intermediate stages, and hence on the value of the rate constant for reduction. The different forms of co-ordination of O_2 to transition-metal compounds have recently been reviewed by Smith and Pilbrow,[64] and the use of N_4 macrocyclic compounds as catalysts has been extensively studied.[65] The metal centres in these compounds have d-electrons available for co-ordination, and it is possible to alter both their energy[66,67] and the co-ordination geometry[65] by axial ligands. These compounds have been used as homogeneous catalysts, as adsorbed redox centres in modified electrodes, as redox centres on modified carbon surfaces, and in polymer electrodes.

Kobayashi et al.[68] have continued previous studies[69] of catalytic oxygen-reduction waves that are produced by water-soluble porphyrins. The complex studied was iron(III) tetrakis(4-N-methylpyridyl)porphyrin (abbreviated to $[Fe^{III}TMP]^{5+}$); the Fe^{II} centre that is generated by reduction reacted very rapidly with oxygen to yield hydrogen peroxide and regenerated the Fe^{III} compound, at potentials where the reduction of oxygen to hydrogen peroxide is kinetically hindered at a glassy carbon electrode. The interesting idea behind this work was to decompose catalytically the H_2O_2 that is formed in the solution, and to achieve a four-electron reduction process in this way. The reaction sequence was:

$$[Fe^{III}TMP]^{5+} + e^- \rightarrow [Fe^{II}TMP]^{4+} \quad (36)$$

$$[Fe^{II}TMP]^{4+} + \tfrac{1}{2}O_2 + H^+ \rightarrow [Fe^{III}TMP]^{5+} + \tfrac{1}{2}H_2O_2 \quad (37)$$

$$H_2O_2 \xrightarrow{catalyst} H_2O + \tfrac{1}{2}O_2 \quad (38)$$

---

[62] G. A. Hamilton, in 'Molecular Mechanisms of Oxygen Activation', ed. O. Hayaishi, Academic Press, New York and London, 1974, Ch. 10, p. 405.
[63] R. A. Sheldon and J. K. Kochi, 'Metal Catalysed Oxidations of Organic Compounds', Academic Press, London, 1981.
[64] T. D. Smith and J. R. Pilbrow, Coord. Chem. Rev., 1981, **39**, 295.
[65] P. D. Smith, B. R. James, and D. H. Dolphin, Coord. Chem. Rev., 1981, **39**, 31.
[66] K. M. Kadish, L. A. Bottomley, S. Kelly, D. Schaeper, and L. R. Shive, Bioelectrochem. Bioenerg., 1981, **8**, 213.
[67] L. A. Bottomley and K. M. Kadish, Inorg. Chem., 1981, **20**, 1348.
[68] N. Kobayashi, M. Fujihira, K. Sunakawa, and T. Osa, J. Electroanal. Chem. Interfacial Electrochem., 1979, **101**, 269.
[69] T. Kuwana, M. Fujihira, K. Sunakawa, and T. Osa, J. Electroanal. Chem. Interfacial Electrochem., 1978, **88**, 299.

and several substituted phthalocyanines were chosen as catalysts for the decomposition of H_2O_2. The compounds (1)—(6) were studied, but little catalytic activity was detected. This is not surprising, since all the compounds that were studied are almost insoluble in acid media, and therefore the experiments must have relied only upon the presence of dispersed solid catalysts close to the surface of the

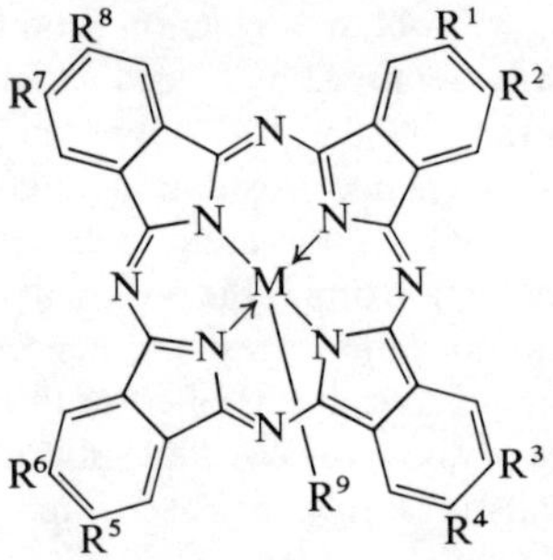

[Fe^{III}OCPc]	(1)	M = Fe^{III}, R^1, ... R^8 = CO$_2$H; R^9 absent
[Fe^{III}OCPc-Im]	(2)	M = Fe^{III}, R^1, ... R^8 = CO$_2$H, R^9 = imidazole
[Fe^{III}TCPc]	(3)	M = Fe^{III}, R^1 = R^3 = R^5 = R^7 = CO$_2$H, R^2 = R^4 = R^6 = R^8 = H; R^9 absent
[Fe^{III}TCPc-Im]	(4)	M = Fe^{III}, R^1 = R^3 = R^5 = R^7 = CO$_2$H, R^2 = R^4 = R^6 = R^8 = H, R^9 = imidazole
[Fe^{III}DAPc]	(5)	M = Fe^{III}, R^1 = R^6 = COC$_6$H$_3$[3,4-(CO$_2$H)$_2$], R^2 = R^3 = R^4 = R^5 = R^7 = R^8 = H; R^9 absent
[Cu^{II}OCPc]	(6)	M = Cu^{II}, R^1, ... R^8 = CO$_2$H; R^9 absent

(Pc = phthalocyanine)

electrode. Four-electron catalytic activity was observed, however, in alkaline solution, when the redox carrier was iron(III) tetra(carboxyphenyl)porphyrin, and iron(III) octacarboxyphthalocyanine was used for decomposing the hydrogen peroxide that was generated. The approach that was followed in this work appears to be of great interest, since it is possible to alter, by chemical means, both the redox potential of the redox carrier and the rate of reaction of oxygen with the transition-metal centre. The relationship between this strategy and the use of metal surfaces that are modified with foreign metal ions, as used by Adzić et al.,[32,33] needs to be investigated further.

Kobayashi et al.[70] studied further the catalytic activity of iron(III) and cobalt(III) tetra-(o-aminophenyl)porphyrins, [Fe^{III}TAPP] and [Co^{III}TAPP]. These complexes were chosen to compare rates of homogeneous catalytic decomposition with those corresponding to electrodes having the TAPP complexes covalently attached to the surface of the electrode. The reduction mechanism that was observed corresponded in both cases to the e.c. catalytic redox mechanism shown in Scheme 4.

The stability of the cobalt-modified electrode was superior to that of the iron

---

[70] N. Kobayashi, T. Matsue, M. Fujihira, and T. Osa, J. Electroanal. Chem. Interfacial Electrochem., 1979, **103**, 427.

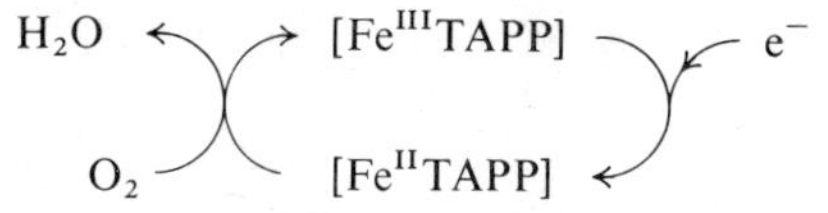

```
H₂O  ←─╲  ╱→  [Fe^III TAPP]  ─╲   ╱─ e⁻
         ╳                       ╳
O₂   ←─╱  ╲→  [Fe^II TAPP]   ←─╱   ╲
```

[TAPP = tetra-(o-aminophenyl)porphyrinato]

Scheme 4

compound, and the results for homogeneous catalysis were similar to those previously observed for $[Fe^{III}TMP]^{5+}$.[68,69] It is interesting to notice, though, that the M^{III}/M^{II} redox potential for $[Co^{III}TAPP]$ was at least 300 mV positive to that of the Fe^{III} compound, but nonetheless, the peak potential for the wave for catalytic reduction of oxygen was not significantly altered. Obviously, the redox potential of the transmitting redox couple is not the only parameter to be considered when modelling these redox-mediated electron-transfer reactions, and the value of the 'chemical' rate constant for the reaction of O_2 with the reduced redox centre must be taken into account. Besides these considerations, for progress to be made by using this strategy, an understanding of the intermediates that are formed in the homogeneous reaction of oxygen with the reduced macrocycle is required. For instance, the possibility of the intermediate formation of μ-peroxo-bridged complexes should be considered.

Catalysis in the reduction of oxygen by a soluble Co^{III} macrocycle was also observed by Geiger and Anson.[71] These authors studied the reaction of O_2 with trans-$[Co([14]ane-N_4)(OH_2)_2]^{3+}$, where $[14]$ane-N_4 stands for 1,4,8,11-tetra-azacyclotetradecane (7). This complex was chosen since it is known to form a

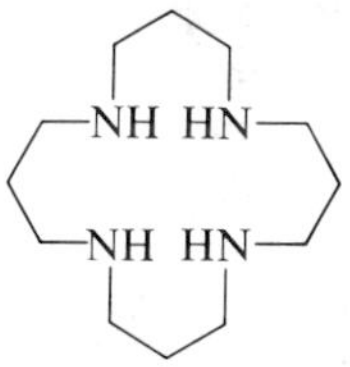

(7)

dinuclear μ-peroxo-bridged complex in solution[72] which could be regarded as comparable to the proposed intermediates formed in the reduction of oxygen on dimeric cofacial porphyrin complexes of cobalt(II).[73,74] Two possible reduction pathways were observed, depending on which reagent, either O_2 or trans-$[Co([14]ane-N_4)(OH_2)_2]^{3+}$ (or $[CoL]^{3+}$ for simplicity), was present in excess. The reason for this is the possibility of a branched mechanism, where either transfer of electrons can occur to the $[CoLO_2]^{2+}$ complex or this can further react with a

---

[71] T. Geiger and F. C. Anson, J. Am. Chem. Soc., 1981, **103**, 7489.
[72] G. L. Wong, J. A. Switzer, B. P. Balakrishnan, and J. F. Endicott, J. Am. Chem. Soc., 1980, **102**, 5511.
[73] J. P. Collman, M. Marrocco, P. Denisevich, C. Koval, and F. C. Anson, J. Electroanal. Chem. Interfacial Electrochem., 1979, **101**, 117.
[74] J. P. Collman, P. Denisevich, Y. Konai, M. Marrocco, C. Koval, and F. C. Anson, J. Am. Chem. Soc., 1980, **102**, 6027.

reduced $[CoL]^{2+}$ molecule, resulting in the intermediate formation of a μ-peroxo-bridged complex prior to the transfer of an electron. The reduction sequence is shown in Scheme 5. Both μ-peroxo and peroxo pathways have been observed, and this work shows the way to analyse oxygen-reduction processes that may involve the formation of μ-peroxo-compounds. An interesting conclusion was that the co-ordination of a macrocyclic cobalt(III) complex to both ends of a peroxide dianion did not result in the activation of the O–O bond to enable a four-electron reduction, and attempts to reduce the μ-peroxo-dimer resulted only in the release of hydrogen peroxide.

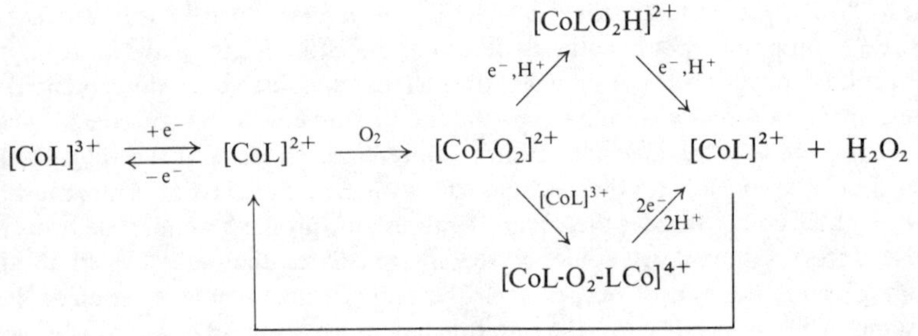

$$[CoL]^{3+} \underset{-e^-}{\overset{+e^-}{\rightleftharpoons}} [CoL]^{2+} \xrightarrow{O_2} [CoLO_2]^{2+} \longrightarrow [CoL]^{2+} + H_2O_2$$

with intermediate $[CoLO_2H]^{2+}$ (via e^-, H^+) and $[CoL\text{-}O_2\text{-}LCo]^{4+}$ (via $[CoL]^{3+}$, then $2e^-/2H^+$)

[L = (7)]

Scheme 5

The peroxo geometry in co-ordination compounds of cobalt has been known for over 70 years,[75] and a great deal of information is available.[64,76] However, it is surprising that the electrochemistry of these complexes has received so little attention, since it would be expected that co-ordination should remove some of the irreversibility that is associated with the transfer of electrons. The kinetic complexities that were observed by Geiger and Anson[71] indicate that the formation of μ-peroxo-complexes is not perhaps the best four-electron-reduction strategy, but the coupling of a μ-peroxo system with a suitable model system for catalase may yield promising results. The complicated features of the electrochemical reduction of these complexes have previously been described by Vlcek[77] and by Barnatt and Charles,[78] and in the latter case, even the evolution of O_2 was observed on reduction of the $[(NH_3)_5Co\text{-}O_2\text{-}Co(NH_3)_5]^{5+}$ complex in acid solutions.

As has been discussed by Adzić,[34] the modification of an electrode surface by specific adsorption of a suitable catalytic centre is an attractive technique to achieve specific electrocatalytic properties. Bettelheim et al.[79] modified a glassy carbon electrode by adsorption of Fe^{III} and Co^{III} tetrapyridylporphines, $[Fe^{III}TPyP]$ and $[Co^{III}TPyP]$. In acid solutions, $[Co^{III}TPyP]$ was strongly adsorbed, and gave catalytic redox waves in the reduction of oxygen, similar to other observations.[68—70]

[75] A. Werner and A. Myelius, Z. Anorg. Chem., 1898, **16**, 245.
[76] G. McLendon and A. E. Martell, Coord. Chem. Rev., 1976, **19**, 1.
[77] A. A. Vlcek, Collect. Czech. Chem. Commun., 1960, **25**, 3037.
[78] S. Barnatt and R. G. Charles, J. Electrochem. Soc., 1962, **109**, 333.
[79] A. Bettelheim, R. J. H. Chan, and T. Kuwana, J. Electroanal. Chem. Interfacial Electrochem., 1979, **99**, 391.

The complex [Fe^{III}TPyP] could not be adsorbed from acid solutions. The reduction product was hydrogen peroxide, as in the previously discussed redox catalytic systems, but the authors reported a significant increase of peak current in linear-sweep voltammetric experiments when catalase was added to the acid solution. The presence of catalase will result in some catalytic decomposition of hydrogen peroxide to oxygen and water, and hence in an increase in the two-electron diffusion-controlled currents.

The effect of adsorbed water-soluble cobalt and iron phthalocyanines on graphite surfaces on the kinetics of reduction of oxygen has been studied by Zagal, Bindra, and Yeager.[80] Cobalt and iron tetrasulphonatophthalocyanines ([Co-TSP] and [Fe-TSP]) were studied as surface catalysts, using the RDE and the RRDE techniques. In the presence of adsorbed [Co-TSP], the reduction of oxygen is almost entirely a two-electron process. The RRDE data for a 0.1M-NaOH solution were analysed, using the diagnostic criteria of Wroblova et al.,[14] and some evidence for the presence of the direct reduction pathway was found, but the rate constant for the direct reaction, k_1, was always at least ten times smaller than k_2.

In contrast, almost no hydrogen peroxide could be detected when the graphite electrode was modified with [Fe-TSP]. In this case, two regions of polarization were observed, and the current showed little dependence on potential at intermediate potentials. Figure 2 shows a schematic diagram of the polarization curves that are

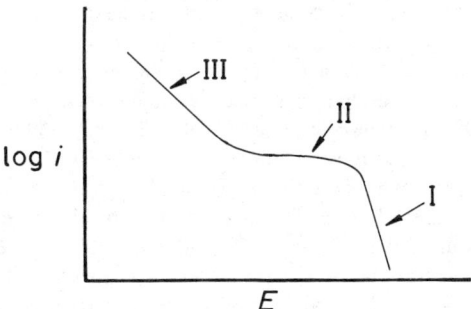

Figure 2 *A schematic representation of the polarization curves for the reduction of oxygen on adsorbed [Fe-TSP]*
(Reproduced by permission from *J. Electrochem. Soc.*, 1980, **127**, 1506)

observed with adsorbed [Fe-TSP]. At low current densities (region I), the reaction order with respect to OH$^-$ ions was -1, and the Tafel slope was -0.030 to -0.035 V per decade. In the intermediate region (region II), the value of the limiting current was independent of pH; at high current densities, the reaction order with respect to OH$^-$ was zero and the Tafel slope was -0.12 V per decade. The polarization behaviour shown in Figure 2 is indicative of a parallel reduction mechanism, with one of the reduction paths having a chemical limiting step, *i.e.*, a parallel *e/e.c.e.* mechanism. The proposed reactions in alkaline solutions were:

$$Fe^{III}OH_{(ads)} + e^- \rightleftharpoons Fe^{II}_{(ads)} + OH^- \tag{39}$$

[80] J. Zagal, P. Bindra, and E. Yeager, *J. Electrochem. Soc.*, 1980, **127**, 1506.

$$Fe^{II}_{(ads)} + O_2 \rightleftharpoons [Fe^{III}\text{-}O_2^-]_{(ads)} \quad (40)$$

$$[Fe^{III}\text{-}O_2^-]_{(ads)} + e^- \rightarrow \text{reduction products} \quad (41)$$

If reaction (41) is the rate-determining step, a limiting Tafel slope of -0.04 V per decade should be observed at low current densities for a transfer coefficient of 0.5 (region I). At more negative potentials, the chemical step (40) becomes rate-determining, and hence no potential dependence is observed (region II), whereas at high current densities a parallel electrochemical mechanism, with the first electron transfer as the rate-determining step, is observed (region III).

It is important to note that the reduction sequence is dependent on the redox properties of the adsorbed phthalocyanine layer, and this gives the possibility of altering the exchange-current density by employing other substituted soluble phthalocyanines. The experimental observations by Zagal et al.[80] and the proposed mechanism are also interesting, owing to the similarity of the reactions studied with the initial stages of activation of the biological hydroxylating cytochrome P-450 system[81] (see later).

The adsorption of metallophthalocyanines on active forms of carbon has been used by van Veen and Visser[82] in the preparation of Teflon-bonded-type electrodes, in a comparative study of the activity of different chelates for oxygen cathodes. The way in which the macrocycle is supported is very important in determining its electrocatalytic activity; in the case of iron phthalocyanine, [FePc], Melendres and Cafasso[83] found that vapour-deposited film electrodes have a significantly lower activity than those prepared by adsorption. There has even been a report that [FePc] film electrodes are unable to support significant oxygen-reduction currents above the background reduction processes of the film electrode.[84]

Another complicating factor in the use of phthalocyanine film electrodes in the study of specific electrocatalytic properties is the poorly understood slow, irreversible transformations of the film. For example, in order to achieve the transfer of electrons across [ZnPc], [NiPc], and H_2Pc films, the electrode has to be cycled between the cathodic and anodic limits;[85] similarly, [FePc] film electrodes[83] have been found to give reproducible voltammetric waves for the reduction of oxygen only by waiting for about 1 hour, at the open-circuit potential, after each scan. These slow processes can be related to the inevitable coupling of redox processes within the film with the migration of ions that is required to maintain the electroneutrality of the film;[86] certainly, the incorporation of ions within electrochromic phthalocyanine films such as $[LuH(Pc)_2]$ results in a profound change in their properties as electrode materials.[86]

A completely different behaviour from that observed for film electrodes was found by Behret et al.[87] when [FePc] was deposited on pyrographite from concentrated H_2SO_4 solutions. The polarization curve was very similar, in this case, to that observed by van Veen and Visser[82] on active carbon, and similar results were obtained by Melendres and Cafasso.[83] The [FePc] electrodes that were obtained

[81] M. J. Coon and R. E. White, in ref. 6, Ch. 2, p. 73.
[82] J. A. R. van Veen and C. Visser, Electrochim. Acta, 1979, 24, 921.
[83] C. A. Melendres and F. A. Cafasso, J. Electrochem. Soc., 1981, 128, 755.
[84] H. Tachikawa and L. R. Faulkner, J. Am. Chem. Soc., 1978, 100, 4379.
[85] F-R. Fan and L. R. Faulkner, J. Am. Chem. Soc., 1979, 101, 4779.
[86] G. C. S. Colins and D. J. Schiffrin, J. Electroanal. Chem. Interfacial Electrochem., in the press.
[87] H. Behret, W. Clauberg, and G. Sandstede, Z. Phys. Chem. (Frankfurt am Main), 1979, 113, 97.

by precipitation contain well-crystallized particles, and it has been suggested that only a fraction of the [FePc] molecules are active.[82]

The above considerations indicate that the study of specific electrocatalytic properties of transition-metal chelates as sites of faradaic processes is best performed under conditions where their solid-state properties do not introduce unforeseeable complications. Therefore, one of the methods used by van Veen and Visser[82] for supporting the catalyst on carbon, its adsorption from solution being monitored spectrophotometrically, can allow a quantitative comparison to be made with other experimental techniques such as RDE studies of adsorbed phthalocyanines. It is interesting to compare results on a molecular basis; for the case of adsorbed [Fe-TSP], the maximum rate of reduction corresponding to processes (39) and (40) is of the order of 2 mA cm^{-2}.[80] This is equivalent to a turnover of ~ 150 electrons per molecule per second for an adsorbed tetrasulphonatophthalocyanine molecule. For the supported [FePc],[82] there is evidence of a limiting process occurring at currents of ~ 10 mA mg^{-1}; taking the surface areas given in ref. 82, a turnover of at least 1000 electrons per molecule per second can be calculated. There are gross oversimplifications in this analysis, but nonetheless it is clear that the state of the active metal centre of a molecule of [Fe-TSP] that is adsorbed from an aqueous solution is quite different from that of [FePc] that is adsorbed on carbon. The redox state of the transition metal in the chelate should affect its spectroscopic properties, and for this reason Melendres and Cafasso[83] studied the Raman spectra of [FePc] that was deposited on glassy carbon. No potential dependence of the Raman spectra was observed, even though, in the potential range studied ($+0.4$ to -1.0 V vs the Hg/Hg_2SO_4 reference electrode), redox processes in the deposited layer did occur. This is a rather surprising result since a change in valency of the central atom should result in significant changes in the spectra of the ligand. The 1500 cm^{-1} band (assigned to the C–N stretching mode) was shifted to 1515 cm^{-1} on immersion in water; this behaviour was assigned to the co-ordination of water to the bridge nitrogen site. The use of X-ray photoelectron spectroscopy (ESCA) showed clearly, however, the change in valency of the central iron atom in [FePc] on oxidation.[88] The $2p$ core-level spectra of iron showed a shift of 1.7 eV when O_2 was admitted to the analysing chamber of the spectrometer. For the di- and poly-iron phthalocyanines it was found that the oxidation state of iron was between $2+$ and $3+$. The ESCA spectra of valence levels indicated the formation of a highly covalent Fe–O_2 bond by interaction of the antibonding molecular orbitals of O_2 with the d_{yz} and d_{xz} orbitals of the iron centre, and the proposed M.O. description for the dimer–oxygen compound is shown in Figure 3.

It was also concluded that dimerization, $i.e.$ conjugated enlargement of the system, leads to delocalization of charge over the whole molecule, and these effects have been exploited in the use of both iron naphthalocyanines[89] and polymeric phthalocyanines[90–93] as catalysts.

[88] S. Maroie, M. Savy, and J. Verbist, $Inorg.\ Chem.$, 1979, **18**, 2560.
[89] G. Magner, M. Savy, and G. Scarbeck, $J.\ Electrochem.\ Soc.$, 1980, **127**, 1076.
[90] H. Behret, H. Binder, G. Sandstede, and G. G. Scherer, $J.\ Electroanal.\ Chem.\ Interfacial\ Electrochem.$, 1981, **117**, 29.
[91] V. E. Kazarinov, M. R. Tarasevich, K. A. Radyushkina, and V. N. Andreev, $J.\ Electroanal.\ Chem.\ Interfacial\ Electrochem.$, 1979, **100**, 225.
[92] L. Kreja and R. Dabrowski, $J.\ Power\ Sources$, 1980, **6**, 35.
[93] A. Kaisheva, $Sov.\ Electrochem.\ (Engl.\ Transl.)$, 1979, **15**, 1326.

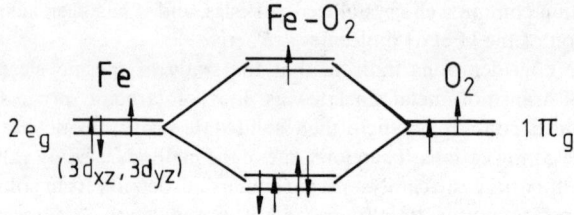

Figure 3 *A molecular-orbital description of the di-[FePc]-O_2 complex* (Reproduced by permission from *Inorg. Chem.*, 1979, **18**, 2650)

One of the reasons for the choice of phthalocyanines as catalyst materials is the possibility of having a catalyst that has a low resistivity, owing to the extended degree of conjugation that is provided by the interaction of the benzene rings of the molecule. Although the porphyrins have a much lower electrical conductivity in the solid state, they have been extensively used as catalysts, owing to their high thermal and chemical stability and the favourable co-ordination geometries assumed by these complexes (square-planar or square-pyramidal). van Veen et al.[94,95] have studied a wide range of central transition-metal atoms (Fe, Co, Mn, Cr, Rh, Ru, Pd, Os, Ir, and Pt) and ligands such as meso-tetra(*para*-substituted phenyl)porphyrin [T(*p*-X)PP], meso-tetra-(2-thionyl)porphyrin (TThP), meso-tetra-(2-pyridyl)-porphyrin (TPyP), octaethylporphyrin (OEP), and tetrabenzoporphyrin (TBP). One of the main objectives of this very substantial piece of research was to establish the existence of a 'volcano'-type relationship between catalytic activity and some property of the chelates that determines the rate of transfer of electrons. As was mentioned before (see p. 141), in the case of platinum alloys,[57] the metal–oxygen bond was used as the relevant variable, similar to the adsorbed intermediates that could be expected to be formed during reduction. For the porphyrin macrocycles, however, van Veen et al.[94] proposed the use of the redox potential of the central transition-metal couple, M^{II}/M^{III}, as the relevant variable. This choice derives from the proposed reduction mechanism, which, in common with that proposed by Zagal et al.[80] for adsorbed [Fe-TSP], involves the reduction of the transition-metal centre as the first reduction step. In acid solutions, the proposed reduction sequence was:

$$M^{III} + e^- \rightleftharpoons M^{II} \qquad (42)$$

$$M^{II} + O_2 + H^+ \xrightarrow{k_{(e)}} M^{III}-O_2H \qquad (43)$$

$$M^{III}-O_2H + e^- \xrightarrow{k_{(f)}} \text{intermediates and products} \qquad (44)$$

By assuming Nernstian equilibrium for reaction (42), the different Tafel slopes and rate constants that are observed could be explained. A volcano-shaped plot was indeed observed, and it was proposed that this resulted from the opposing effects of decreasing the number of active sites and increased charge-transfer constants as the

---

[94] J. A. R. van Veen, J. F. van Baar, C. J. Kroese, J. G. F. Coolegem, N. de Wit, and H. A. Colijn, *Ber. Bunsenges. Phys. Chem.*, 1981, **85**, 693.
[95] J. A. R. van Veen, J. F. van Baar, and C. J. Kroese, *Ber. Bunsenges. Phys. Chem.*, 1981, **85**, 700.

redox potential of the chelate was made more negative. The order of activity in 4M-H_2SO_4 was $Ir \gg Co$, $Rh > Fe > Ru > Os$, Pd, Pt, Cr, Mn, H_2; in 6M-KOH it was $Ru > Co > Fe > Ir$.

para-Substitution in the phenyl rings had a surprisingly small effect, and the order of activity for other ligands was (for cobalt) [CoPc] < [CoTBP] < [CoTPP] < [CoOEP]; for iron it was [FeTBP] < [FeOEP] < [FeTPP] < [FeTPyP] < [FeT(perfluoro)PP], [FePc], and it was concluded that the main effects are due to changes at the metal centre. Kazarinov et al.[91] reached a similar conclusion, and also found that [CoTPP] was more active than [FeTPP]. Behret, Clauberg, and Sandstede[9] studied the kinetics of reduction of O_2 by these two chelates, using the RRDE technique and discs of platinum, gold, pyrographite, and glassy carbon as supports for the chelate. The comparison of behaviour with different supports is very important for establishing if the observed effects are due solely to the catalyst. It was found, in accordance with the results obtained by other authors,[5,91] that hydrogen peroxide is the main reduction product with a [CoTPP]-covered electrode and that a parallel and sequential mechanism was operative. The ratio k_1/k_2 was 0.6—1.5 and 0.4 for the iron and cobalt complexes, respectively, in acid solutions, whereas the values observed in alkaline solutions were 3.4—6.7 and ~ 1.0, respectively, in the potential range 0.5—0.2 V *vs* a hydrogen electrode in the same solution. The Tafel slopes that were observed were very different from those reported by van Veen et al.[94] For instance, for [CoTPP] and [FeTPP] in acid solutions, the values were -0.04 and -0.065 V per decade, respectively,[94] when porous Teflon-bonded Norit BRX stationary carbon electrodes were used, whereas the slopes from the work with rotating-disc electrodes were -0.17 and -0.15 V per decade, respectively,[90] which would indicate that the transfer of the first electron is the rate-determining step. The discrepancy between these results needs clarification; the differences between these results must be related to the experimental approach taken. van Veen et al.[94] used a carbon with a large surface area as the support for the catalyst, whereas a planar disc electrode was used as the support in the work of Behret et al.[90] In the latter case, the time taken for diffusion of reduction products from the catalytic surface under convective conditions is very small, and, considering the ring currents observed by these authors, very little catalytic decomposition of H_2O_2 occurred in their experiments. In the experiments with active-carbon-supported electrodes, the electrode material was active in the catalytic decomposition of H_2O_2.[94] Reaction orders of 1 with respect to the concentrations of both hydrogen peroxide and catalyst were found. Moreover, it is very significant that there was a clear correlation between the rate of reduction of oxygen and the catalytic activity for decomposition of H_2O_2. It is tempting to suggest, therefore, that a redox equilibrium similar to that discussed in relation to the reduction of O_2 on carbon (see p. 130), but coupled in this case to the catalytic decomposition of H_2O_2, prevails, and the role of the catalyst is also to decompose the hydrogen peroxide that is formed. It is unlikely that the only function of the catalyst is to decompose H_2O_2, since, for instance, no correlation between the efficiency of the catalyst and heterogeneous rate constants for decomposition of H_2O_2 was observed for alkaline solutions.

The catalytic activity of some carbon-supported N_4 chelates is greatly increased by heat treatment. The complexes [CoPc] and [RuPc] showed great improvements in activity after heating the supported catalyst at 650 °C, whereas no effect was observed for [FePc], [MnPc], [ZnPc], and the Norit BRX support itself. In the case of the porphyrins, heat treatment resulted in an improvement of activity for all the chelates studied; e.g., by a factor of 5 for the cobalt and by two orders of magnitude for the ruthenium and iron porphyrins for use with 4M-H_2SO_4. Enhanced activity was also found in KOH for [CoTPP], [FeTPP], and [IrTPP] by a factor of three, one, and two orders of magnitude, respectively.[94]

Heat treatment of carbon-supported porphyrins also resulted in a large enhancement of the long-term stability of the oxygen electrodes prepared with these materials.[95] The changes occurring on pyrolysis of cobalt tetra(p-methoxyphenyl)-porphyrin, [TMeOPPCo], were studied by Fuhrmann et al.[96] by measuring the specific conductivity, X-ray diffraction, composition, catalytic activity towards the decomposition of hydrogen peroxide, BET area, and electrocatalytic properties of the pyrolysed product. Maximum activity was obtained for heat treatment at about 550 °C, and the X-ray results indicated that β-cobalt is formed. The overall chemical composition of the pyrolysis product of the porphyrin was $\sim C_{40}H_{15}N_3Co$ in this temperature range, and it is not clear at present what is the chemical nature of the products that give the increased performance for reduction of O_2. The formation of new chemical compounds between the chelate and the carbon surface by heat treatment has been proposed by van Veen.[95] In these, the essential features of the environment of the transition-metal centre and its surrounding N_4 chelate that are necessary for electrocatalysis remain intact after heat treatment.

The use of polymeric phthalocyanines offers the attraction of combining a high electrical conductivity in the solid state with the presence of reactive metal centres at the surface that can act as specific anchoring co-ordinating points for oxygen and its reduction products. Kreja and Dabrowski[92] used the activation energy of the electrical conductivity of pellets of polyphthalocyanines to characterize the degree of polymerization. Polymers were prepared from pyromellitic dianhydride, from mixtures of tetracyanobenzene and dicyanobenzene, and also from mixtures of tetracyanobenzene and dicyanonaphthalene. Substitution in the ring was introduced by polycondensation of iron octamethoxyphthalocyanine. Since the methoxy-group is electron-donating, substitution will result (in this case) in an increased electron density at the central atom. The polymers were supported on active carbon and made into oxygen electrodes; the corresponding monomers were supported in a similar way, for comparison purposes. A linear dependence between electrochemical activity and the activation energy of conductivity was found. This, it was argued, caused a decrease in the ionization energy of the catalyst, and hence aided the formation of a charge-transfer complex between the oxygen molecule and the redox centres of the polymer. Electrical conductivity is, however, not the only parameter that is useful in determining the catalytic activity of the polyphthalocyanine, and Kreja and Rozploch[97] found a relationship between the e.s.r. linewidth, the concentration of paramagnetic centres in the presence of oxygen, and the electrochemical activity. The e.s.r. linewidth of the polyphthalocyanines followed the order Co > Fe > Cu > Ni > Mg, and a linear relationship with current density was not observed (note that the cobalt polymers are usually less active than the iron

---

[96] A. Fuhrmann, K. Wiesener, I. Iliev, S. Gamburzev, and A. Kaisheva, *J. Power Sources*, 1981, **6**, 69.
[97] L. Kreja and F. Rozploch, *Bull. Acad. Pol. Sci.*, 1979, **27**, 809.

compounds). A linear relationship between current density and the spin concentration was, however, demonstrated.

Behret et al.[90] studied the kinetics of reduction of oxygen in alkaline solutions on polyphthalocyanines, using the RRDE technique. The activity found was in the order Fe > Co > Ni. In the case of both the cobalt and the nickel polymers, the final reduction product was H_2O_2, but, in all of the polymers that were studied, the RRDE diagnostic criteria that were used showed evidence for the simultaneous occurrence of the direct and sequential reduction pathways. The mechanism proposed by these authors can be summarized as follows: the initial step involves the formation of a $M-O_2^-$ centre (or MO_2H), and the presence of simultaneous direct and sequential mechanisms is accounted for by assuming that the metal–superoxo complex can either be reduced further to hydrogen peroxide (and partially diffuse in the solution, thus giving a sequential contribution to the measured current) or can decompose according to:

$$MO_2H^- \rightarrow M^+O^- + OH^- \qquad (45)$$

with M^+O^- undergoing further rapid reduction. This O–O reductive splitting of the adsorbed peroxide will appear as a direct contribution of the reduction of O_2 to the measured current, since this pathway cannot result in the desorption of peroxide in the solution. This common pathway of the 'direct' and sequential reaction can explain the similarity of the E–log i dependence for both processes, as discussed in the case of the reduction of O_2 on platinum in alkaline solutions.[30,52]

The electrochemical activity of iron naphthalocyanine in alkaline solutions as a catalyst for the oxygen cathode has been investigated by Magner, Savy, and Scarbeck.[89] The rationale behind the choice of this macrocycle is to use a π-electron system that is larger than that of the phthalocyanines, since previous work had indicated[88] that the free energy of adsorption of oxygen on the iron centre could be optimized by delocalization of π-electrons. The generation of H_2O_2 from the reduction of O_2 was suppressed by the catalyst. Linear plots of N or of I_D/I_R (N is the collection efficiency; I_D and I_R are disc and ring currents respectively) versus $\omega^{+\frac{1}{2}}$ were observed, instead of the usual dependence on $\omega^{-\frac{1}{2}}$,[8] and this was considered to result from a mechanism involving control by the desorption of peroxide. The iron naphthalocyanine–carbon electrode appeared to be attractive for use as an air cathode in the zinc–air battery.

The idea of attaching active centres to a polymeric structure that is supported on an electrode, and using them to mediate in charge-transfer reactions, has been extensively studied in recent years. These electrode modifications can be made by electropolymerization in situ[98] and by simple exposure of electrodes to the polymer solution,[99] etc. However, polymer electrodes have not been extensively studied as oxygen cathodes. Hiratsuka et al.[100] prepared polytetracyanoethylene (PTCNE)–metal films by heating tetracyanoethylene and the metal in vacuum. The metals studied were iron, copper, and nickel. The Fe–PTCNE films could sustain fast transfer of electrons for redox processes in solution, and oxygen could be reduced. The ohmic drop across the film appears to be significant for the copper

[98] J. E. Dubois, P. C. Lacaze, and M. Pham, J. Electroanal. Chem. Interfacial Electrochem., 1981, 117, 233.
[99] N. Oyama and F. C. Anson, J. Am. Chem. Soc., 1979, 101, 3450.
[100] K. Hiratsuka, H. Sasaki, and S. Toshima, Chem. Lett., 1979, 751.

complex. A different approach was taken by Kikuchi et al.,[101] who used polystyrene to immobilize [CoPc] on a glassy carbon electrode surface. In 6M-KOH, a similar activity to that observed for adsorbed [CoPc] on glassy carbon was reported, but the stability was greatly improved by the presence of the polymer support.

Haas and Zumbrunnen[102] have studied a variety of electrochemical processes at a glassy carbon electrode that is modified by an adsorbed layer of 1-hydroxyphenazine. This acts as an electron mediator when fixed to an electrode surface, and the rate constant of the overall transfer of electrons for this type of electrode is determined by the second-order reaction between the attached redox species and the redox couple in solution. The effect of the adsorbed hydroxyphenazine layer was to catalyse the reduction of oxygen to H_2O_2. In a 1M-H_2SO_4 solution, the half-wave potential that was observed was -0.13 V vs the S.C.E., but the observed limiting current (7.5×10^{-4} A cm^{-2}) was lower than the value (3.7×10^{-3} A cm^{-2}) that can be calculated from the theory of the RDE at the rotation rate that was studied. This and other evidence presented indicates that, in the presence of a discrete number of redox-active centres, only part of the surface can be regarded as being electro-active.

Besides the phthalocyanines and porphyrin chelates, other macrocyclic complexes have been studied as catalysts for the reduction of oxygen. In acidic solutions, the dibenzotetra-aza-annulenes (TAA) are more active catalysts than the tetraphenylporphyrins or the phthalocyanines.[91] Similar results were obtained by Behret et al.[87] for [CoTAA] that was supported on platinum, gold, and pyrolytic graphite. Similarly to the behaviour of [CoTPP],[9] the main reduction product was H_2O_2, but from the RRDE results there was evidence for a contribution to the reduction by a direct mechanism. Values of k_1/k_2 of 0.1 and 0.3 were found in acidic and alkaline solutions, respectively, at 0.5 V vs the dynamic hydrogen electrode. The Tafel slopes that were found were -0.04 and -0.11 V per decade respectively, indicating that the transfer of the second electron was rate-determining in acidic solutions, whereas in alkaline solutions the transfer of the first electron was the rate-determining step. A cobalt hexa-azacyclotetradecine complex was found to increase the rate of reduction of O_2 on gold,[103] although the ratio of disc current to ring current (and hence the reduction mechanism) remains unaltered.

[FeTAA] was a better catalyst for the direct reduction process,[87] and the ratio k_1/k_2 was 0.4 in acidic media. This greater contribution by direct reduction for iron chelates is in agreement with the behaviour observed for [FeTPP],[87] [FePc],[87] and [FeTSP][80] in comparison with the corresponding cobalt compounds. It was proposed[87] that the simultaneous occurrence of the direct and sequential processes for the TAA chelates is similar to the mechanism observed for the TPP and Pc ligands, i.e., the splitting of the O–O bond of adsorbed peroxide occurs in parallel with its desorption.

The all-important question of catalysing the direct four-electron reduction of O_2 to water has been approached by Collman, Marrocco, Denisevich, Konai, Koval, and Anson[73,74] by studying two transition-metal porphyrin catalysts. It is very

---

[101] T. Kikuchi, H. Sasaki, and S. Toshima, *Chem. Lett.*, 1980, 5.
[102] O. Haas and H. R. Zumbrunnen, *Helv. Chim. Acta*, 1981, **64**, 854.
[103] H. Behret, G. Sandstede, G. G. Scherer, and H. G. Schomann, *J. Electroanal. Chem. Interfacial Electrochem.*, 1981, **117**, 339.

significant that the terminal enzyme of the electron-transport chain in mitochondria, the cytochrome oxidase, is an enzyme that contains two haems and two copper centres, some of these centres probably acting as 'reservoirs' for electrons and some others as co-ordination centres for O_2. It appears that no H_2O_2 is formed during reduction, and that a direct process is operative.[104, 105] The systems studied by Collman et al.[73,74] were dimeric face-to-face cobalt porphyrins that were kept in position by two linkages. The separation between metal centres was altered by using a four-, five-, or six-atom linkage to fix the relative position of the porphyrin moieties, and it was found that the dicobalt complex in which the chelating rings were separated by a four-atom linkage catalysed the reduction of O_2 without causing significant generation of H_2O_2. The distance between centres was found to be critical, and the compounds having more than a four-atom linkage catalysed the reduction mainly to hydrogen peroxide, as was also the case with the monomeric cobalt chelates.

The disproportionation of H_2O_2 was not catalysed by the dimers studied, and as a consequence, for the complex leading to a four-electron reduction, a two-centre co-ordination mechanism was proposed. The reduction sequence that was proposed involved the initial formation of Co^{II} centres, followed by chemical reaction with O_2 to give the corresponding μ-peroxo-complex. Contrary to the behaviour in solution of other μ-peroxo-complexes, this could easily be reduced to H_2O, thus leading to an overall direct mechanism. The work of Collman, Anson, and co-workers[73,74] is of fundamental importance in studies of the chemistry and electrochemistry of O_2, since it has shown the possibility of relating modern architectural synthetic chemical procedures to recent advances in the understanding of the mechanisms of the electrochemistry and biochemistry of oxygen.

5 Oxides and Mixed Oxides of the Transition Metals

These compounds are of interest owing to the possibility of combining a low bulk resistivity with active transition-metal centres on the surface. Vilinskaya et al.[106] have studied the kinetics of reduction of O_2 on the cobaltites of cobalt, magnesium, manganese, and cadmium that were supported on carbon black, using the RRDE technique. These compounds have the spinel structure, and catalyse the direct four-electron reduction to water. For instance, at 0.75 V vs the hydrogen electrode, the ratio k_1/k_2 was 4.0, 5.2, 5.9, and 7 for $CoCo_2O_4$, $CdCo_2O_4$, $MgCo_2O_4$, and $MnCo_2O_4$, respectively, and very small amounts of H_2O_2 could be detected at the ring. Trunov and Verenikina[107] continued previous work on the electrocatalytic properties of Ni–Co mixed oxides and studied the surface coverage of chemisorbed oxygen molecules by linear-sweep voltammetry. Adsorption of O_2 was carried out in air and reduction of the chemisorbed layer was followed in an oxygen-free 7M-KOH solution. A good correlation was found between the ratio of the concentrations of reducible sites to cations at the surface and electrical conductivity.

---

[104] B. G. Malmström, *Biochim. Biophys. Acta*, 1979, **549**, 281.
[105] B. G. Malmström, in ref. 6, Ch. 5, p. 181.
[106] V. S. Vilinskaya, N. G. Bulavina, V. Ya. Shepelev, and R. Kh. Burshtein, *Sov. Electrochem. (Engl. Transl.)*, 1979, **15**, 805.
[107] A. M. Trunov and N. N. Verenikina, *Sov. Electrochem. (Engl. Transl.)*, 1981, **17**, 115.

It was proposed, therefore, that the numbers of electrochemically active sites in a series of mixed oxides of the same metals is a function of the conductivity of the oxide. Another spinel-type oxide, Fe_3O_4, has also been studied as a catalyst for reduction of O_2;[108] in 0.1M-KOH the exchange-current density depended on the degree of oxidation of the spinel, changing from $(0.2\text{---}0.6) \times 10^{-11}$ to 1.6×10^{-11} A cm^{-2} on reduction. It is interesting to notice that the Tafel slope that was observed was -70 to -80 mV per decade, which differs significantly from the results observed for passive iron.[20] In the latter case, the outer layer of the passive film contains the γ-Fe_2O_3 oxide, which is a much weaker electrocatalytic material than Fe_3O_4.

Ternary oxides of platinum, palladium, rhodium, chromium, and cobalt, of composition $PtCoO_2$, $PdCoO_2$, $PdRhO_2$, and $PdCrO_2$, have been studied by Carcia et al.[109] The ternary oxides that were studied had the delafossite structure. These are layered compounds, and one of the interesting properties originally observed with the $PtCoO_2$ compound was its extremely low resistivity in the basal plane, similar to that of copper.[110] The structure of this compound, which is characteristic of this series, consists of alternating layers of cobalt and platinum atoms. The platinum atoms are linearly co-ordinated to two oxygen atoms whereas cobalt is octahedrally co-ordinated by oxygen atoms. The authors prepared these compounds by a simple gas-phase sputtering technique and found that electrocatalytic activity was only related to the presence of platinum or palladium in the lattice and was nearly independent of the transition-metal cation. In particular, if the current densities that could be obtained with the most active compound, i.e. $PtCoO_2$, were related to unit area of catalyst, the behaviour was similar to that observed for pure platinum. Octahedral cobalt is not a favoured co-ordination geometry for electrocatalysis, and the reduction of O_2 occurs on the platinum atoms in these mixed oxides; indeed, the Pt–Pt spacing of the (100) and (110) surfaces was found to be similar to those of bulk platinum metal, and it was suggested that the mechanism of reduction of O_2 is probably similar on both surfaces.

Nguyen Cong and Brenet[111] studied the reduction of O_2 on a series of mixed oxides of composition $A_xB_yMn_{3-x-y}O_4$ with A = Zn, Cr, or Al, B = Ni or Cu, $x > 0$, and $y < 1$. The main purpose of this work was to elucidate the role of Mn^{4+} and Mn^{3+} ions, and it was shown that the simultaneous presence of both of these species in the lattice is a necessary condition for electrocatalysis by these manganites. An e.c. reduction mechanism was proposed, with the Mn^{4+}/Mn^{3+} acting as a redox shuttle:

$$Mn^{4+} + e^- \rightarrow Mn^{3+} \qquad (46)$$

$$Mn^{3+} + O_2 \rightarrow Mn^{4+} + O_2^- \rightarrow \text{products} \qquad (47)$$

Bronoel et al.[112] compared the electrocatalytic properties of ferrites that have the composition $La_{1-x}Sr_xFeO_{3-y}$ with those of the copper manganite $Cu_{1.4}Mn_{1.6}O_y$, lanthanum strontium cobaltite ($La_{0.5}Sr_{0.5}CoO_3$), and surface oxides that are formed

---

[108] N. F. Razina, Izv. Akad. Nauk Kaz. SSR, Ser. Khim., 1980, No. 1, p. 75.
[109] P. F. Carcia, R. D. Shannon, P. E. Bierstedt, and R. B. Flippen, J. Electrochem. Soc., 1980, **127**, 1974.
[110] R. D. Shannon, D. B. Rogers, and C. T. Prewitt, Inorg. Chem., 1971, **10**, 713.
[111] H. Nguyen Cong and J. Brenet, J. Appl. Electrochem., 1980, **10**, 433.
[112] G. Bronoel, J. C. Grenier, and J. Reby, Electrochim. Acta, 1980, **25**, 1015.

on nickel and on nickel–cobalt alloys in 1M-KOH. All of the oxides were reduced in the same potential range as that for reduction of O_2, and when both the reduction currents of the oxides and the surface areas were taken into account it was found that the activity per unit area was inferior to that of nickel and its alloys for all of the mixed oxides studied. The nickel–cobalt alloy was the most active compound, as expected from the known good catalytic properties of $NiCo_2O_4$.

Perovskite-type oxides show good electrical conductivity and catalytic activity for the reduction of oxygen. Typical oxides that have this structure are the lanthanum nickel oxides, which are non-stoicheiometric compounds that have Ni^{2+} and Ni^{3+} ions as well as oxygen vacancies in the lattice.[113] The high electronic conductivity of these compounds and their catalytic activity in the reduction of O_2 are related to the presence of a partially filled σ^* band in the oxide.[113] Matsumoto and Sato[114] continued previous work on mixed oxides with the perovskite-type structure as electrode materials and studied the electrocatalytic behaviour of $La_{0.7}Pb_{0.3}MnO_3$. It was found that, in alkaline solutions, the reduction of O_2 did not occur through the diffusion of O^{2-} ion in the lattice, as had previously been proposed by Kudo et al.[115] for the perovskites, but as an electrochemical process at active sites at the oxide/solution interface. The exchange-current density was $i_0 = 2 \times 10^{-9}$ A cm^{-2} and the Tafel slope was -80 mV per decade on the region of low current density. The value of i_0 was greater than that observed for other perovskites; if the rate of transfer of electrons is determined only by the width of the σ^* band, the value of i_0 should have been lower than that of $La_{1-x}Sr_xMnO_3$,[116] owing to the narrowness of the σ^* band of $La_{0.7}Pb_{0.3}MnO_3$. The authors proposed specific surface effects of the manganese cation to account for the increased catalytic activity.

Titanium dioxide has been proposed as an interesting semiconducting oxide for use in the photoelectrolysis of water, but the behaviour of this oxide as an electrode material for reduction of oxygen has not been extensively studied. Since oxygen is generated at the surface of the TiO_2 during photoelectrolysis, it is important to understand its electrochemical properties on this semiconductor, and Parkinson et al.[117] have studied its reduction for this reason, using a RRDE technique. In alkaline solutions, reduction occurs at potentials close to the flat-band potential, and H_2O_2 was detected at the ring. In the absence of O_2, linear-sweep voltammetry showed the occurrence of surface redox processes in the same potential range as that at which reduction of O_2 takes place. The authors proposed that these surface sites acted as mediators in electron transfer, not only in the case of O_2 but also for the reduction of $[Fe(CN)_6]^{3-}$. The number of surface species was calculated as 3×10^{14} cm^{-2}: if the density of rutile is taken as 4.26 g cm^{-3},[23] the number of surface species that was calculated is very close to the number of titanium atoms present per unit area of oxide surface. The peak potential showed a similar dependence on pH (-59 mV per unit of pH) to that observed for the flat-band potential, suggesting that there is a protonation equilibrium. A similar dependence is obtained, however, by

---

[113] Y. Matsumoto, H. Yoneyama, and H. Tamura, *J. Electroanal. Chem. Interfacial Electrochem.*, 1977, **79**, 319.
[114] Y. Matsumoto and E. Sato, *Electrochim. Acta*, 1980, **25**, 585.
[115] T. Kudo, H. Obayashi, and M. Yoshida, *J. Electrochem. Soc.*, 1977, **124**, 321.
[116] Y. Matsumoto and E. Sato, *Electrochim. Acta*, 1979, **24**, 421.
[117] B. Parkinson, F. Decker, J. F. Julião, and M. Abramovich, *Electrochim. Acta*, 1980, **25**, 521.

considering the changes in the inner potential difference across the oxide/solution interface that are caused by the equilibrium:

$$O^{2-}_{(oxide)} + H_2O \rightleftarrows 2\,OH^-_{(aq)} \tag{48}$$

and no specific interaction with the transition-metal centres needs to be invoked. This is a general problem in the study of reaction orders with respect to the hydroxide ion for oxide-covered metals, since the specific participation of OH^- in the rate-determining step must be deconvoluted from the dependence on pH of the inner potential difference across the oxide/solution interface. The general question of a tunnelling mechanism to surface states from the bulk of the semiconductor is one of practical importance, and the authors suggested that, for TiO_2, the relatively high-energy surface states that are present at less than 0.5 eV below the edge of the conduction band could explain the easy transfer of charge that is observed for the redox process at the surface.

6 The Electrochemistry of Oxygen in Molten Salts and in Fuel Cells at High Temperatures

The main interest in this area is in studies related to the molten carbonate fuel cell. Appleby and Nicholson[118] investigated the reduction of O_2 on gold in a Li/K (53:47 atom %) carbonate melt. Two reduction peaks were observed in linear-sweep voltammetry. The first peak resulted from the reduction of the peroxide ion that is homogeneously formed by the reaction of oxygen with carbonate ions:

$$O_2 + 2\,CO_3^{2-} \rightarrow 2\,O_2^{2-} + 2\,CO_2 \tag{49}$$

and a reductive splitting of the O–O bond of the peroxide was proposed as the rate-determining step:

$$O_2^{2-} + e^- \rightarrow O^{2-} + [O^-] \quad (r.d.s.) \tag{50}$$

followed by:

$$[O^-] + e^- \rightarrow O^{2-} \tag{51}$$

The proposed second reduction process was the reduction of superoxide ion, in equilibrium in the melt:

$$O_2 + O_2^{2-} \rightleftarrows 2\,O_2^- \tag{52}$$

$$O_2^- + e^- \rightarrow O_2^{2-} \tag{53}$$

The same authors also studied the apparent catalytic effects that are observed when silver is used as the material for the oxygen electrode in carbonate melts.[119] Current densities that are an order of magnitude above those of gold were observed if the silver electrode was pre-oxidized with H_2O_2 or HNO_3. It was concluded that the effects were due to the dissolution of O_2 in the silver cathodes, and the electrode could therefore act as an oxygen-transfer medium. Sabbatini et al.[120] tried to correlate the nature of the species that were formed on nickel electrodes when they

---

[118] A. J. Appleby and S. B. Nicholson, *J. Electroanal. Chem. Interfacial Electrochem.*, 1980, **112**, 71.
[119] A. J. Appleby and S. B. Nicholson, *J. Electrochem. Soc.*, 1980, **127**, 759.
[120] L. Sabbatini, B. Morelli, P. Zambonin, and B. A. DeAngelis, *J. Chem. Soc., Faraday Trans. 1*, 1979, **75**, 2628.

were used as cathodes in carbonate melts with their potentiostatic behaviour. An ESCA technique was used and the formation of higher oxidation states on the nickel surface could be detected. Irreversibility at this electrode was attributed to the involvement of this surface species in the rate-determining step.

It should be added, finally, that most of the work published on the molten carbonate system has dealt with the technological aspects of the development of a system that is competitive with the phosphoric acid fuel cell.

There has recently been interest also in solid oxide electrolytes for use in fuel cells that operate at high temperatures.[121] The process that occurs at this cathode is the reduction of oxygen to oxide. This has been studied by Wang and Nowick by a current-interruption technique,[122] using an electrolyte of doped ceria. On platinum, the rate-determining step that was proposed was the transfer of an O^- ion that is adsorbed on the platinum electrode to the electrolyte, followed by electron transfer; Tafel behaviour was observed at high oxygen pressures. The same authors extended previous transient-technique studies to potentials above the cathodic limiting current[123] in doped ceria solid electrolytes.

Sasaki et al.[124] used a.c. impedance techniques and proposed that the rate-determining step on platinum in an yttria-stabilized zirconia electrolyte is either the dissociation of weakly adsorbed oxygen or the subsequent diffusion of adsorbed oxygen atoms over the surface and their reduction to oxide.

7 Non-aqueous Solvents and Reactions of Superoxide Ion

The superoxide ion is very unstable in aqueous solutions, but it is now well recognized that it is sufficiently stable in some non-aqueous solvents to be regarded as a useful reagent.[125] The electrochemical generation of O_2^- in non-aqueous solvents by the reduction of O_2 is now well documented, and for instance, in solvents such as pyridine,[125] DMF,[126] or DMSO,[127] the one-electron reduction of O_2 yields well-defined cyclic voltammetric waves, with a ratio of cathodic to anodic peak currents of unity. Thus, the simplicity of generation of O_2^- opens interesting possibilities for organic electrosynthetic processes. Gritzner and Rechberger[128] studied 2,2'-thiodiethanol as a solvent for non-aqueous electrochemistry. Dioxygen was reduced on a dropping mercury electrode and an anodic reverse peak was observed on the cyclic voltammogram. Although the use of a dropping mercury electrode introduces serious difficulties in the quantitative analysis of voltammograms, the presence of a reverse peak must be indicative of the stability of the O_2^- ion in this solvent.

The superoxide ion can enter into a wide range of homogeneous reactions, such as acid–base equilibria, nucleophilic attack, and reduction and oxidation reactions, and it can act as a complexation agent.[125] The O_2^- ion also occurs in living

---

[121] G. B. Barbi and C. M. Mari, *Mater. Chem.*, 1981, **6**, 35.
[122] D. Y. Wang and A. S. Nowick, *J. Electrochem. Soc.*, 1979, **126**, 1155.
[123] D. Y. Wang and A. S. Nowick, *J. Electrochem. Soc.*, 1980, **127**, 113.
[124] J. Sasaki, J. Mizusaki, S. Yamauchi, and K. Fueki, *Bull. Chem. Soc. Jpn.*, 1981, **54**, 1688.
[125] D. T. Sawyer and M. J. Gibian, *Tetrahedron*, 1979, **35**, 1471.
[126] C. L. Hussey, T. M. Laher, and J. M. Achord, *J. Electrochem. Soc.*, 1980, **127**, 1484.
[127] S. Kudo and A. Iwase, *Bull. Chem. Soc. Jpn.*, 1979, **52**, 908.
[128] G. Gritzner and P. Rechberger, *J. Electroanal. Chem. Interfacial Electrochem.*, 1980, **109**, 333.

organisms as part of the mechanism of utilization of oxygen, and there is currently a great deal of interest and argument regarding its role in biochemical processes.[6,7,129,130]

Hussey, Laher, and Achord[126] have studied the reaction between electrogenerated $O_2^{\bar{\cdot}}$ and 2-, 3-, and 4-nitroanilines and also its reaction with NN-dimethyl-4-nitroaniline, using DMF as the solvent. The cyclic voltammetry of the $O_2/O_2^{\bar{\cdot}}$ couple was used to follow the reaction between $O_2^{\bar{\cdot}}$ and the nitroanilines. Both 2- and 4-nitroaniline reacted with the electrochemically generated $O_2^{\bar{\cdot}}$ ions, giving voltammograms that are characteristic of an *e.c.* process. However, the $O_2^{\bar{\cdot}}$ species acted in this case only as a base, as verified by comparing the visible—u.v. spectra of the products of long-term electrolysis with that obtained for the nitroanilines with added KOH/18-crown-6 ether or MeONa. The *e.c.e.* mechanism that was proposed involved the intermediate protonation of $O_2^{\bar{\cdot}}$ in the reactions:

$$O_2 + e^- \rightleftharpoons O_2^{\bar{\cdot}} \qquad (54)$$

$$O_2^{\bar{\cdot}} + RNH_2 \rightarrow O_2H\cdot + RNH^- \qquad (55)$$

$$O_2H\cdot + e^- \rightarrow O_2H^{\bar{\cdot}} \qquad (56)$$

Since 3-nitroaniline is more basic than the 2- and 4-derivatives, and NN-dimethyl-4-nitroaniline has no available protons, the lack of reactivity with $O_2^{\bar{\cdot}}$ gives support to the proposed mechanism.

Roberts and Sawyer[131] studied the reactivity of electrochemically generated $O_2^{\bar{\cdot}}$ as a nucleophile in substitution reactions with chlorinated substrates. The rate constants of the reactions between $O_2^{\bar{\cdot}}$ and the chlorinated substrates were derived from cyclic voltammetric measurements. Rate constants calculated for reactions with CH_3Cl, CH_2Cl_2, $CHCl_3$, CCl_4, and p,p'-DDT were 80, 9, 460, 1300, and 130 dm^3 mol^{-1} s^{-1}, respectively. Complete loss of chlorine from the chloro-derivatives was observed in all cases, and the authors suggested the possibility of disposing of chlorinated wastes by allowing them to react with electrochemically generated superoxide ion.

The reaction between $O_2^{\bar{\cdot}}$ and 1,2-phenylenediamine in DMF has been studied by Hussey *et al.*,[132] using cyclic voltammetry and chronoamperometry. An *e.c.* mechanism was observed in the reduction of oxygen, and homogeneous transfer of electrons and of hydrogen atoms was proposed:

$$O_2^{\bar{\cdot}} + RNH_2 \rightarrow O_2H^- + RNH\cdot \qquad (57)$$

The versatility of $O_2^{\bar{\cdot}}$ as a reagent is highlighted by the work of Sagae, Fujihira, Lund, and Osa,[133] who were able to cause *oxidation* of the side-chains of methylpyridine and methylpyridine N-oxides with $O_2^{\bar{\cdot}}$ ions that were generated by the *reduction* of oxygen. All the compounds studied had reduction potentials that were more negative than that of dioxygen. Both 2- and 3-methylpyridine and 3-

---

[129] I. Fridovich, in 'Advances in Inorganic Biochemistry', ed. G. L. Eichhorn and L. G. Marzilli, Elsevier North Holland, New York, 1979, Ch. 2, p. 67.
[130] H. A. O. Hill, in 'New Trends in Bio-Inorganic Chemistry', ed. R. J. P. Williams and J. R. R. Fausto da Silva, Academic Press, London, 1978, p. 173.
[131] J. L. Roberts and D. T. Sawyer, *J. Am. Chem. Soc.*, 1981, **103**, 712.
[132] C. L. Hussey, T. M. Laher, and J. M. Achord, *J. Electrochem. Soc.*, 1980, **127**, 1865.
[133] H. Sagae, M. Fujihira, H. Lund, and T. Osa, *Heterocycles*, 1979, **13**, 321.

methylpyridine N-oxide did not show significant reactivity to yield the corresponding carboxylic acids, but 4-methylpyridine, 2-methylpyridine N-oxide, and 4-methylpyridine N-oxide gave yields of 11, 33, and 28%, respectively, of the corresponding pyridinecarboxylic acids. In a different context, the insertion of oxygen atoms (under reductive conditions for the O_2 molecule) is the hydroxylating mode of action of the enzyme cytochrome P-450.[6,7] The proposed reaction mechanism involved the reaction of both the electrochemically generated superoxide ions and of molecular oxygen with the radical intermediates:

$$O_2 + e^- \rightleftharpoons O_2^{\cdot -} \tag{58}$$

$$RCH_3 + O_2^{\cdot -} \rightarrow RCH_2{\cdot} + O_2H^- \tag{59}$$

$$RCH_2{\cdot} + O_2 \rightarrow RCH_2OO{\cdot} \tag{60}$$

$$RCH_2OO{\cdot} + O_2^{\cdot -} \text{ (or } e^-\text{)} \rightarrow RCH_2OO^- + O_2 \tag{61}$$

$$RCH_2OO^- \rightarrow RCHO + OH^- \tag{62}$$

$$RCHO + O_2^{\cdot -} \rightarrow RCO{\cdot} + O_2H^- \tag{63}$$

$$RCO{\cdot} + O_2H^- \rightarrow [RCOOOH]^{\cdot -} \tag{64}$$

$$[RCOOOH]^{\cdot -} \rightarrow RCOO{\cdot} + OH^- \tag{65}$$

$$RCOO{\cdot} + O_2^{\cdot -} \text{ (or } e^-\text{)} \rightarrow RCOO^- + O_2 \tag{66}$$

There are several uncertainties regarding this proposed reaction sequence; for instance, reaction (59) would probably require a degree of reactivity of the methylpyridine that does not appear to be warranted by the lack of electrochemical activity of these compounds which the authors observed.

Superoxide ion can act as an oxidizing or a reducing agent, and Stanbury, Mulac, Sullivan, and Taube[134] studied both processes in the reaction of $O_2^{\cdot -}$ with (isonicotinamide)pentammineruthenium-(II) and -(III) $\{[Ru(NH_3)_5(isn)]^{2+}$ and $[Ru(NH_3)_5(isn)]^{3+}\}$, using pulse-radiolysis techniques. The interest of this work is the possibility of obtaining kinetic information on the intermediates in the reduction of oxygen that can be compared with the Marcus theory of transfer of electrons. In particular, it has been shown that the self-exchange rates for the HO_2/HO_2^- couple range over six orders of magnitude when comparing data obtained with the $[Ru(NH_3)_5(isn)]^{2+}$ system, $[Fe(CN)_6]^{4-}$, and $[Mo(CN)_8]^{4-}$; the origin of this discrepancy is not clear. The understanding of these questions is very relevant to the mechanism of dismutation of superoxide ion, which is catalysed in living organisms that utilize oxygen[130] by the enzyme superoxide dismutase (SOD).[129,130]

The superoxide ion can co-ordinate to transition-metal centres,[125] and it has been proposed that, in the case of the Cu–Zn superoxide dismutases, complexation to the Cu^{II} centre occurs *prior* to electron exchange.[129] The co-ordination of electrogenerated $O_2^{\cdot -}$ with bis(acetylacetonato)cobalt(II), $[Co(acac)_2]$, in acetone and DMSO has been studied by Kudo and Iwase.[127] A catalytic current in the region of the reduction of O_2 to $O_2^{\cdot -}$ was observed and long-term electrolysis resulted in the

---

[134] D. M. Stanbury, W. A. Mulac, J. C. Sullivan, and H. Taube, *Inorg. Chem.*, 1980, **19**, 3735.

eventual formation of the corresponding μ-peroxo-complex. This reaction was, however, rather slow, taking several hours to reach completion. The chemical transformations proposed are shown in Scheme 6.

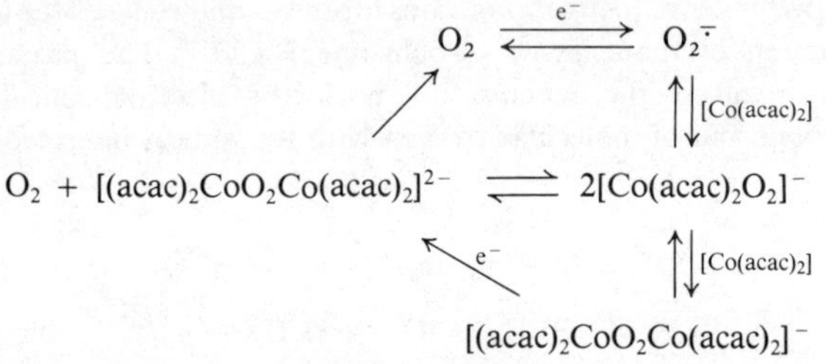

$$O_2 \xrightleftharpoons{e^-} O_2^-$$

$$O_2 + [(acac)_2CoO_2Co(acac)_2]^{2-} \rightleftharpoons 2[Co(acac)_2O_2]^-$$

$$[(acac)_2CoO_2Co(acac)_2]^-$$

with vertical equilibria involving [Co(acac)$_2$] and electron transfer.

Scheme 6

It is interesting to note that the μ-peroxo-dicobalt(II) complex could not be reduced in the accessible potential range. These results are in marked contrast with the work of Geiger and Anson[71] on the reduction of the μ-peroxo-complex of *trans*-[Co([14]ane-N$_4$)(OH$_2$)$_2$]$^{3+}$ from aqueous solutions (see Scheme 6). In this case, the use of a non-aqueous solvent such as DMSO must have a remarkable influence on the stability of these O$_2$-containing co-ordination compounds. The formation of the μ-peroxo-dicobalt(II) complex from [Co(acac)$_2$O$_2$]$^-$ is the equivalent of the homogeneous dismutation of O$_2^-$ ion. It is interesting to note that co-ordination to a [Co(acac)$_2$] complex does not result in a large increase in rates of dismutation as compared with electrogenerated O$_2^-$ in aprotic solvents. The rate constant for the homogeneous reaction:

$$O_2^- + O_2H\cdot \rightarrow O_2 + O_2H^- \tag{67}$$

is $\sim 9 \times 10^7$ dm^3 mol^{-1} s^{-1} in water, but, unfortunately, this rate has little bearing on the expected rates in aprotic solvents, where the formation of a peroxo (O$_2^{2-}$) species in the complete absence of a proton donor will be an unfavourable process.

Tomat and Rigo[135] generated OH$\cdot$ radicals by the electrochemical reduction of O$_2$ to H$_2$O$_2$ in the presence of Fe^{II} and Ti^{III}. The transition-metal cations in the low oxidation state were concurrently generated by reduction, and a reaction similar to that of the Fenton reagent generated the OH$\cdot$ radicals according to:

$$H_2O_2 + M_{red} \xrightarrow{k_D} OH\cdot + OH^- + M_{ox} \tag{68}$$

with k_D equal to 500 and 60 dm^3 mol^{-1} s^{-1} for Ti^{III} and Fe^{II}, respectively. The oxidation of cyclohexane to cyclohexanone in the presence of a high concentration of HCl gave a maximum yield of 55% when using Fe^{III} as the catalyst for decomposition of H$_2$O$_2$. The oxidation mechanism that was proposed involved the abstraction of hydrogen:

$$C_6H_{12} + OH\cdot \rightarrow C_6H_{11}\cdot + H_2O \tag{69}$$

---

[135] R. Tomat and A. Rigo, *J. Appl. Electrochem.*, 1980, **10**, 549.

followed by the insertion of O_2:

$$C_6H_{11}\cdot + O_2 \rightarrow C_6H_{11}O_2\cdot \quad (70)$$

and reduction of the peroxyl radical by Fe^{II}:

$$C_6H_{11}O_2\cdot + Fe^{II} \rightarrow C_6H_{10}O + Fe^{III} + OH^- \quad (71)$$

A large decrease in the yield of cyclohexanone was observed if the Cl^- ion was replaced by HSO_4^- or ClO_4^-, and it was proposed that, in the presence of a large excess of Cl^- ions, chlorine atoms were produced according to:

$$OH\cdot + H^+ + Cl^- \rightarrow Cl\cdot + H_2O \quad (72)$$

The $Cl\cdot$ atoms are more reactive than $OH\cdot$ for abstraction of hydrogen from aliphatic C–H bonds, and as a consequence, the yield of oxidation product was significantly higher in the presence of Cl^- ions.

The catalytic effect of the specific co-adsorption of iron(III) deuteroporphyrin and tri-n-butyl phosphate (TBP) on the reduction of oxygen at a mercury electrode has been studied by Sohr and Wienhold.[136] The adsorption of TBP on mercury is known to stabilize the O_2^- ion and to split into two the first polarographic wave of the reduction of O_2. The catalytic mechanism that has been proposed involves the intermediate formation of an iron(II)–TBP–μ-superoxo intermediate. The interaction of the superoxide ion with iron porphyrins and the electrochemical properties of these complexes are of great relevance in the understanding of the hydroxylating mechanism of cytochrome P-450.[6] In particular, the reduction of the $Fe^{III}O_2^-$ complex, i.e. the transfer of the second electron, is the least understood reaction in the functioning of cytochrome P-450. Welborn, Dolphin, and James[137] studied this process, using ferrous octaethylporphyrin, $[Fe^{II}(OEP)]$. The reduction of the superoxo-complex of $[(Fe^{II})(OEP)]$ was carried out at $-25\,^\circ C$ in MeCN–DMSO; the peroxo-derivative $[Fe^{II}(OEP)O_2]^-$ was formed and found to be identical with that produced by the reaction of $[Fe^{I}(OEP)]$ with O_2 or of $[Fe^{II}(OEP)]$ with O_2^- radical ion. The electrochemical generation of this peroxo-complex is an important step in attempting to perform an electrochemical hydroxylation reaction: it was argued that the electronic configuration of the peroxo-complex must be similar to the intermediate in the P-450 cycle prior to cleavage of the O–O bond and the insertion of oxygen (see Scheme 7). Unfortunately, attempts to epoxidize styrene with this system were not successful.

8 The Biochemistry of Oxygen

Comprehensive reviews of the biochemical reactions that utilize oxygen have been published;[6,7,138] the purpose of this section is merely to point out some of the possibilities that are emerging and to establish the relevance of electrochemistry in this area.

Traditionally, the biological function of oxygen was considered to be restricted to serving as the terminal acceptor of electrons in cellular respiration. The first

---

[136] H. Sohr and K. Wienhold, *Z. Phys. Chem. (Leipzig)*, 1980, **261**, 633.
[137] C. H. Welborn, D. Dolphin, and B. R. James, *J. Am. Chem. Soc.*, 1981, **103**, 2869.
[138] O. Hayaishi, 'Molecular Mechanisms of Oxygen Activation', Academic Press, New York and London, 1974.

observations of enzymes that are capable of catalysing insertion reactions of oxygen on organic substrates were made in 1950; since then, many oxygenases have been studied and characterized.[138] There are two broad classes of oxygenases; the *dioxygenases* catalyse reactions where both oxygen atoms are incorporated into the substrate whereas the *mono-oxygenases* catalyse reactions whereby one oxygen atom is inserted while the other is reduced to the OH^- ion or to water. Of the latter class of enzymes, cytochrome *P*-450 has been extensively studied. In what follows, we shall briefly discuss this enzyme, as an example of a system that shows important electrosynthetic possibilities, and also cytochrome aa_3 (or cytochrome oxidase), as the biochemical analogue of a catalyst for an air cathode.

Cytochrome *P*-450.—The reaction that is catalysed by this enzyme is:

$$RH + O_2 + NADPH + H^+ \rightarrow ROH + H_2O + NADP^+ \qquad (73)$$

The transfer of electrons from the reduced nucleotide occurs through a series of intermediates, and the inserted oxygen atom comes from the oxygen molecule. The prosthetic group is a ferriprotoporphyrin-IX,[139] and the iron centre plays a key role in the hydroxylating reaction. There is, at present, general agreement regarding the overall sequence of reactions, which are shown in Scheme 7.[6,7] The binding of the substrate to the resting Fe^{3+} enzyme is the trigger of the whole reduction sequence. Binding alters the co-ordination environment of the transition-metal centre and the redox potential of the enzyme, thus enabling the transfer of electrons to occur (Step 2). This triggering mechanism by the substrate that is to be hydroxylated is probably

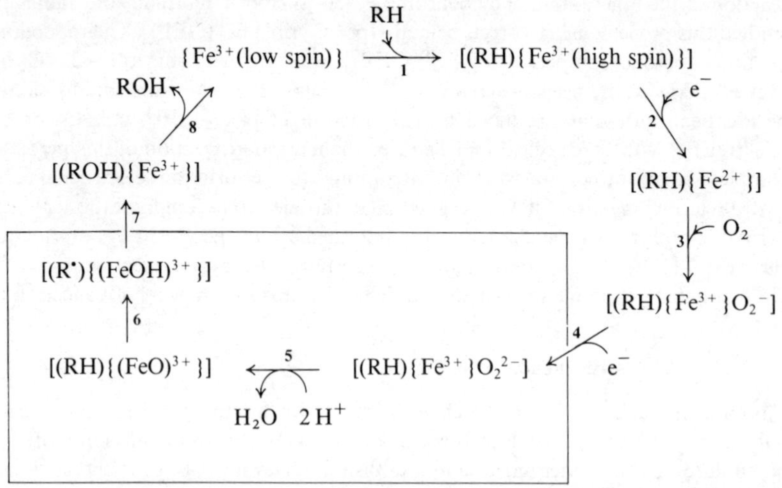

$$\{Fe^{3+}(\text{low spin})\} \xrightarrow[1]{RH} [(RH)\{Fe^{3+}(\text{high spin})\}]$$

$$ROH \xrightarrow{8} [(ROH)\{Fe^{3+}\}] \qquad \xrightarrow{2, e^-} [(RH)\{Fe^{2+}\}]$$

$$\uparrow 7 \qquad \qquad \downarrow 3, O_2$$

$$[(R^\bullet)\{(FeOH)^{3+}\}] \qquad [(RH)\{Fe^{3+}\}O_2^-]$$

$$\uparrow 6 \qquad \qquad \downarrow 4, e^-$$

$$[(RH)\{(FeO)^{3+}\}] \xleftarrow[H_2O \; 2H^+]{5} [(RH)\{Fe^{3+}\}O_2^{2-}]$$

No definite evidence is yet available for the proposed steps within the box

Scheme 7

---

[139] R. L. Tsai, I. C. Gunsalus, and K. Dus, *Biochem. Biophys. Res. Commun.*, 1971, **45**, 1300.

an essential feature of the *P*-450 system, since, in the absence of a starting instruction when only substrate is available, the cyclic behaviour can result in the continuous generation of hydrogen peroxide. The formation of H_2O_2 in the absence of substrate has certainly been observed.[6] The transfer of electrons and the formation of a superoxo-complex by reaction with O_2 (Steps **2** and **3**) have been well characterized. It is important to stress that there are many similar electrochemical reactions, equivalent to these steps, that have been observed,[80] and some of them have been discussed in this Report. The difficulty in finding an electrochemical or chemical model system lies in the reaction whereby the O–O bond is split; the mechanism of this process is not known at present, and the steps **4** to **7** in Scheme 7 have been proposed by analogy with carbene-insertion reactions.[140] In this model, the true hydroxylating agent is the ferryl radical, FeO^{3+}, which can be regarded as an oxygen atom that is co-ordinated to an Fe^{3+} centre. No firm spectroscopic evidence exists for this intermediate in biological systems, but an analysis of the products seems to require that the two-step hydroxylation reaction given by steps **6** and **7** is present.[141] If the FeO^{3+} intermediate is the hydroxylating species, a very special stabilization mechanism must be operative in the enzyme, in order to protect this strongly oxidizing agent from the transfer of electrons, since the enzyme functions in a reducing environment.

If FeO^{3+} is a necessary intermediate which must have a lifetime at least as long as the time of a molecular vibration, it would appear to be difficult to reproduce the function of *P*-450 electrochemically on any reasonable model compound that is attached to a metallic electrode, since the transfer of electrons to what is essentially an oxygen *atom* that is co-ordinated to an Fe^{3+} centre would in all likelihood be extremely fast.

The reduction of oxygen on transition-metal macrocycles shows, however, some interesting possibilities. As previously discussed, in general, for the reduction of O_2, the use of cobalt complexes results in the generation of H_2O_2, but the use of iron redox centres leads to a significant proportion of direct four-electron reduction to water. It is obvious, considering the corresponding molecular dimensions of the chelates studied, that the possibility of a two-centre co-ordination, as proposed for the face-to-face cobalt diporphyrins,[73,74] is not the origin of the direct four-electron reduction; therefore this process must occur on a single redox centre. As discussed for the polyphthalocyanines, the reductive splitting of the O–O bond of the transition-metal–superoxo intermediate in the reduction of oxygen could account for the observed direct reduction pathway. The similarity between this reduction sequence and the redox changes in cytochrome *P*-450 is striking. Whether or not it is possible to stabilize the intermediates of this process so that an insertion reaction can occur is an open question, but it would seem to be an attractive electrosynthetic approach, in particular if it is remembered that some of the likely reactions, such as the epoxidation of double bonds, are undoubtedly of great commercial relevance.

Cytochrome Oxidase.—Cytochrome oxidase (or cytochrome aa_3) is the terminal enzyme of the respiratory chain; its function is to reduce oxygen to water, catalysing

---

[140] G. A. Hamilton, *J. Am. Chem. Soc.*, 1964, **86**, 3391.
[141] J. T. Groves, in ref. 6, Ch. 3, p. 125.

the oxidation of cytochrome c and generating adenosine triphosphate, ATP. Several review articles on the properties of this enzyme have been published recently.[6,105] Cytochrome oxidase contains two haem groups and two copper atoms, and the functional relationship between the different redox centres is a matter of controversy. There is, however, general agreement that, in the normal reduction process, no hydrogen peroxide is generated. This is an important feature that is of relevance for air cathodes, where the generation of H_2O_2 is detrimental for the carbon support. Petty, Welch, Wilson, Bottomley, and Kadish[142] have recently proposed the initial formation of a μ-peroxo-complex between an iron and a copper centre to account for the available spectroscopic evidence; recent EXAFS data[143] indicate an [Fe–Cu] separation of ~0.3 nm, which is sufficient to accommodate a μ-peroxo bridge. It was proposed that further reduction occurs via the formation of a μ-oxo bridge. Mixed dinuclear complexes of known geometry offer some attractive features as catalysts for the reduction of oxygen; apart from the work on the face-to-face cobalt porphyrins,[73,74] these possibilities have not been exploited.

Hill, Walton, and Higgins[144] were recently able to drive the reduction of oxygen via the terminal oxidase of *Pseudomonas aeruginosa* electrochemically. This was achieved by coupling the transfer of electrons from a gold electrode to the oxidase through a layer of specifically adsorbed 1,2-bis-(4-pyridyl)ethene. Hill and co-workers[145,146] had previously used 4,4'-bipyridyl to form an adsorbed layer that catalysed the transfer of electrons to cytochrome c from a gold electrode, and this elegant work shows the possibility of coupling biochemical systems with electrode processes.

9 Applications

Some of the recent applications of the electrochemistry of O_2 are described here. The original electrolytic method for the manufacture of hydrogen peroxide has been replaced by a chemical route; however, the generation of bleaching agents *in situ* for the pulp industry is an interesting area of application which has been studied by Oloman and Watkinson.[147] These authors used a simple parallel-plate cell with a bed cathode made of inexpensive graphite particles. The cathode bed was separated from the anode plate by a porous diaphragm which allowed the diffusion of the oxygen that was generated to the cathode; thus a separate anode compartment was not required. The process was carried out in alkaline solutions, and the operating conditions of this simple cell are given in Table 3. The authors estimated the capital cost of a plant to produce 500 kg of H_2O_2 per day at about \$150 000 at 1978 prices. Balej and Spalek[148] proposed the use of porous gas electrodes, made of carbon black and Teflon, for the generation of H_2O_2 from the reduction of oxygen with reported current efficiencies of 92—100%.

[142] R. H. Petty, B. R. Welch, L. J. Wilson, L. A. Bottomley, and K. M. Kadish, *J. Am. Chem. Soc.*, 1980, **102**, 611.
[143] W. E. Blumberg and J. Peisach, *Biophys. J.*, 1979, **25**, 34a.
[144] H. A. O. Hill, N. J. Walton, and I. J. Higgins, *FEBS Lett.*, 1981, **126**, 282.
[145] M. J. Eddowes and H. A. O. Hill, *J. Am. Chem. Soc.*, 1979, **101**, 4461.
[146] M. J. Eddowes, H. A. O. Hill, and K. Uosaki, *J. Am. Chem. Soc.*, 1979, **101**, 7113.
[147] C. Oloman and A. P. Watkinson, *Sven. Papperstidn.*, 1980, **83**, 405.
[148] J. Balej and O. Spalek, Czech. P. 184 557.

Table 3 *Operating conditions of a generator of hydrogen peroxide that has a graphite bed cathode*[147]

Current density	400—1600 A m^{-2}
Oxygen pressure	0.1—2.2 MPa
Concentration of NaOH	1—3 mol dm^{-3}
Current efficiency	53—83%
Power consumption	12 kW h per kg
Cross-sectional flow rate	1—3 dm^3 dm^{-2} min^{-1}
Thickness of cathode	0.3 cm
Dimensions of the cathode	78 × 5 cm

The application of oxygen cathodes that were developed for use in fuel cells to the generation of pure oxygen from air has been studied by Tseung and Jasem.[149] Portable generators of oxygen are of interest for a wide range of applications where the use of bottled gas is either not possible or uneconomic. The electrolysis of water has been used for portable oxygen generators, but the concurrent generation of hydrogen represents a potential hazard; from this point of view, the replacement of the cathodic evolution of hydrogen by the reduction of oxygen is attractive. Tseung and Jasem[149] took this idea a step further, recognizing the difficulties in the use of efficient air cathodes regarding the generation of hydrogen peroxide, and therefore used this intermediate partially as a redox carrier of oxygen. Hydrogen peroxide was thus generated at the cathodic compartment and was then decomposed catalytically, using Teflon-bonded $CoFe_2O_4$ and $NiCo_2O_4$ in the compartment where oxygen was evolved. The anode was a Teflon-bonded $NiCo_2O_4$ electrode, and the best power consumption that was achieved was of 2.69 kW h per dm^3 of oxygen, which was significantly lower than conventional electrochemical oxygen extractors (4.38 kW h per dm^3 of oxygen).

The possibility and the problems in the use of air electrodes in the chlor-alkali industry have been discussed by Yeager and Bindra.[2] For a membrane cell, operating with anolyte and catholyte feeds of 15% NaCl and 30% NaOH, respectively, at 85 °C and at 300 mA cm^{-2}, a reduction in cell voltage of 1 V is possible which represents a direct saving in electrical energy of ~28%. This figure, however, does not reflect the true difference in energy requirements between cells that operate with hydrogen-evolution cathodes and with air electrodes, since the thermal value of the hydrogen that is generated has to be taken into account. This, it was argued, reduces the total energy saving to 16%, and therefore the choice of air cathodes must be set against other considerations, such as the cost of replacement of existing plant, *etc*. An alternative strategy is to keep existing chlor-alkali technology and to use the hydrogen in a fuel cell to generate electricity. The advantage of this approach is in making use of developing technologies,[9] while maintaining the main features of a well-tested production procedure such as chlor-alkali cells. Yeager and Bindra[2] also reviewed the different mechanisms for reduction of O_2 and the possibilities of using transition-metal chelates as catalysts for the reduction of O_2 or for the decomposition of H_2O_2. The use of several air cathodes for the chlor-alkali

---

[149] A. C. C. Tseung and S. M. Jasem, *J. Appl. Electrochem.*, 1981, **11**, 209.

industry,[150,151] as well as air electrodes for batteries[152] and other applications,[153] has been described in the patent literature.

10 Final Remarks

The mechanistic aspects of the reduction of oxygen on different substrates are not yet clearly understood. Part of the difficulty is in the different possible pathways and intermediates involved; electrochemical techniques alone are perhaps not sufficiently discriminating, and the use of spectroelectrochemical methods for the elucidation of mechanisms for the reduction of O_2 is necessary.

There is a correlation between modes of co-ordination of the oxygen molecule and the reduction pathway, as has been pointed out by Yeager,[154] but a clear understanding of the relationship between the electronic properties of a transition-metal centre and the kinetics of reduction is not yet available.

Perhaps one of the most promising new areas of research in the reduction of oxygen will be in the modelling of biochemical reactions which utilize oxygen, not only for the intrinsic interest in the understanding of the basic processes of life, but also for the possible synthetic applications that can be envisaged.

---

[150] H. B. Johnson and R. D. Chamberlin, US P. 4 244 793 (1981).
[151] R. L. La Barre, US P. 4 221 644 (1980).
[152] A. Winsel, Ger. Offen. 2 924 181 (1980).
[153] I. Morcos, Can. P. 1 080 204 (1980).
[154] E. Yeager, in 'Electrocatalysis on Non-Metallic Surfaces', NBS Special Publication No. 455, 1976, p. 203.

5
Organic Electrochemistry – Synthetic Aspects

BY J. GRIMSHAW AND D. PLETCHER

PART I: General Topics and Reductions by *J. Grimshaw*

1 General Topics

In the extended period that is covered in this Chapter, books have appeared on organic electrosynthesis,[1] on the electrochemistry of biological molecules,[2] and on the electrochemistry of alcohols, phenols, and ethers.[3] English editions of textbooks on the rotating disc electrode[4] and on the kinetics of electrode processes[5] have been published and the engineering aspects of the design of electrochemical reactions have been covered.[6] The most substantial of a great number of review articles deal with recent developments in organic electrochemistry,[7] anodic substitution,[8] electrochemistry of biological systems,[9] and structure and mechanism in organic electrochemistry.[10]

Reference electrodes for use in propylene carbonate[11] and in N-methylpyrrolidone[12] have been described. A platinum–hydrogen anode for use in non-aqueous solvent systems has been proposed.[13] Protons are generated in the anode reaction and the cell is operated without a diaphragm, so that any bases that are formed at the cathode are neutralized. So far, only modest current densities have been achieved.

A number of workers have reported on the electrochemical properties of chemically modified surfaces of solid electrodes. Glassy carbon can be subjected to surface oxidation in order to generate covalently bound carbonyl functions, which can be used to form bonds with other organic molecules that possess a suitable

---

[1] A. P. Tomilov, M. Ya. Fioshin, and V. A. Smirnov, 'Electrosynthesis of Organic Compounds', Khimiya, Leningrad, 1976.
[2] G. Dryhurst, 'Electrochemistry of Biological Molecules: Purines, Pyrimidines, Pteridines, Flavins, Pyrroles, Porphyrins and Pyridines', Academic Press, New York, 1977.
[3] 'Encyclopaedia of Electrochemistry of the Elements, Volume II', ed. H. Lund, Dekker, New York, 1978.
[4] Yu. V. Pleskov and V. Yu. Filinovskii, 'The Rotating Disc Electrode', Consultants Bureau, New York, 1976.
[5] T. Erdey-Gruz, 'Kinetics of Electrode Processes', Hungarian Academy of Sciences, Budapest, 1975.
[6] D. J. Pickett, 'Electrochemical Reactor Design', Elsevier, Amsterdam, 1977.
[7] M. M. Baizer, *Top. Pure Appl. Electrochem.*, 1975, 185; J. Q. Chambers, *Int. Rev. Sci., Org. Chem. Series 2*, 1975, **10**, 317.
[8] L. Eberson and K. Nyberg, *Tetrahedron*, 1976, **32**, 2185.
[9] 'ACS Symposium Series Volume 38: Electrochemical Studies of Biological Systems', ed. D. T. Sawyer, American Chemical Society, Washington, 1977.
[10] L. Eberson and K. Nyberg, *Adv. Phys. Org. Chem.*, 1976, **12**, 1.
[11] N. H. Cuong, A. Maiornikoff, and H. D. Hurwitz, *J. Electroanal. Chem. Interfacial Electrochem.*, 1976, **72**, 107.
[12] M. Breant and M. Lavergne, *Bull. Soc. Chim. Fr.*, 1976, 28.
[13] C. P. Andrieux, J. M. Dumas-Bouchiat, and J. M. Saveant, *J. Electroanal. Chem. Interfacial Electrochem.*, 1977, **83**, 355.

functional group.[14] A surface that was modified with (−)-camphoric anhydride showed no asymmetric induction for the oxidation of *p*-nitrobenzyl methyl sulphide to the sulphoxide.[15] Molecules have also been covalently bonded to the surfaces of platinum[16] and ruthenium oxide.[17] In another approach to the problem of modification of a surface, polymer films have been coated onto platinum from organic solutions and found to be sufficiently stable for electrochemistry to occur at the modified surface. Terephthalate ester polymers, poly(4-nitrostyrene),[18] poly(vinylferrocene,[19] and poly(acryloyl chloride) that is coupled to dopamine[20] have been used for surface modification.

Interest still continues in improving the technique for following electrochemical reactions at transparent electrodes by u.v. spectroscopy,[21] and methods have been developed for producing optically transparent carbon and mercury–carbon surfaces.[22] Raman spectroscopy has also been applied to the characterization of intermediates at electrode surfaces.[23]

A new development in the study of rates of electrochemical reactions has been the application of fast Fourier-transform procedures.[24] The convolution-integral technique has been applied to chronopotentiometry.[25] Convolution linear-sweep voltammetry has been shown to give results at sweep rates in excess of 10^3 V s^{-1}, so that the kinetics of the decomposition of the radical-anion of benzaldehyde in ethanol could be studied.[26] The kinetics of decomposition of acetophenone radical-anion have also been studied by this method.[27] The mathematical processes of semi-integration and convolution-integration have been shown to be identical.[28] Studies of this kind only yield information about the slow reaction step, and much theoretical discussion and experiment has been expended in showing whether transfer of electrons following the slow reaction step occurs from the electrode [the e.c.e. mechanism; (1)] or by transfer from a radical-anion [the d.i.s.p. mechanism;

[14] C. M. Elliott and R. W. Murray, *Anal. Chem.*, 1976, **48**, 1247; J. C. Lennox and R. W. Murray, *J. Electroanal. Chem. Interfacial Electrochem.*, 1977, **78**, 395; M. Fujihira, A. Tamura, and T. Osa, *Chem. Lett.*, 1977, 361; J. F. Evans, T. Kuwana, M. T. Henne, and G. P. Royer, *J. Electroanal. Chem. Interfacial Electrochem.*, 1977, **80**, 409; M. Sharp, *Electrochim. Acta*, 1978, **23**, 287.
[15] B. E. Firth and L. L. Miller, *J. Am. Chem. Soc.*, 1976, **98**, 8272.
[16] J. R. Lenhard and R. W. Murray, *J. Electroanal. Chem. Interfacial Electrochem.*, 1977, **78**, 195.
[17] P. R. Moses and R. W. Murray, *J. Electroanal. Chem. Interfacial Electrochem.*, 1977, **77**, 393.
[18] L. L. Miller and M. R. Van de Mark, *J. Am. Chem. Soc.*, 1978, **100**, 639, 3223; *J. Electroanal. Chem. Interfacial Electrochem.*, 1978, **88**, 437; J. B. Kerr, L. L. Miller, and M. R. Van de Mark, *J. Am. Chem. Soc.*, 1980, **102**, 3383.
[19] A. Mertz and A. J. Bard, *J. Am. Chem. Soc.*, 1978, **100**, 3222.
[20] C. Degrand and L. L. Miller, *J. Am. Chem. Soc.*, 1980, **102**, 5728.
[21] E. Steckhan and D. A. Yates, *Ber. Bunsenges. Phys. Chem.*, 1977, **81**, 369.
[22] T. P. de Angelis, R. W. Hurst, A. W. Yacynych, H. B. Mark, and W. R. Heineman, *Anal. Chem.*, 1977, **49**, 1395.
[23] M. Fleischmann, P. J. Hendra, A. J. McQuillan, R. C. Paul, and E. S. Reid, *J. Raman Spectrosc.*, 1976, **4**, 269; M. R. Suchanski and R. P. Van Duyne, *J. Am. Chem. Soc.*, 1976, **98**, 250; R. P. Van Duyne, *J. Phys. (Paris) Colloq.*, 1977, 239.
[24] R. J. Schwall, A. M. Bond, and D. E. Smith, *Anal. Chem.*, 1977, **49**, 1805; R. J. O'Halloran and D. E. Smith, *ibid.*, 1978, **50**, 1391; P. R. Griffiths, *Transform. Tech. Chem.*, 1978, 355.
[25] L. Nadjo and J. M. Saveant, *J. Electroanal. Chem. Interfacial Electrochem.*, 1977, **75**, 181.
[26] J. M. Saveant and D. Tessier, *J. Electroanal. Chem. Interfacial Electrochem.*, 1977, **77**, 225; *J. Phys. Chem.*, 1978, **82**, 1723.
[27] C. Amatore, L. Nadjo, and J. M. Saveant, *J. Electroanal. Chem. Interfacial Electrochem.*, 1978, **90**, 321.
[28] K. B. Oldham, *J. Electroanal. Chem. Interfacial Electrochem.*, 1976, **72**, 371; K. B. Oldham and M. Goto, *Anal. Chem.*, 1976, **48**, 1671.
[29] C. Amatore and J. M. Saveant, *J. Electroanal. Chem. Interfacial Electrochem.*, 1977, **85**, 27; 1978, **86**, 227; 1979, **102**, 21.

(2)].[29] Reductions of naphthalene and of anthracene in DMF that contains phenol as the proton donor are considered to follow the d.i.s.p. mechanism.[30]

$$A + e^- \rightleftharpoons A^{\bar{\cdot}} \xrightarrow{+H^+} AH\cdot \xrightarrow{+e^-} AH^- \quad (1)$$

$$A + e^- \rightleftharpoons A^{\bar{\cdot}} \xrightarrow{+H^+} AH\cdot \xrightarrow{+A^{\bar{\cdot}}} AH^- + A \quad (2)$$

The application of cyclic differential pulse voltammetry to electrochemical problems is also being developed,[31] and a theoretical study of thin-layer sweep voltammetry continues.[32]

The theory of homogeneous redox catalysis of fast irreversible electrochemical reactions that are promoted by a mediator which undergoes a reversible one-electron electrode reaction has been developed in detail.[33,34] This technique has been used to estimate the rate of transfer of electrons from catalyst radical-anion to substrate (with chlorobenzene as the substrate)[35] and the rate of decomposition for 2-chloroquinoline radical-anion (in DMF, $k = 6 \times 10^5 \, s^{-1}$) as well as for other aryl halides[36] whose radical-anions are very unstable. The standard redox potentials for these radical-anions can also be determined by the method.[37]

2 Reductions of Aliphatic and Aromatic Compounds

General.—Radical-anions have been shown to undergo alkylation reactions in the presence of alkyl halides or dimethyl sulphate. Thus the reduction of anthraquinone[38] (Scheme 1) in DMF that contains methyl bromide leads to the quinol dimethyl ether; benzil[39] shows a related reaction. The radical-anion of phenazine[40] (generated in acetonitrile) reacts with ethyl bromide or dimethyl sulphate in a similar way to give the 9,10-dialkyl-9,10-dihydrophenazine. Cyclic compounds[41] can be formed in good yield by the reduction of appropriate substrates in the presence of α,ω-dibromo-alkanes (Scheme 1). In the reaction between radical-anions and t-butyl chloride, the rate-determining step is thought to be transfer of an electron between the two species, followed by radical substitution by the resulting t-butyl radical. Ultimately, t-butyl derivatives of the dihydro-substrate are formed. Reactions between t-butyl chloride and quinoline,[42]

---

[30] C. Amatore and J. M. Saveant, *J. Electroanal. Chem. Interfacial Electrochem.*, 1980, **107**, 353.

[31] K. F. Drake, R. P. Van Duyne, and A. M. Bond, *J. Electroanal. Chem. Interfacial Electrochem.*, 1978, **89**, 231.

[32] E. Laviron, *J. Electroanal. Chem. Interfacial Electrochem.*, 1978, **87**, 31; C. P. Andrieux, J. M. Dumas-Bouchiat, and J. M. Saveant, *ibid.*, p. 39.

[33] J. M. Saveant and S. K. Binh, *J. Electroanal. Chem. Interfacial Electrochem.*, 1978, **91**, 35.

[34] C. P. Andrieux, J. M. Dumas-Bouchiat, and J. M. Saveant, *J. Electroanal. Chem. Interfacial Electrochem.*, 1980, **113**, 1; C. P. Andrieux, C. Blockman, J. M. Dumas-Bouchiat, F. M'Halla, and J. M. Saveant, *ibid.*, p. 19.

[35] C. P. Andrieux, J. M. Dumas-Bouchiat, and J. M. Saveant, *J. Electroanal. Chem. Interfacial Electrochem.*, 1978, **87**, 55.

[36] C. P. Andrieux, C. Blockman, J. M. Dumas-Bouchiat, F. M'Halla, and J. M. Saveant, *J. Am. Chem. Soc.*, 1980, **102**, 3806.

[37] C. P. Andrieux, C. Blockman, J. M. Dumas-Bouchiat, and J. M. Saveant, *J. Am. Chem. Soc.*, 1979, **101**, 3431.

[38] R. A. Misra and A. K. Yadar, *Electrochim. Acta*, 1980, **25**, 1221.

[39] T. Troll, H. Leffler, and W. Elbe, *Electrochim. Acta*, 1979, **24**, 969.

[40] D. K. Root, R. O. Pendarvis, and W. H. Smith, *J. Org. Chem.*, 1978, **43**, 778.

[41] C. Degrand, P-L. Compagnon, G. Belot, and D. Jacquin, *J. Org. Chem.*, 1980, **45**, 1189.

[42] C. Degrand and H. Lund, *Acta Chem. Scand., Ser. B*, 1977, **31**, 593.

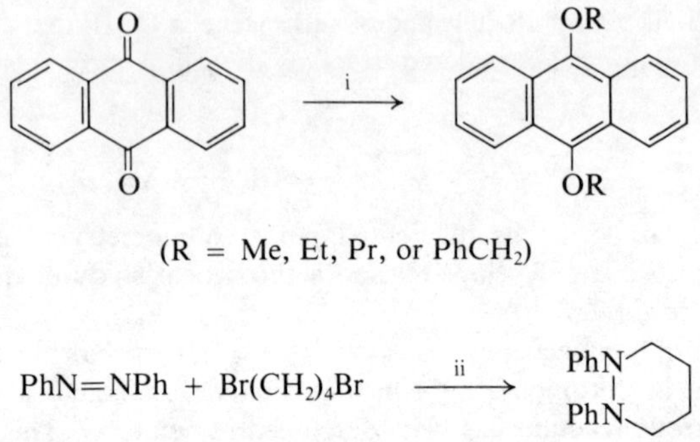

(R = Me, Et, Pr, or PhCH$_2$)

PhN=NPh + Br(CH$_2$)$_4$Br $\xrightarrow{ii}$ PhN–(CH$_2$)$_4$–NPh

Reagents: i, 2e$^-$, RBr, LiCl, DMF; ii, 2e$^-$

Scheme 1

isoquinoline,[42] pyrene,[43] and benzophenone[44] have been studied; in some reactions, 1-bromoadamantane[45] replaces the t-butyl chloride (Scheme 2). 1-Ethyl-4-methoxycarbonylpyridinium iodide[46] undergoes similar electron-transfer reactions towards alkyl halides, but only from the stage of two-electron addition.

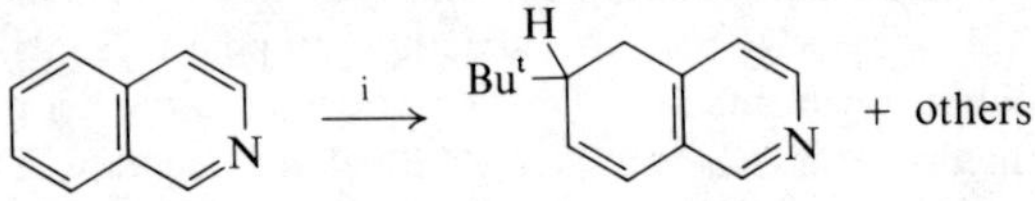

isoquinoline $\xrightarrow{i}$ Bu^t-dihydroisoquinoline + others

Reagents: i, 2e$^-$, Bu^{t}Cl, DMF

Scheme 2

Radical-anions are also acylated by acetic anhydride in DMF. Ethyl cinnamate[47] gives the product shown in Scheme 3, and other activated olefins, 2- and 4-styrylpyridine, and styrene take part in similar reactions.[48] Anthracene[49] gives the enol ester product shown. Cathodic acylation of both aliphatic and aromatic nitro-compounds gives the corresponding N,O-diacyl-hydroxylamine.[50]

Electrogenerated bases have been used to effect the formation of a carbanion from fluorene. The carbanion is captured by carboxylation with carbon dioxide, so that the process is a synthesis of fluorene-9-carboxylic acid.[51] Electrogenerated bases, in acetonitrile, form the cyanomethyl carbanion, which will condense with methyl benzoate and NN-dimethylthiobenzamide (Scheme 4), but, since the base is itself

---

[43] P. E. Hansen, A. Berg, and H. Lund, *Acta Chem. Scand., Ser. B*, 1976, **30**, 267.
[44] L. H. Kristensen and H. Lund, *Acta Chem. Scand., Ser. B*, 1979, **33**, 735.
[45] U. Hess, D. Huhn, and H. Lund, *Acta Chem. Scand., Ser. B*, 1980, **34**, 413.
[46] H. Lund and L. H. Kristensen, *Acta Chem. Scand., Ser. B*, 1979, **33**, 495.
[47] H. Lund and C. Degrand, *Tetrahedron Lett.*, 1977, 3593.
[48] H. Lund and C. Degrand, *Acta Chem. Scand., Ser. B*, 1979, **33**, 57.
[49] H. Lund, *Acta Chem. Scand., Ser. B*, 1977, **31**, 424.
[50] L. Christensen and P. E. Iversen, *Acta Chem. Scand., Ser. B*, 1979, **33**, 352.
[51] R. C. Hallcher and M. M. Baizer, *Liebigs Ann. Chem.*, 1977, 737; R. C. Hallcher, M. M. Baizer, and D. A. White, US P. 4 072 583, 1978.

generated by reduction of the substrate, the product yield cannot exceed 50%. These two reactions do not proceed with azobenzene as the pro-base whereas the corresponding condensation with NN-dimethylbenzamide will take place if azobenzene is used as the pro-base.[52]

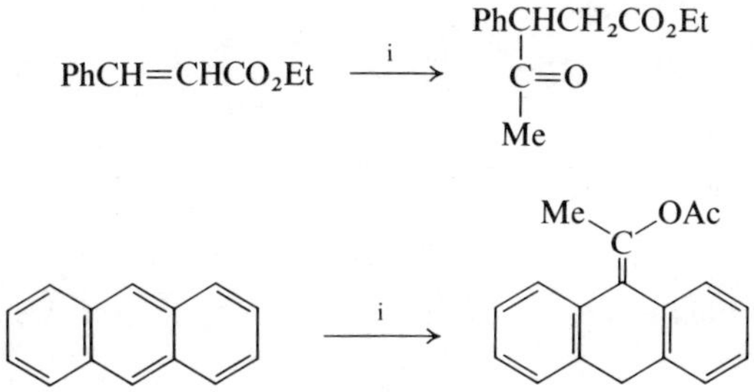

$$PhCH=CHCO_2Et \xrightarrow{i} \underset{Me}{\underset{|}{\overset{PhCHCH_2CO_2Et}{\overset{|}{C}}=O}}$$

[anthracene] $\xrightarrow{i}$ [9-(1-acetoxyethylidene)-9,10-dihydroanthracene, Me\C/OAc]

Reagents: i, $2e^-$, Ac_2O, DMF

Scheme 3

$$PhCO_2Me + MeCN \xrightarrow{e^-} PhCOCH_2CN + PhCO_2H + MeOH$$

$$\underset{S}{\overset{\|}{PhCNMe_2}} + MeCN \xrightarrow{e^-} \underset{NMe_2}{\underset{|}{PhC}=CHCN} + PhCH=CHPh + PhCHO$$

Scheme 4

Several workers have examined the catalytic hydrogenation of substrates with hydrogen that is generated electrochemically from an acid solution at a cathode made of the hydrogenation catalyst. Smooth platinum[53] was used to convert phenol into cyclohexanol, but after a systematic investigation it has been suggested that rhodium black that has been deposited on carbon[54] is a better catalyst. Platinized platinum cathodes[55] were used to reduce allyl and crotyl alcohols to propene and butene respectively.

Protecting groups which can be removed in an electrochemical step form a useful addition to synthetic chemistry, and this subject has been reviewed.[56] N-Tosyl and N-benzyloxycarbonyl groups[57] can be removed electrochemically from protected peptides without racemization. Cathodic cleavage of the tritylone group from protected alcohols in the presence of other alcohol-protecting groups has been demonstrated (Scheme 5).[58a] 4-Picolyl has been recommended[58b] as a protecting

---

[52] L. Kirstenbrügger, P. Mischke, J. Voss, and G. Wiegand, *Liebigs Ann. Chem.*, 1980, 461.
[53] R. A. Misra and B. L. Sharma, *Electrochim. Acta*, 1979, **24**, 727.
[54] L. L. Miller and L. Christensen, *J. Org. Chem.*, 1978, **43**, 2059.
[55] G. Horanyi, G. Inzelt, and K. Torkas, *J. Electroanal. Chem. Interfacial Electrochem.*, 1979, **101**, 101.
[56] V. G. Mairanovskii, *Angew. Chem.*, 1976, **88**, 283.
[57] V. G. Mairanovskii and N. F. Loginova, *Bioorg. Khim.*, 1976, **2**, 1497.
[58] (a) C. van der Stouwe and H. J. Schafer, *Tetrahedron Lett.*, 1979, 2643; (b) A. Gosden, R. Macrae, and G. T. Young, *J. Chem. Res. (S)*, 1977, 22.

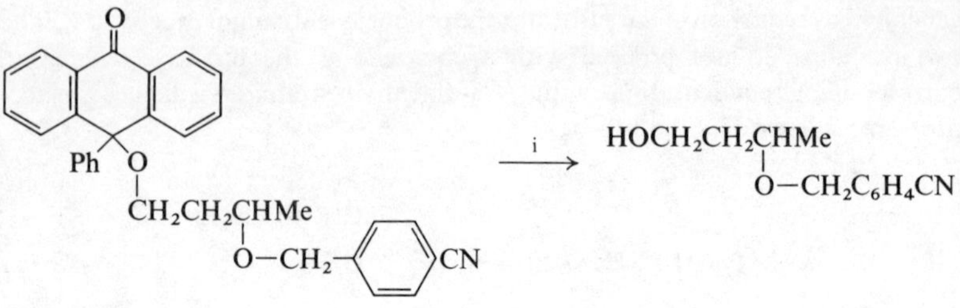

Reagents: i, MeOH, −1.3 V vs Ag/AgCl

Scheme 5

group for OH and SH functions, being removable by electroreduction in aqueous acidic solution.

Hydrocarbons.—Tetrakis(trifluoromethyl)ethene forms a stable anion-radical that has been characterized by polarography and by e.s.r. spectroscopy.[59] Conjugated olefins can be reduced to dihydro-compounds at a mercury cathode. Vitamin D_2 undergoes 1,6-addition of hydrogen to give a mixture of two stereoisomers.[60] β-Carotene gives a dihydro-derivative that has been claimed by one group[61] to be 7,7'- and by another group[62] to be 15,15'-dihydro-β-carotene. The hydrocarbon axerophthene, which is related to Vitamin A, is also reduced to a dihydro-compound.[61] The reaction of cyclopentadiene, in DMF that contains lithium bromide, between an iron anode and a nickel cathode affords ferrocene in good yield.[63]

Two groups[64,65] have demonstrated the existence of the radical-anions of cis- and trans-stilbene by spectroscopic means and shown that, in solution, the cis-form is rapidly converted into the trans-form. The kinetics of this conversion in HMPT were determined by a flash photolysis technique. Isomerism occurs by disproportionation of the radical-anion to give the stilbene dianion. This exists in a skew form, rather than a planar form, and collapses with the donation of one electron to an acceptor, to give mainly the trans-radical-anion with some cis-radical-anion.[65] The radical-anion of cis-azobenzene also rapidly isomerizes to the trans-form at room temperature.[66]

The radical-anions that are derived from styrenes will attack the solvent in a nucleophilic manner, and the reaction can be made of preparative value.[67] α-Methylstyrene adds twice to DMF (Scheme 6) and with acetonitrile it adds once to form 4-phenylpentan-2-one.

---

[59] E. A. Polenov, V. V. Minin, S. R. Sterlin, and L. M. Yagupol'skii, *Izv. Akad. Nauk SSSR, Ser. Khim.*, 1978, 482.
[60] N. A. Bogoslovskii, I. A. Titova, and V. G. Mairanovskii, *Khim. Farm. Zh.*, 1977, **11**, 30.
[61] V. G. Mairanovskii, L. A. Vakulova, N. T. Ioffe, A. A. Engovatov, E. I. Korunova, T. I. Maksakova, and G. I. Samokhvalov, *Khim. Farm. Zh.*, 1977, **11**, 111.
[62] S-M. Park, *J. Electrochem. Soc.*, 1978, **125**, 216.
[63] H. Lehmkuhl and W. Eisenbach, Ger. Offen. 2 720 165, 1978 (*Chem. Abstr.*, 1978, **89**, 50 600).
[64] B. S. Jensen, R. Lines, P. Pagsberg, and V. D. Parker, *Acta Chem. Scand., Ser. B*, 1977, **31**, 707.
[65] H. C. Wang, G. Levin, and M. Szwarc, *J. Am. Chem. Soc.*, 1977, **99**, 2642.
[66] E. Laviron and Y. Mugnier, *J. Electroanal. Chem. Interfacial Electrochem.*, 1978, **93**, 69.
[67] R. Engels and H. J. Schafer, *Angew. Chem.*, 1978, **90**, 483.

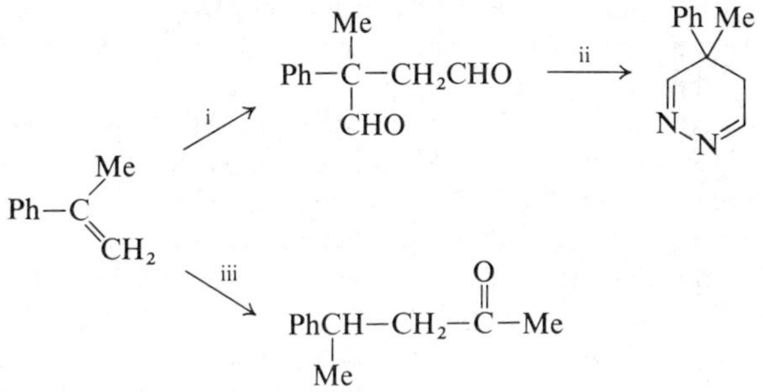

Reagents: i, e⁻, DMF; ii, NH$_2$NH$_2$; iii, e⁻, MeCN

Scheme 6

Chlorobenzene has been suggested as a solvent for the determination of reversible reduction potentials of aromatic hydrocarbons.[68] The solvent can be obtained rigorously anhydrous, so that reversible addition of two electrons can be observed. A micro-working electrode is necessary, so that currents are small, and tetrahexylammonium perchlorate serves as the soluble salt. Another investigation of aromatic hydrocarbon anion–dianion redox systems uses dimethoxyethane as the solvent and achieves reversible behaviour at very fast sweep rates.[69] An electrochemical cell[70] for the preparation of stable solutions of radical-anions in DMF or acetonitrile has been designed so that the last traces of impurities can be removed by circulating the electrolyte over alumina prior to electrolysis.

The radical-cations and radical-anions of alternant aromatic hydrocarbons show equal solvation energies, and the mean of the two redox potentials is -0.31 V vs the S.C.E.[71] This common mid-point has been proposed as the zero point against which to measure the change in energy of an electrode reaction. For aromatic hydrocarbons, the separation of the first two reduction potentials is about 0.5 V, regardless of the substrate. In contrast, this separation is much less for Hückel non-aromatic cyclophanes, and it depends on the solvent.[72,73] The reduction of some cyclophanes (e.g. [2$_4$]-paracyclophanetetraene) in DMF proceeds by a two-electron reversible step to form the dianion, although in dimethoxyethane the radical-anion can be detected by e.s.r. spectroscopy.[73]

Carbo-cations, cyclopropenyl ion, and tropylium ion and its benzo-derivatives show two reduction waves in aqueous sulphuric acid.[74] The free radical that is

---

[68] R. Lines and V. D. Parker, *Acta Chem. Scand., Ser. B*, 1977, **31**, 369.
[69] T. Saji and S. Aoyagui, *J. Electroanal. Chem. Interfacial Electrochem.*, 1979, **78**, 163.
[70] R. Lines, B. S. Jensen, and V. D. Parker, *Acta Chem. Scand., Ser. B*, 1978, **32**, 510.
[71] V. D. Parker, *J. Am. Chem. Soc.*, 1976, **98**, 98.
[72] H. Kojima, A. J. Bard, H. N. C. Wong, and F. Sondheimer, *J. Am. Chem. Soc.*, 1976, **98**, 5560.
[73] K. Ankner, B. Lamm, B. Thulin, and O. Wennerstrom, *Acta Chem. Scand., Ser. B*, 1978, **32**, 155; F. Gerson, W. Huber, and O. Wennerstrom, *Helv. Chim. Acta*, 1978, **61**, 2763; K. Ankner, B. Lamm, B. Thulin, and O. Wennerstrom, *J. Chem. Soc., Perkin Trans. 2*, 1980, 1301.
[74] M. R. Feldman and W. C. Flythe, *J. Org. Chem.*, 1978, **43**, 2596.

obtained by one-electron reduction of the heptaphenyltropylium ion in acetonitrile is stable on the time-scale of cyclic voltammetry, and it slowly dimerizes, probably by reaction on the phenyl rings.[75]

The preparative-scale reduction of substituted naphthalenes to their dihydro-compounds[76] in acetonitrile that contains water has been described and a number of steroidal naphthalene compounds have been reduced in liquid ammonia. Equilenin 3-methyl ether (1)[77] is reduced to the dihydro-ring-A compound, while the corresponding phenol gives a mixture of the two stereoisomeric alcohols that are obtained by addition of two moles of hydrogen to ring A. The vinylbenzene-type steroid (2) affords two isomeric dihydro-derivatives by saturation of the vinyl group.[78] Reduction of pyrene in DMF gives a complex mixture of di-, tetra-, and hexa-hydro-derivatives.[79] The reduction of pyrene in the presence of t-butyl chloride gives 1-t-butylpyrene (52%) together with minor amounts of other t-butylated products, probably by homogeneous reduction of the alkyl chloride by the radical-anion of pyrene followed by aromatic substitution of the radical by the t-butyl radical that is thus formed.[80] The reduction of crystalline graphite at very negative potentials, while simultaneously exposing the surface to various electrophiles such as organic bromides, sulphones, or carbon dioxide, results in modification of the surface of the crystals.[81]

Halogen-containing Compounds.—Several groups of workers have examined the reduction of alkyl halides. The value of $E_{\frac{1}{2}}$ for this reduction in DMSO varies with the size of the tetra-alkylammonium ion and with the anion in the supporting electrolyte,[82] moving to more negative potentials for larger anions. Evidence has

---

[75] J. M. Leal, T. Tcherani, and A. J. Bard, *J. Electroanal. Chem. Interfacial Electrochem.*, 1978, **91**, 275.
[76] A. Misono, T. Naga, and T. Yamagishi, Jpn. Kokai 77 45 707 (*Chem. Abstr.*, 1978, **88**, 105 009).
[77] K. Junghans, G. A. Hoyer, and G. Cleve, *Chem. Ber.*, 1979, **112**, 2631.
[78] K. Junghans, *Chem. Ber.*, 1976, **109**, 395.
[79] P. E. Hansen, O. Blaabjerg, and A. Berg, *Acta Chem. Scand.*, Ser. B, 1978, **32**, 720.
[80] P. E. Hansen, A. Berg, and H. Lund, *Acta Chem. Scand.*, Ser. B, 1976, **30**, 267.
[81] G. Bernard and J. Simonet, *J. Electroanal. Chem. Interfacial Electrochem.*, 1980, **112**, 117.
[82] A. J. Fry and R. L. Krieger, *J. Org. Chem.*, 1976, **41**, 54.

been gathered for the formation of alkyl radical intermediates. These can be trapped during the reduction of 1-bromodecane in DMF by the addition of N-methylformamide (ca 0.05%), when N-decyl-N-methylformamide is produced.[83] The cyclized compound (3) is one of the products[84] from reduction of 6-iodo-1-phenylhexyne, arising by intramolecular trapping of the primary alkyl radical. The ratio of products formed is dependent on the reduction potential, and the corresponding bromo-compound gives mainly RH and R_2Hg products.[84]

$$PhC\equiv C(CH_2)_4-I \xrightarrow[DMF]{Hg\ cathode,} R-H + R_2Hg +$$
[R—I]

(3)

Dialkylmercury compounds are commonly found as products from the reduction of alkyl halides; they may arise by the reaction of the alkyl radical with the surface of the mercury cathode.[82—85] It has also been shown, for perfluorohexyl iodide, that the cathode becomes covered with the alkyl-mercury iodide in a chemical reaction.[86] Adsorbed alkyl-mercury iodide is reduced to the dialkyl-mercury at more negative potentials. Radical intermediates also disproportionate, to give alkene and alkane. The value of n for reduction of 1-iododecane depends[87] upon the duration of electrolysis and the water content of the DMF solvent. For controlled-potential electrolysis, n approaches 1 in moist DMF and 1.28 in dry DMF, at all potentials. In short-sampling pulse polarography, two waves can be seen, each with $n=1$, so that a value of 2 for n is obtained by this technique at negative potentials. The reactions of 1-bromo- and of 1-iodo-norbornane are much cleaner, yielding norbornane and no mercury compounds;[88] again, the half-wave potential changes with the identity of the supporting electrolyte.

The stereochemistry of the reduction of alkyl halides in which the halogen is attached to an asymmetric carbon atom to alkanes has also been re-examined. All groups agree that the optical yield depends on the solvent and on the supporting electrolyte. The reduction of a tertiary alkyl chloride, i.e. 6-chloro-2,6-dimethyloctane, proceeded with inversion of configuration, but the optical yield could not be assessed because the required specific rotations are not known with certainty.[89] The reduction of 1-deuterio-1-bromo-1-phenylethane in DMF favours retention with Li^+ cation and inversion with Et_4N^+, while Li^+ in t-butyl alcohol favours inversion.[90] The reduction of 1-bromo-1-methoxycarbonyl-2,2-diphenyl-cyclopropane also gives either retention or inversion, depending on the conditions,

---

[83] G. M. McNamee, B. C. Willett, D. M. LaPerriere, and D. G. Peters, J. Am. Chem. Soc., 1977, **99**, 1831.
[84] B. C. Willett, W. M. Moore, A. Salajegheh, and D. G. Peters, J. Am. Chem. Soc., 1979, **101**, 1162.
[85] L. A. Avaca, E. R. Gonzalez, and N. R. Stradiotto, Electrochim. Acta, 1977, **22**, 225.
[86] P. Calas, P. Moreau, and A. Commeyras, J. Electroanal. Chem. Interfacial Electrochem., 1977, **78**, 271; 1978, **89**, 373.
[87] D. M. La Pierre, B. C. Willett, W. F. Carroll, E. C. Torp, and D. G. Peters, J. Am. Chem. Soc., 1978, **100**, 6293.
[88] W. F. Carroll and D. G. Peters, J. Org. Chem., 1978, **43**, 4633.
[89] T. Nonaka, T. Ota, and K. Odo, Bull. Chem. Soc. Jpn., 1977, **50**, 419.
[90] R. B. Yamasaki, M. Tarle, and J. Casanova, J. Org. Chem., 1979, **44**, 4519.

while the optical yield depends on the reduction potential.[91] The reduction of 1,1-dibromo-2,2-diphenylcyclohexane in buffered aqueous ethanol in the presence of strychnine, emetine, or yohimbine, all of which were strongly adsorbed at mercury, gives rise to optically active 1-bromo-2,2-diphenylcyclopropane.[92]

Allyl halides, which show multiple waves on reduction at mercury or platinum electrodes, give a single reduction wave on vitreous carbon, with an apparent value of 1 for n.[93] This arises because there is rapid nucleophilic substitution of the allyl carbanion that is formed on a second molecule of allyl halide. Contrary to previous assertions, allyl halides are reduced in a two-electron step, and reduction of the allyl radical to the anion cannot be seen as a separate wave.

Intermediates from the reduction of benzyl chlorides can be trapped with acyl chlorides to yield ketones[94] and with imines to yield amines[95] (Scheme 7) in reactions which have value in synthesis. The reduction of α-halogeno-ketones gives the ketone enolate ion, and this can be trapped by its Michael addition to acrylonitrile.[96] A further useful synthesis is the conversion of cyclic β-diketones into the corresponding β-chloro-enones by reaction with a phosphorus halide and then reductive removal of chlorine to yield the enone.[97]

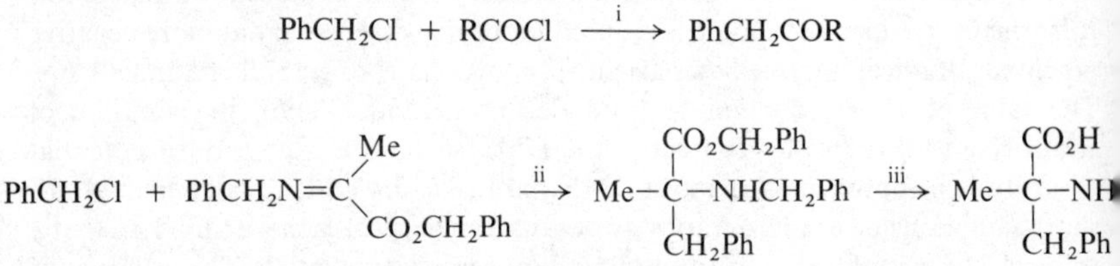

$$PhCH_2Cl + RCOCl \xrightarrow{i} PhCH_2COR$$

$$PhCH_2Cl + PhCH_2N=C\begin{matrix}Me\\CO_2CH_2Ph\end{matrix} \xrightarrow{ii} Me-\underset{CH_2Ph}{\underset{|}{\overset{CO_2CH_2Ph}{\overset{|}{C}}}}-NHCH_2Ph \xrightarrow{iii} Me-\underset{CH_2Ph}{\underset{|}{\overset{CO_2H}{\overset{|}{C}}}}-NH_2$$

Reagents: i, e⁻; ii, e⁻, DMF; iii, H₂, Pd

Scheme 7

Electrogenerated square-planar nickel(I) complexes interact with an alkyl halide to re-form nickel(II) and an alkyl radical. The radical is now generated remote from the electrode surface, where it can interact with other components of the solution.[98] Typical products are the dimer and the alkane and alkene that are formed by transfer of hydrogen atom(s). In the presence of acrylonitrile or ethyl acrylate, the radical is trapped by addition to the activated olefin.[99] Acetonylacetone complexes of iron similarly catalyse the conversion of 1-bromo-octane into hexadecane in 59% yield.[100] 2,2'-Bipyridyl complexes of cobalt have been used in this way to catalyse the conversion of allyl chloride into hexa-1,5-diene.[101]

[91] S. Jaouannet, R. Hazard, and A. Tallec, *J. Electroanal. Chem. Interfacial Electrochem.*, 1980, **111**, 397.
[92] R. Hazard, S. Jaouannet, and A. Tallec, *Tetrahedron Lett.*, 1979, 1105.
[93] A. J. Bard and A. Mertz, *J. Am. Chem. Soc.*, 1979, **101**, 2959.
[94] T. Shono, I. Nishiguchi, and O. Ohmizu, *Chem. Lett.*, 1977, 1021.
[95] T. Iwasaki and K. Harada, *J. Chem. Soc., Perkin Trans. 1*, 1977, 1730.
[96] N. Yamashina and H. Miura, Jpn. Kokai 77 151 122 (*Chem. Abstr.*, 1978, **88**, 169 669).
[97] R. Liesenberg, H. Matschiner, and H. Schick, Ger. (East) P. 126 997 (*Chem. Abstr.*, 1978, **88**, 169 666).
[98] K. P. Healy and D. Pletcher, *J. Organomet. Chem.*, 1978, **161**, 109; C. Gosden, K. P. Healy, and D. Pletcher, *J. Chem. Soc., Dalton Trans.*, 1978, 972.
[99] C. Gosden and D. Pletcher, *J. Organomet. Chem.*, 1980, **186**, 401.
[100] J. L. Hall, R. D. Greer, and P. W. Jennings, *J. Org. Chem.*, 1978, **43**, 4364.
[101] S. Margel and F. C. Anson, *J. Electrochem. Soc.*, 1978, **125**, 1232.

Carbenes can be generated by the reduction of *gem*-dichloro- or -dibromo-compounds and trapped by their reaction with olefins at 0 to −5 °C. Dichlorocarbene was best generated by the reduction of CCl_4 or of $CHCl_3$, in chloroform or dichloromethane as the solvent. Difluorocarbene resulted from the reduction of CF_2Br_2 in dichloromethane.[102] The reduction of the trichloromethyl group in aqueous dioxan affords good yields of the corresponding dichloromethyl compound, and this reaction has been used in some syntheses of derivatives of amino-acids.[103] The electrochemical reduction of 1,2-bis(dibromomethyl)benzene yields the same product (4) as had previously been obtained by chemical reduction.[104] A study of the reduction and rearrangement of 1,1-diaryl-2,2-

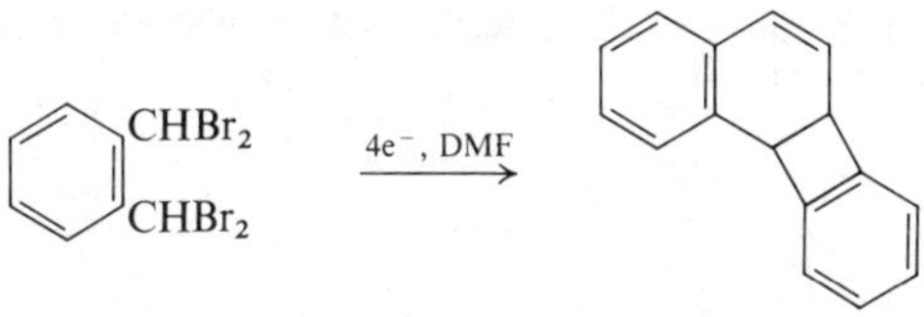

$$C_6H_4(CHBr_2)_2 \xrightarrow{4e^-,\ DMF} \text{(4)}$$

(4)

dichloroethenes to diaryl-ethynes has shown that the rearrangement only occurs under conditions of high current density, and preferably with lithium bromide (in DMF) as the supporting electrolyte (see Scheme 8). The addition of 3 to 5% of water completely suppresses the rearrangement and the formation of 1,1-diaryl-2-chloroethene.[105]

$$Ar_2C=CCl_2 \xrightarrow[DMF]{e^-} ArC\equiv CAr\ +\ Ar_2C=CHCl$$

Scheme 8

The reduction potential for less-alkylated alkene dibromides occurs at more negative potentials, so that selective debromination of diene tetrabromides is possible, and it affords a means for protecting the less-alkylated double-bond in the diene.[106] An example of the use of this protecting group is given in Scheme 9. The reduction of 2-bromo-1-iodotrinorbornane afforded the unstable bridgehead olefin $\Delta^{1,2}$-bornene, trapped as its furan adduct.[107] Co-elimination of the halide and the sulphonyl group to give an alkene occurs during the reduction of β-halogenoethyl sulphones.[108] A study of related co-elimination reactions in a group of substrates (5) when they were reduced in DMF showed the preferred formation of the olefin when R is Cl, $MeSO_3$, or $C_7H_7SO_3$ and the preferred replacement of chlorine by hydrogen

---

[102] H. P. Fritz and W. Kornrumpf, *Liebigs Ann. Chem.*, 1978, 1416; *J. Electroanal. Chem. Interfacial Electrochem.*, 1979, **100**, 217.
[103] T. Iwasaki, Y. Urabe, Y. Ozaki, M. Miyoshi, and K. Matsumoto, *J. Chem. Soc., Perkin Trans. 1*, 1976, 1019.
[104] L. Rampazzo, A. Inesi, and R. M. Bettolo, *J. Electroanal. Chem. Interfacial Electrochem.*, 1977, **83**, 341.
[105] A. Mertz and G. Thumm, *Liebigs Ann. Chem.*, 1978, 1526.
[106] U. Husstedt and H. J. Schafer, *Synthesis*, 1979, 964; *J. Chem. Res. (S)*, 1977, 131.
[107] E. Stamm, L. Walder, and R. Keese, *Helv. Chim. Acta*, 1978, **61**, 1545.
[108] E. A. Berdnikov, S. B. Fedorov, and Yu. M. Kargin, *Zh. Obshch. Khim.*, 1978, **48**, 875.

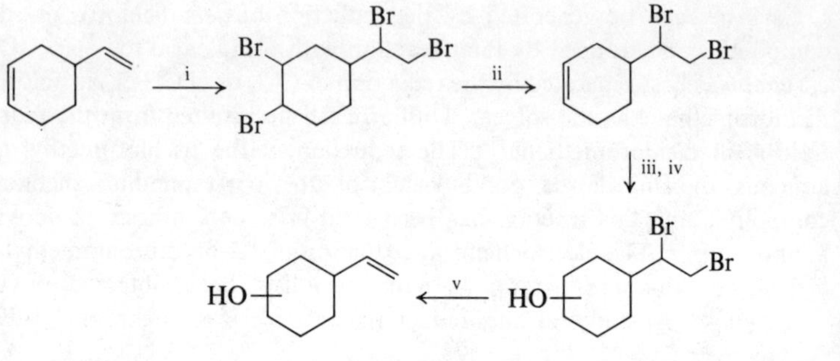

Reagents: i, Br_2; ii, at -1.4 V vs the S.C.E., in DMF; iii, B_2H_6; iv, oxidation; v, at -1.8 V vs the S.C.E., in DMF

Scheme 9

$$PhCH-CCl_3 \xrightarrow{e^-, DMF} PhCH=CCl_2 + PhCH-CHCl_2$$
         | |
         R   R

(5) R = H, Cl, OMe, $MeSO_3$, p-$C_7H_7SO_3$, or p-$XC_6H_4CO_2$

when R is H or OMe.[109] Mixed products were obtained when R is a substituted benzoate group. The diphenyl phosphate ester group also behaves as a leaving group in related co-elimination reactions.[110] Examples are available of the formation of olefins by reduction of bromohydrins in DMF.[111] The reduction of derivatives of trichloroethanol (Scheme 10) appears to result in elimination (to form the olefin) in DMF[112] and in the replacement of one chlorine by hydrogen in aqueous dioxan.[103] trans-1,4-Dibromobut-2-ene and the corresponding dichlorobutene are reduced to butadiene in DMF.[113]

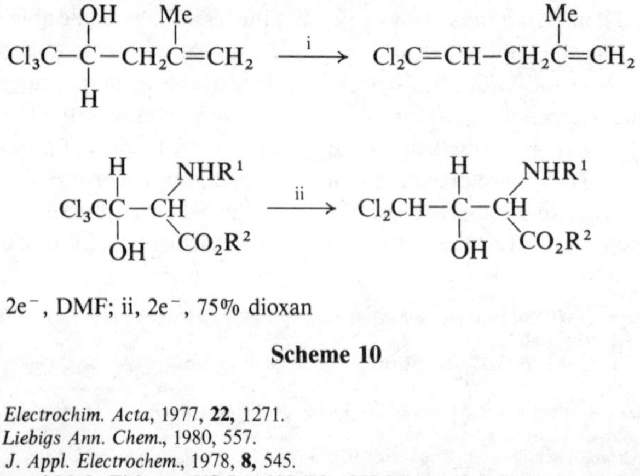

$$Cl_3C-\underset{H}{\underset{|}{\overset{OH}{\overset{|}{C}}}}-CH_2\underset{}{\overset{Me}{\overset{|}{C}}}=CH_2 \xrightarrow{i} Cl_2C=CH-CH_2\overset{Me}{\overset{|}{C}}=CH_2$$

$$Cl_3C\underset{OH}{\overset{H}{\overset{|}{C}}}-\underset{CO_2R^2}{\overset{NHR^1}{\overset{|}{CH}}} \xrightarrow{ii} Cl_2CH-\underset{OH}{\overset{H}{\overset{|}{C}}}-\underset{CO_2R^2}{\overset{NHR^1}{\overset{|}{CH}}}$$

Reagents: i, $2e^-$, DMF; ii, $2e^-$, 75% dioxan

Scheme 10

---

[109] A. Mertz, *Electrochim. Acta*, 1977, **22**, 1271.
[110] J. Engels, *Liebigs Ann. Chem.*, 1980, 557.
[111] D. Cipris, *J. Appl. Electrochem.*, 1978, **8**, 545.
[112] M. Alvarez and M. L. Fishman, US P. 4 022 672, 1977 (*Chem. Abstr.*, 1977, **87**, 31 177).
[113] E. Brillas and J. M. Costa, *J. Electroanal. Chem. Interfacial Electrochem.*, 1976, **69**, 435.

Organic Electrochemistry – Synthetic Aspects: Reductions

The reduction of 1,3-di-iodo-compounds to the corresponding cyclopropane proceeds in good yield for the complex nucleoside derivative (6).[114] This type of ring-closure has given useful yields from the reduction of dimethyl 2,(ω−1)-dibromoalkane-1,ω-dioates (7).[115] The reduction of 1,4-dihalogenotrinorbornanes in DMF was expected to yield [2.2.1]propellane, and, although the hydrocarbon could not be isolated, it is presumed to be an intermediate in the formation of trinorbornane and bis-(1-trinorbornyl)mercury.[116]

	n =	3	4	5	6	7
	% yield =	60	32	52	60	20

Aryl halides will undergo homogeneous transfer of an electron from the electrochemically generated radical-anion of a suitable arene or heteroarene redox catalyst.[117] The rate of this transfer has been used to estimate the standard redox potential for those compounds where the value cannot be obtained directly: e.g., fluorobenzene, −2.97; chlorobenzene, −2.78; bromobenzene, −2.44 V vs the S.C.E. Radical-anions of aryl halides decompose to give the aryl radical, which can undergo a number of competing reactions: abstraction of hydrogen from the solvent, reduction to the carbanion by the electrode, and transference of an electron from another radical-anion. This competition has been examined in a quantitative manner.[118] The abstraction of hydrogen from DMSO can lead to complex products by the interation of the aryl radical with products arising from decomposition of the solvent.[119] Radical-anions can undergo nucleophilic substitution by a process of breaking and then re-forming of bonds:

$$Ar\text{—}X^{1\,\overline{}} + Y^- \rightarrow Ar\text{—}Y^{1\,\overline{}} + X^{-\,\cdot}$$

and, provided that $E^\ominus$ for the product ArY is more negative than $E^\ominus$ for the substrate ArX, we have a situation where the substitution

$$ArX + Y^- \rightarrow ArY + X^-$$

is catalysed by the radical-anion of ArX. Such reactions proceed well in liquid

---

[114] T. Adachi, T. Iwasaki, M. Muneji, and I. Inoue, *J. Chem. Soc., Chem. Commun.*, 1977, 248.
[115] S. Satoh, M. Itoh, and M. Tokuda, *J. Chem. Soc., Chem. Commun.*, 1978, 481.
[116] W. F. Carroll, Jr. and D. G. Peters, *J. Am. Chem. Soc.*, 1980, **102**, 4127.
[117] C. P. Andrieux, C. Blockman, J. M. Dumas-Bouchiat, and J. M. Saveant, *J. Am. Chem. Soc.*, 1979, **101**, 3431; C. P. Andrieux, C. Blockman, and J. M. Saveant, *J. Electroanal. Chem. Interfacial Electrochem.*, 1979, **105**, 413.
[118] F. M'Halla, J. Pinson, and J. M. Saveant, *J. Am. Chem. Soc.*, 1980, **102**, 4120.
[119] F. M'Halla, J. Pinson, and J. M. Saveant, *J. Electroanal. Chem. Interfacial Electrochem.*, 1978, **89**, 347.

ammonia, which is a poor hydrogen-atom-donor solvent, so that aryl radicals are not lost in a competing reaction.[120]

The sequence of reactions of cleavage of aryl halide radical-ions and then abstraction of hydrogen by the aryl radical can be a useful method for the replacement of halogen by hydrogen (Scheme 11).[121] The *N*-methylbenzanilides (8) have the correct conformation to undergo intramolecular radical substitution as a competing reaction. A study of substituted benzanilides showed that this cyclization competes very successfully for chloro-compounds (and progressively less successfully for bromo- and iodo-compounds) with the replacement of halogen by hydrogen. It is suggested that, where the replacement of hydrogen occurs to a large extent, this is because the aryl radical is generated close to the surface of the electrode, where it undergoes further addition of electrons. Radical cyclization of the *N*-methylbenzanilides leads to two products, as shown, in relative proportions which depend on the number of *ortho*-substituents in the amine ring.[122] *ortho*-Substituents favour the diphenyl product. Thus the process was unsuccessful for the synthesis of the benzophenanthridine system from (9).[123] The reaction has been used in a synthesis of the aporphine ring system (Scheme 12).[124]

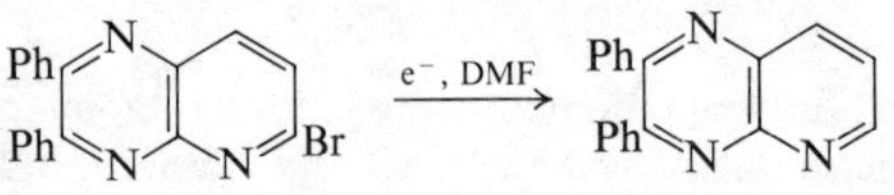

Scheme 11

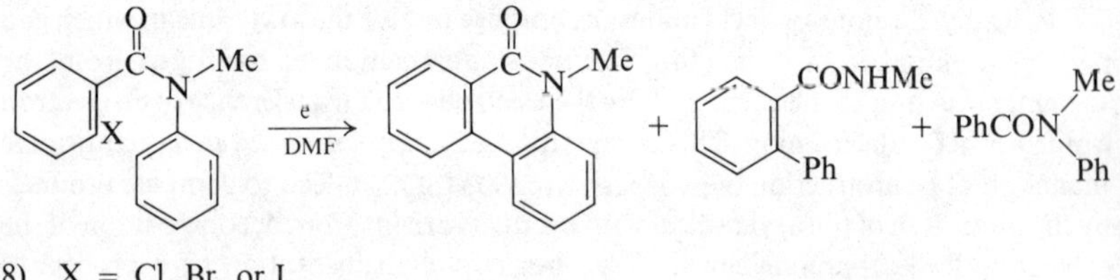

(8) X = Cl, Br, or I

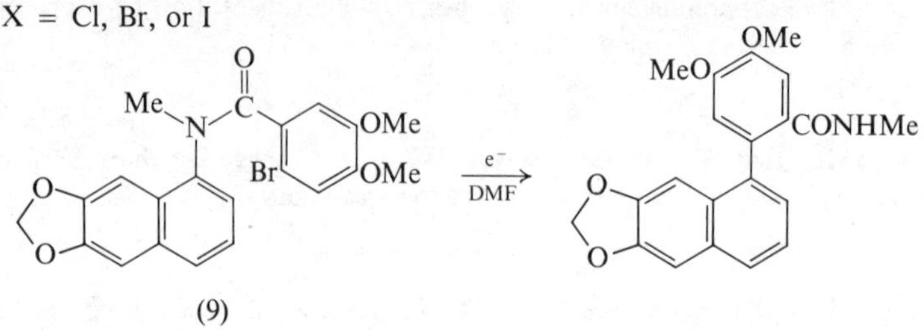

(9)

[120] J. Pinson and J. M. Saveant, *J. Am. Chem. Soc.*, 1978, **100**, 1506; J. M. Saveant and A. Thiebault, *J. Electroanal. Chem. Interfacial Electrochem.*, 1978, **89**, 335; C. Amatore, J. Chaussard, J. Pinson, J. M. Saveant, and A. Thiebault, *J. Am. Chem. Soc.*, 1979, **101**, 6012.
[121] J. Armand, K. Chekir, and J. Pinson, *Can. J. Chem.*, 1978, **56**, 1804.
[122] J. Grimshaw, R. J. Haslett, and J. Trocha-Grimshaw, *J. Chem. Soc., Perkin Trans. 1*, 1977, 2448; J. Grimshaw and D. Mannus, *ibid.*, p. 2456; J. Grimshaw and R. J. Haslett, *J. Chem. Soc., Perkin Trans. 2*, 1980, 657.
[123] W. J. Begley and J. Grimshaw, *J. Chem. Soc., Perkin Trans. 1*, 1977, 2324.
[124] R. Gottlieb and J. L. Neumeyer, *J. Am. Chem. Soc.*, 1976, **98**, 7108.

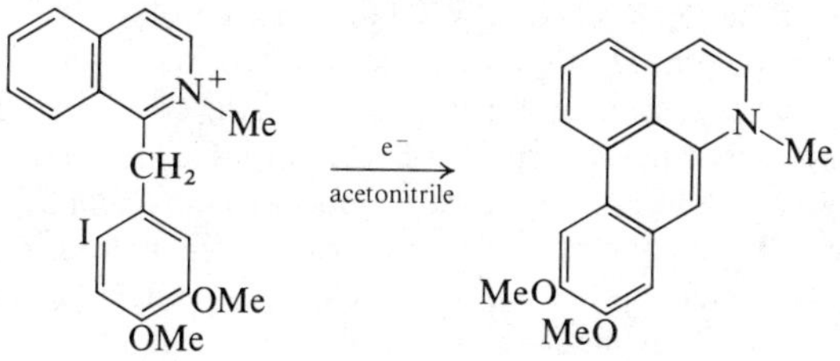

Scheme 12

Carbonyl Compounds.—Carbonyl groups can be used to trap the carbanions that are formed as intermediates in the reduction of tetrachloromethane and ethyl trichloroacetate in DMF (Scheme 13), and, where ethyl trichloroacetate is used, further rearrangement of the initial product occurs to give a keto-ester.[125] The carbanion intermediates from two-electron reduction of aliphatic ketones in liquid ammonia react with tetraethylammonium ions in the supporting electrolyte to give the C-ethylated tertiary alcohol along with the expected secondary alcohol.[126] The stereochemistry of the reduction of acyclic[127] and cyclic ketones[128—130] to the

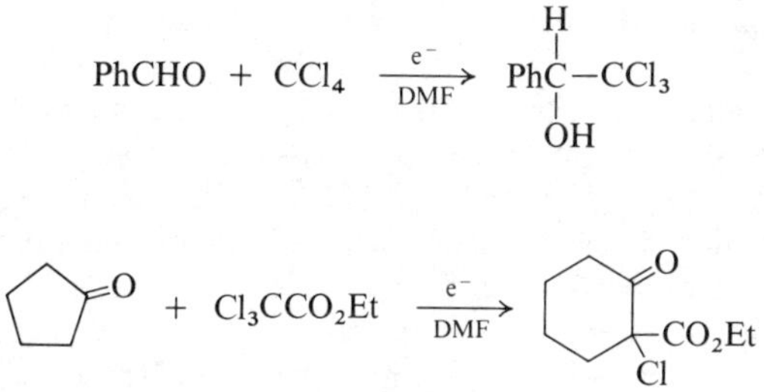

$$PhCHO + CCl_4 \xrightarrow[DMF]{e^-} PhC(H)(OH)-CCl_3$$

Scheme 13

corresponding secondary alcohol in protic solvents has been investigated by several groups. Ketones with a chiral centre in the α-position are reduced to a mixture of *threo*- and *erythro*-alcohols in a ratio which is changed to a small degree if the pH of the solution is altered. For conformationally rigid but unhindered cyclohexanones, the ratio of axial to equatorial alcohol depends markedly on the charge/ionic radius of the supporting cation. The *eq*-isomer is favoured in the presence of Li^+ or Bu_4N^+ but the *ax*-isomer is strongly favoured in the presence of Mg^{2+} or Zn^{2+}, so that the method becomes of some preparative value for the production of axial alcohols. Increasing steric hindrance adjacent to the carbonyl function leads to the replacement of CO by CH_2, particularly when $Zn(ClO_3)_2$ is the supporting

---

[125] F. Karrenbrock and H. J. Schafer, *Tetrahedron Lett.*, 1978, 1521.
[126] E. M. Abbot and A. J. Bellamy, *J. Chem. Soc., Perkin Trans. 2*, 1978, 254.
[127] T. Nonaka, Y. Kusayanagi, and T. Fuchigami, *Electrochim. Acta*, 1980, **25**, 1679.
[128] J. P. Coleman, R. J. Holman, and J. H. P. Utley, *J. Chem. Soc., Perkin Trans. 2*, 1976, 879; R. J. Holman and J. H. P. Utley, *ibid.*, p. 884.
[129] G. Le Guillanton and M. Lamant, *Nouv. J. Chim.*, 1978, **2**, 157.
[130] G. Le Guillanton, *Electrochim. Acta*, 1980, **25**, 1351.

electrolyte. This reduction has been applied to 7-oxo-steroids,[131] where reaction in dioxan–D_2O that contains D_2SO_4 gives good yields for the replacement of CO by CD_2. The reduction of 2-cyanocyclohexanone gives the cis-alcohol in good yields when LiCl is the supporting electrolyte whereas Et_4NI favours the elimination of cyanide ion to give cyclohexanone, which is reduced to the alcohol.[129] Temperature has a profound effect on the ratio of cis- to trans-isomers from reduction of 2-ethoxycarbonylcyclopentanone and 2-ethoxycarbonylcyclohexanone; a temperature of $-6\,°C$ favours the cis-isomer while one of $80\,°C$ favours the thermodynamically preferred trans-isomer.[130] No very dramatic difference in the cis/trans ratio was found on changing the supporting cation from Li^+ to Mg^{2+}.

α-Keto-acids are converted into α-amino-acids by reduction at mercury in the presence of ammonia and ammonium chloride. Current densities have to be low in order that the working electrode is not too negative, otherwise reduction of ketone to secondary alcohol occurs.[132] Reductive amination can also be carried out at cathodes of platinum or palladium black, but there is no advantage over a mercury cathode.

Studies on the reduction of acetophenone to optically active 1-phenylethanol in protic solvents, using an optically active conducting salt, continue.[133] No optical induction was observed when acetophenone, 4-acetylpyridine, or ethyl phenylglyoxylate was reduced at a modified carbon electrode that was prepared with optically active groups attached to the graphite.[134] Reduction of acetophenone in acetonitrile that contains a chiral electrolyte gives asymmetric induction in both the 1-phenylethanol and the ($\pm$)-pinacol that is formed. Asymmetric induction in the pinacol is destroyed by the presence of water, which presumably breaks up intimate ion-pairs between the cation and the ketyl radical-anion.[135] The degree of asymmetric induction in 1-phenylethanol depends on the cathode potential. Both 2- and 4-acetylpyridine give substantial asymmetric induction in the secondary alcohol that is formed by reduction at mercury in the presence of a catalytic amount of strychnine. 3-Acetylpyridine showed no induction under similar conditions.[136] Phenylglyoxylic acid and its oxime are reduced at mercury, in the presence of catalytic amounts of alkaloids that are strongly adsorbed at the electrode, to give optically active mandelic acid and phenylglycine, respectively. The optical yield varies with pH and with the working potential of the electrode, being highest in acidic solution. It is suggested that the optical yield depends on a competition between protonation from the electrode side or from the solution side of a complex that is formed between the reduced ketone and the adsorbed alkaloid.[137] Reduction of (−)-menthyl phenylglyoxylate occurs with the generation of asymmetry at the

---

[131] G. Phillipou, C. J. Seaborn, and I. A. Blair, *Aust. J. Chem.*, 1979, **32**, 2767.
[132] E. A. Jeffery and A. Meisters, *Aust. J. Chem.*, 1978, **31**, 73; E. A. Jeffery, O. Johansen, and A. Meisters, *ibid.*, p. 79.
[133] D. Brown and L. Horner, *Liebigs Ann. Chem.*, 1977, 77; L. Horner and W. Brich, *ibid.*, 1978, 710; *Chem. Ber.*, 1978, **111**, 574.
[134] L. Horner and W. Brich, *Liebigs Ann. Chem.*, 1977, 1354.
[135] W. J. M. Tilborg and C. J. Smit, *Recl. Trav. Chim. Pays-Bas*, 1978, **97**, 89.
[136] J. Kopilov, S. Shatzmiller, and E. Kariv, *Electrochim. Acta*, 1976, **21**, 535; J. Kopilov, E. Kariv, and L. L. Miller, *J. Am. Chem. Soc.*, 1977, **99**, 3450; J. Hermolin, J. Kopilov, and E. Gileadi, *J. Electroanal. Chem. Interfacial Electrochem.*, 1976, **71**, 245.
[137] M. Jubault, E. Raoult, and D. Peltier, *Electrochim. Acta*, 1980, **25**, 1359.

secondary alcohol centre that is formed; the optical yield is also highest in acidic solution and at low current densities.[138]

Stereoselectivity in the pinacolization of 4-hydroxybenzaldehyde has been determined, using h.p.l.c. as an analytical tool. Reduction in aqueous media at either alkaline or neutral pH gives a ratio of ($\pm$)- to *meso*-isomers of 0.77—0.94:1, contrary to previous suggestions that the *meso*-form predominates in alkaline solution.[139] The ($\pm$):*meso* ratio from pinacolization of acetophenone in DMF is strongly influenced by the water content of the solvent. Under anhydrous conditions, with lithium as the cation, the ratio is either 14:1 or 19:1, because dimerization occurs through cation–radical-anion pairs. The ion-pairs are broken up by traces of water, so the ratio then falls. Tetrabutylammonium ions give rise to a ratio of ($\pm$)-:*meso*-pinacol of about 4:1, and in protic solvents the ratio is 1.8—2.6:1.[140]

Reduction of polyenecarbonyl compounds in DMF gives good yields of the pinacol in the presence of $CrCl_3 \cdot 6H_2O$. The method is applicable to a number of polyene pinacols that are otherwise difficult to prepare (Scheme 14).[141] A chromium(III) complex of the ketone is the species which undergoes reduction. For benzaldehyde, the ratio of ($\pm$)-:*meso*-pinacol in the presence of chromium(III) is 1.3—3.1:1, as opposed to *ca* 14:1 in the same solvent in the absence of chromium ions. Reduction of 1,4-diacetylbenzene and benzene-1,4-dialdehyde gives polymeric pinacols as well as dihydro- and tetrahydro-products.[142]

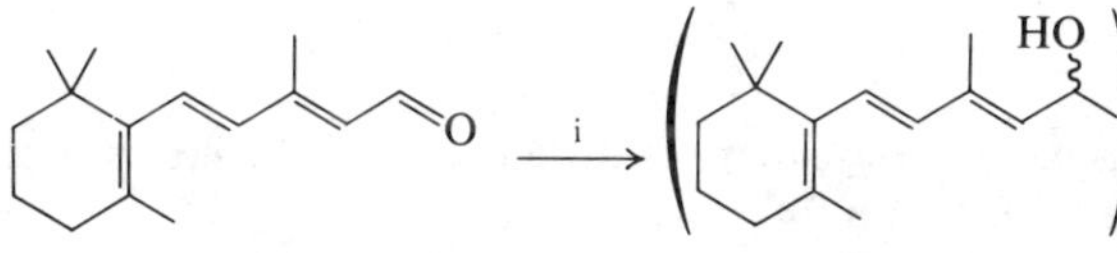

Reagents: i, e⁻, DMF, $CrCl_3 \cdot 6H_2O$

[25% yield]

Scheme 14

α-Hydroxyacetophenone is reduced exclusively to acetophenone in protonic solvents at pH 3.4, but at pH 7—10 there is competition between this reaction and the formation of (1,2-dihydroxyethyl)benzene.[143] This removal of a hydroxy-group that is adjacent to a carbonyl function has been used in a recommended process for converting xylose into 2-deoxyxylitol (Scheme 15).[144]

Phenyl cyclopropyl ketone is reduced, in aqueous ethanol and at pH 3, to the expected mixture of carbinol and pinacol. However, when the cyclopropane ring also bears a phenyl substituent, the ring undergoes reductive cleavage.[145] A related

---

[138] M. Jubault, E. Raoult, and D. Peltier, *Electrochim. Acta*, 1977, **22**, 67.
[139] D. F. Tomkins and J. H. Wagenknecht, *J. Electrochem. Soc.*, 1978, **125**, 372.
[140] A. Bewick and D. J. Brown, *J. Chem. Soc., Perkin Trans. 2*, 1977, 99.
[141] D. W. Sopher and J. H. P. Utley, *J. Chem. Soc., Chem. Commun.*, 1979, 1087.
[142] L. Horner and H. Hoenl, *Liebigs Ann. Chem.*, 1978, 1.
[143] S. Letellier, *Electrochim. Acta*, 1980, **25**, 105.
[144] H. Berbalk, K. Eichinger, and P. Fricko, *Synthesis*, 1978, 42.
[145] L. Mandel, J. C. Johnston, and R. A. Day, *J. Org. Chem.*, 1978, **43**, 1616.

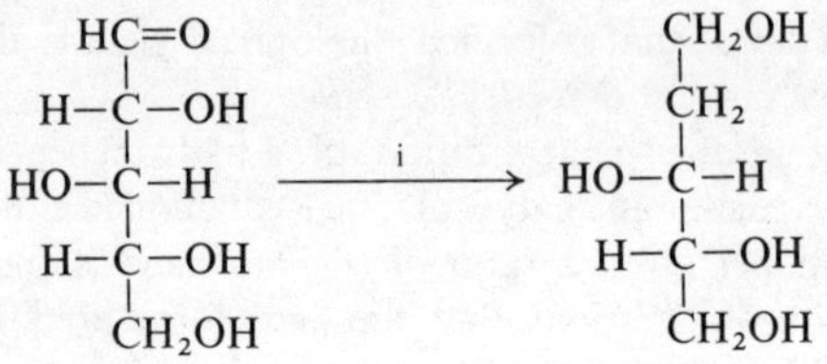

$$\begin{array}{c}\text{HC=O}\\\text{H}-\text{C}-\text{OH}\\\text{HO}-\text{C}-\text{H}\\\text{H}-\text{C}-\text{OH}\\\text{CH}_2\text{OH}\end{array}\xrightarrow{\text{i}}\begin{array}{c}\text{CH}_2\text{OH}\\\text{CH}_2\\\text{HO}-\text{C}-\text{H}\\\text{H}-\text{C}-\text{OH}\\\text{CH}_2\text{OH}\end{array}$$

Reagents: i, e$^-$, H$_2$O, Na$_2$SO$_4$, at pH 11

Scheme 15

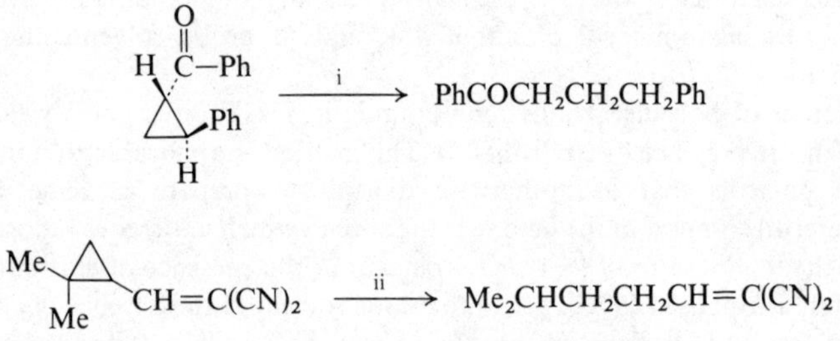

[cyclopropane with H, C(=O)Ph, Ph, H substituents] $\xrightarrow{\text{i}}$ PhCOCH$_2$CH$_2$CH$_2$Ph

[Me,Me-cyclopropane–CH=C(CN)$_2$] $\xrightarrow{\text{ii}}$ Me$_2$CHCH$_2$CH$_2$CH=C(CN)$_2$

Reagents: i, e$^-$, EtOH, H$_2$O, at pH 3; ii, e$^-$, acetonitrile

Scheme 16

reductive cleavage of 1-cyclopropylethylidenemalononitriles in acetonitrile has been described[146] (Scheme 16).

Reactive species from the reduction of the carbonyl function in DMF can be trapped intramolecularly by a neighbouring olefin group, thus giving rise to a useful ring-closure reaction (Scheme 17).[147] Intermolecular coupling between reduction products of benzophenone and vinyl acetate has also been found (Scheme 17), and the yield from this type of reaction is reduced by the introduction of substituents onto the vinyl substrate.[148]

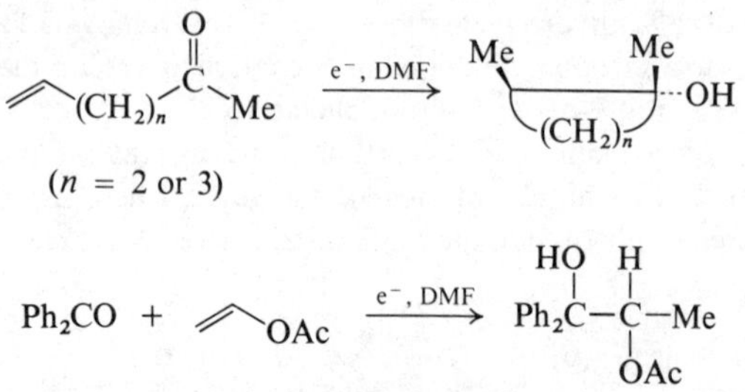

$$\text{CH}_2=\text{CH}-(\text{CH}_2)_n-\text{C}(=\text{O})\text{Me}\xrightarrow{e^-,\text{DMF}}\text{[cyclic product with Me, Me, OH, (CH}_2)_n\text{]}$$

(n = 2 or 3)

$$\text{Ph}_2\text{CO}+\text{CH}_2=\text{CH-OAc}\xrightarrow{e^-,\text{DMF}}\text{Ph}_2\text{C}(\text{OH})-\text{CH}(\text{OAc})-\text{Me}$$

Scheme 17

---

[146] A. J. Bellamy and J. B. Kerr, *Acta Chem. Scand., Ser. B*, 1979, **33**, 370.
[147] T. Shono, I. Nishiguchi, H. Ohmizu, and M. Mitani, *J. Am. Chem. Soc.*, 1978, **100**, 545.
[148] T. Shono, H. Ohmizu, and S. Kowakami, *Tetrahedron Lett.*, 1979, 4091.

The reduction of α-alkyl-acetoacetic esters in aqueous ethanolic sulphuric acid at a lead cathode proceeds with rearrangement to form a straight-chain hydrocarbon (the Tafel rearrangement). Evidence has been presented to show that this reaction proceeds via cyclopropane intermediates.[149]

Carbon monoxide reacts with methoxide ions that are generated at a steel cathode in methanol that contains some sodium methoxide to form methyl formate. In the presence of amines, the corresponding formamide can be isolated, and the method shows possibilities for large-scale synthesis.[150] Reduction of carbon monoxide (at a pressure of 230 atm) in DMF that contains tetrabutylammonium bromide yields 30—48% of the salt of squaric acid.[151]

Results from a number of studies on the reduction of carbon dioxide indicate the complex nature of this reaction. The products that have been isolated[152,153] are carbon monoxide, formate, oxalate, glycolate, and malate. Formate is found in aqueous solvents, and indium seems to be the best cathode material for obtaining this product. The other products are formed in DMF or DMSO, and a mechanistic proposal (Scheme 18) has been made to account for the observations.[152] In anhydrous DMF, $E^{\ominus}$ for the first step is -2.21 V vs the S.C.E. and the second-order rate constant for deactivation of the radical ion has been estimated[154] as 10^7 dm^3 mol^{-1} s^{-1}.

$$CO_2 + e^- \rightleftharpoons CO_2^{\cdot -}$$

$$CO_2^{\cdot -} + H_2O \longrightarrow HO^- + HCO_2^{\cdot} \xrightarrow{e^-} HCO_2^-$$

$$2\ CO_2^{\cdot -} \longrightarrow {}^-O_2C-CO_2^-$$

$$CO_2^{\cdot -} + CO_2 \longrightarrow O=\overset{\cdot}{C}-O-C\underset{O^-}{\overset{O}{\diagup\!\!\!\diagdown}} \xrightarrow{e^-} CO + CO_3^{2-}$$

Scheme 18

The influence of steric and electronic effects of methyl substituents on the redox potential of the benzophenone radical-anion system has been examined.[155] The squaric acid derivatives (10) and (11) each show two reversible one-electron-reduction steps in aprotic solvents.[156]

The reduction of carboxylic acids to the primary alcohol has received attention. Reduction of benzoic acid proceeds best on a lead cathode, owing to specific adsorption of the substrate.[157] Reduction in ethanol as solvent proceeds best in the

---

[149] S. Wawzonek and J. E. Durham, *J. Electrochem. Soc.*, 1976, **123**, 500.
[150] D. Cipris, *J. Electrochem. Soc.*, 1980, **127**, 1045.
[151] G. Silvestri, S. Gambino, G. Filardo, G. Spadaro, and L. Palmisano, *Electrochim. Acta*, 1978, **23**, 413.
[152] J. C. Gressin, D. Michelet, L. Nadjo, and J. M. Saveant, *Nouv. J. Chim.*, 1979, **3**, 545.
[153] F. Wolf and J. Rollin, *Z. Chem.*, 1977, **17**, 337; K. Ito, T. Murata, and S. Ikeda, *Nagoya Kogyo Daigaku Gakuho*, 1975, **27**, 209 (*Chem. Abstr.*, 1977, **87**, 5347).
[154] E. Lamy, L. Nadjo, and J. M. Saveant, *J. Electroanal. Chem. Interfacial Electrochem.*, 1977, **78**, 403.
[155] J. Grimshaw and R. Hamilton, *J. Electroanal. Chem. Interfacial Electrochem.*, 1980, **106**, 339.
[156] S. Hunig and H. Putter, *Chem. Ber.*, 1977, **110**, 2524; K. Komatsu and R. West, *J. Chem. Soc., Chem. Commun.*, 1976, 570.
[157] I. Taniguchi, A. Yoshiyama, and T. Sekine, *Denki Kagaku Oyobi Kogyo Butsuri Kagaku*, 1977, **45**, 442 (*Chem. Abstr.*, 1978, **88**, 6464).

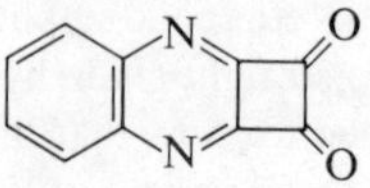

(10)

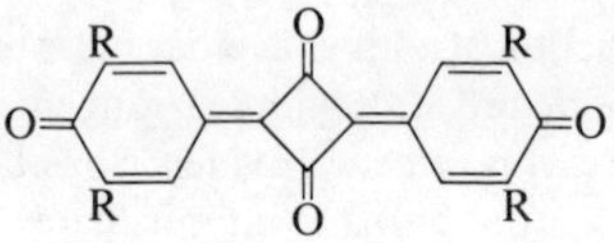

(11) R = Bu^t

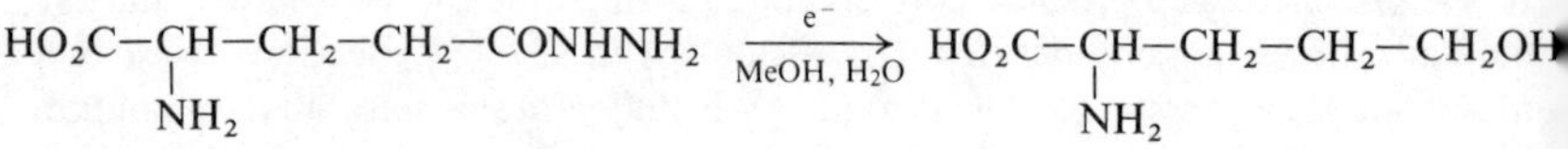

HO$_2$C—CH—CH$_2$—CH$_2$—CONHNH$_2$ $\xrightarrow[\text{MeOH, H}_2\text{O}]{e^-}$ HO$_2$C—CH—CH$_2$—CH$_2$—CH$_2$OH
     | |
   NH$_2$ NH$_2$
(12)

presence of acetic acid, which suppresses hydrogenation of the benzene ring.[158] The best yields are obtained from reduction of phenyl esters in ethanol solution at a mercury cathode. Phenyl benzoates are reduced quantitatively to the benzyl alcohol while phenyl esters of fatty acids give acceptable yields of the corresponding alcohols. Reduction of S-alkyl esters of thioacids also gives acceptable yields of the primary alcohol.[158] Applications include the selective reduction of glutaric acid hydrazide (12)[159] and the conversion of ribono-γ-lactone into a mixture of ribose and arabinose.[160]

Activated Olefins.—β-Keto-esters and β-diketones can be regarded as activated olefins in their enol form. The reduction of 1,3-diphenylpropane-1,3-dione in aqueous ethanol at pH 4.2 gives a mixture of (±)- and *meso*-pinacols, isolated as the products from further aldol-like condensations of the remaining carbonyl functions.[161a] Ethyl benzoylacetates also undergo reductive coupling in DMF and then further cyclization under the basic conditions which prevail, and only the *trans*-

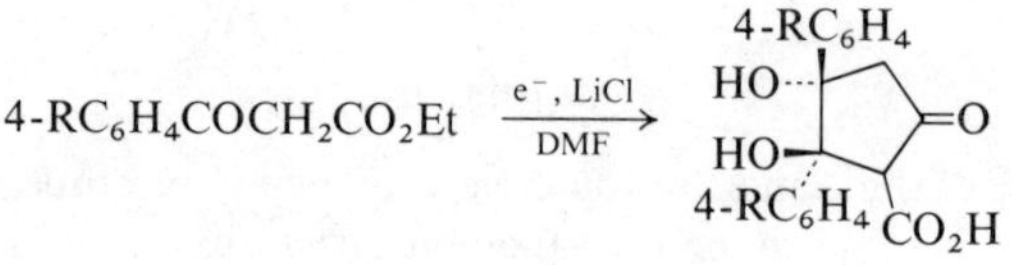

4-RC$_6$H$_4$COCH$_2$CO$_2$Et $\xrightarrow[\text{DMF}]{e^-, \text{LiCl}}$ (13)

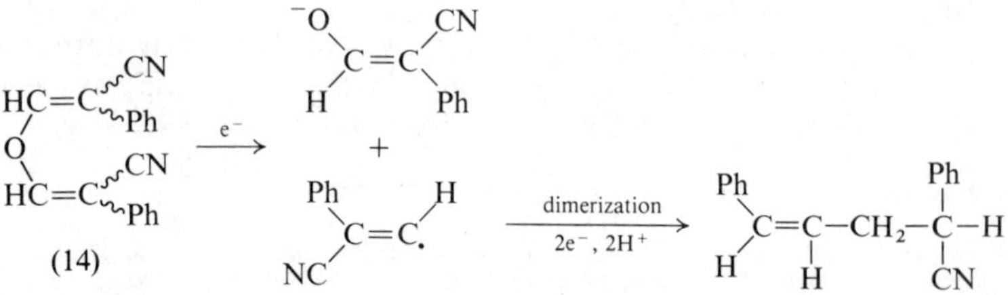

(14)

[158] L. Horner and H. Hoenl, *Liebigs Ann. Chem.*, 1977, 2036.
[159] M. Matsuoka and Y. Kokusenya, Jpn. Kokai 77 153 915 (*Chem. Abstr.*, 1978, **88**, 191 468).
[160] A. Korczynski, L. Piszczek, J. Swiderski, and A. Donjec, *Przem. Chem.*, 1977, **56**, 589 (*Chem. Abstr.*, 1978, **88**, 8818).
[161] A. J. Klein and D. H. Evans, *J. Org. Chem.*, 1977, **42**, 2560; Y. Brun and G. Mousset, *C.R. Hebd. Seances Acad. Sci., Ser. C*, 1978, **286**, 525.

diol (13) is isolated.[161b] In contrast, the related ether (14) undergoes reductive cleavage followed by elimination of the radicals that are formed.[162]

α,ω-Diketopolymers of the carotenoid series have been shown to undergo reductive acetylation in a mixture of methylene chloride and acetic anhydride. In the case of astacene (15), the initially formed acetylated polyene can be reduced further at a more negative potential.[163]

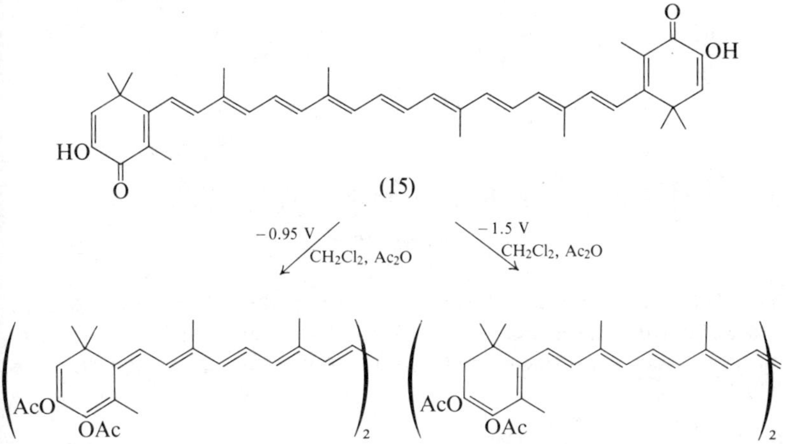

(15)

−0.95 V / CH_2Cl_2, Ac_2O −1.5 V / CH_2Cl_2, Ac_2O

Reduction of dibenzalsuccinic acid gives only di- and tetra-hydro-derivatives, with no indication of intramolecular cyclization.[164] The stereochemistry of related reductions of cycloalkene-1,2-dicarboxylic acids has been discussed.[165] Reductive cyclization of the bis-activated olefins (16) does occur, but the introduction of further β-phenyl substitution on the olefin group suppresses the cyclization in favour of hydrogenation of the olefin.[166] Compound (17) undergoes bimolecular reductive coupling in aqueous DMF; no monomeric cyclization products are found in the reaction mixture.[167] Intramolecular reductive coupling to form six-membered rings has been demonstrated for two bis-activated olefins (Scheme 19).[168a] The stereochemistry of the hydro-dimers from cyclohexenone, isophorone, and (±)-piperitone has been elucidated.[168b]

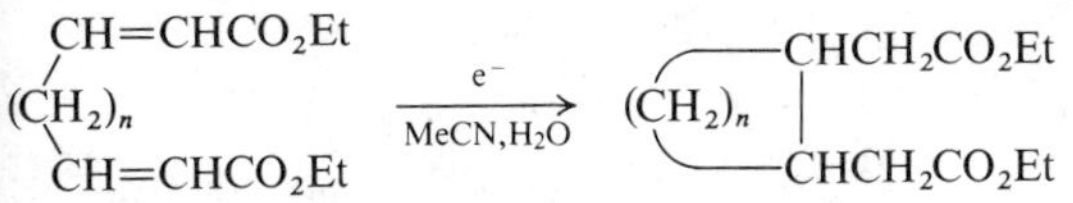

(16) n = 3 or 4

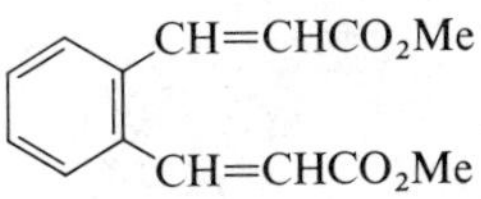

(17)

---

[162] G. Mabon and G. Le Guillanton, *C.R. Hebd. Seances Acad. Sci., Ser. C*, 1976, **282**, 319.
[163] E. A. H. Hall, G. P. Moss, J. H. P. Utley, and B. C. L. Weedon, *J. Chem. Soc., Chem. Commun.*, 1978, 387; 1976, 586.
[164] J. Anderson and L. Eberson, *Nouv. J. Chim.*, 1977, **1**, 413.
[165] R. Hazard, J. Sarrazin, and A. Tallec, *Electrochim. Acta*, 1979, **24**, 239; 1980, **25**, 1071.
[166] C. P. Andrieux, D. J. Brown, and J. M. Saveant, *Nouv. J. Chim.*, 1977, **1**, 157.
[167] J. Anderson, L. Eberson, and C. Svensson, *Acta Chem. Scand., Ser. B*, 1978, **32**, 234.
[168] (a) B. Terem and J. H. P. Utley, *Electrochim. Acta*, 1979, **24**, 1081; L. Mandell, R. F. Daley, and R. A. Day, *J. Org. Chem.*, 1976, **41**, 4087; (b) J. Grimshaw and R. J. Haslett, *J. Chem. Soc., Perkin Trans. I*, 1979, 395.

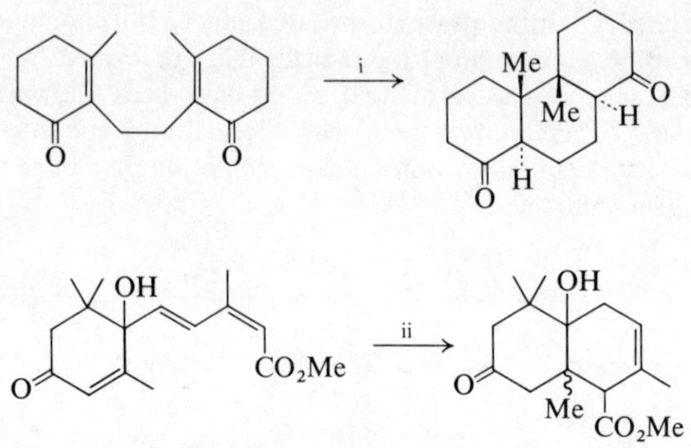

Reagents: i, e$^-$, acetonitrile, H$_2$O; ii, e$^-$, acetonitrile

Scheme 19

The reduction of compound (18) at the first wave (-1.67 V vs the S.C.E.) gives a mixture of dihydro-dimer products, whereas reduction at the second wave in the presence of acetic anhydride gives the intramolecular cyclization product that is shown.[169]

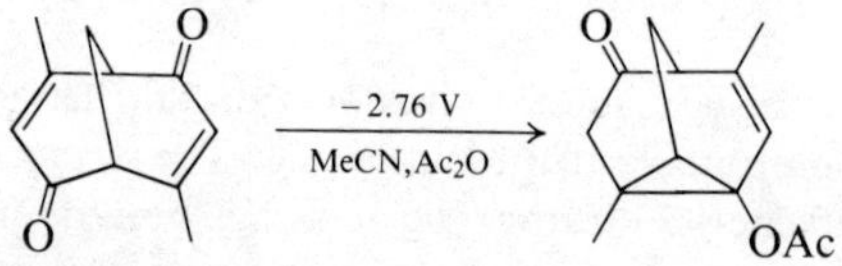

Several $\alpha\beta$-unsaturated aldehydes have been shown to undergo both head-to-head dimerization, to form the pinacol, and head-to-tail dimerization in aqueous ethanol (Scheme 20).[170a] Retinal gives satisfactory yields of the pinacol on reduction in acetonitrile, in the presence of diethyl malonate as the proton donor, while the α- and β-ionones can be pinacolized satisfactorily under ordinary conditions in

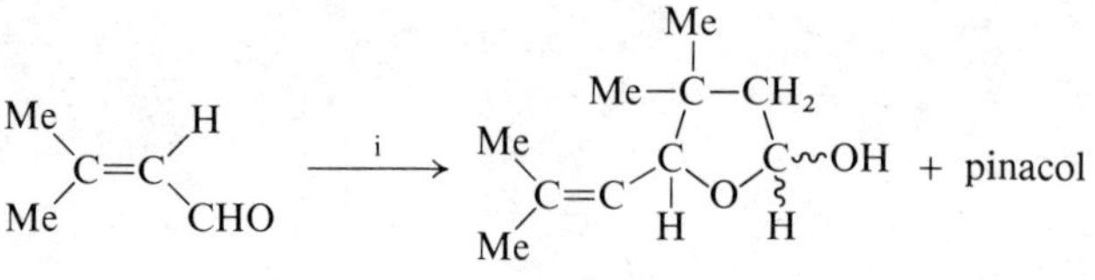

Reagents: i, e$^-$, EtOH, H$_2$O

Scheme 20

[169] J. M. Mellor, B. S. Pons, and J. H. A. Stibbard, *J. Chem. Soc., Chem. Commun.*, 1979, 759; *J. Chem. Soc., Perkin Trans. 1*, 1981, 3092.
[170] (a) J. C. Johnston, J. D. Faulkner, L. Mandell, and R. A. Day, *J. Org. Chem.*, 1976, **41**, 2611; (b) L. A. Powell and R. A. Wightman, *J. Am. Chem. Soc.*, 1979, **101**, 4412; R. E. Sioda, B. Terem, J. H. P. Utley, and B. C. L. Weedon, *J. Chem. Soc., Perkin Trans. 1*, 1976, 561.

DMF.[170b] The yield of hydro-dimer from reduction of N-ethylmaleimides is very dependent on the substituents that are attached to the olefin. The parent yields hydro-dimer in aqueous ethanol at neutral pH; the introduction of one methyl substituent results in a mixture of dihydro-compound and the hydro-dimer; the introduction of two methyl substituents results in the dihydro-compound being formed exclusively.[171]
Investigations on the trapping of radical-anions by t-butyl chloride and by acetic anhydride have been mentioned previously. In what appear to be related reactions, the reduction of acrylonitrile and methyl acrylate in the presence of aldehydes gives the mixed hydro-dimer product (Scheme 21).[172] Activated olefins that can more easily be reduced than carbon dioxide itself are carboxylated if they are reduced in DMF that is saturated with carbon dioxide (Scheme 21).[173] The mixed coupling (Scheme 21) of an activated olefin and a diene is possible in acceptable yields if the half-wave potentials for the two components do not differ by more than 0.6 V; radical–radical coupling of the corresponding two radical-ions is the probable mechanism in this case.[174]

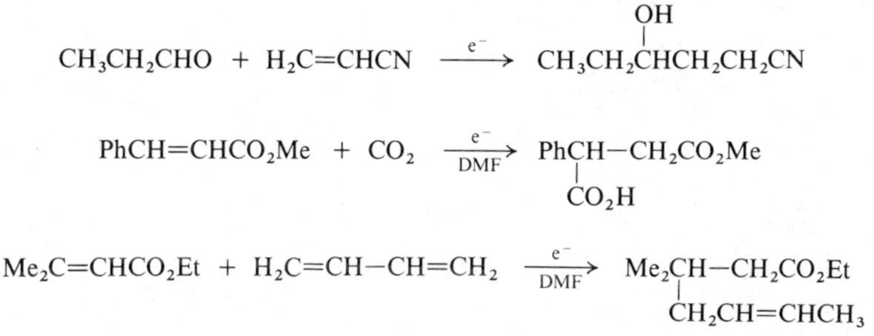

$$CH_3CH_2CHO + H_2C{=}CHCN \xrightarrow{e^-} CH_3CH_2\overset{OH}{\underset{|}{C}}HCH_2CH_2CN$$

$$PhCH{=}CHCO_2Me + CO_2 \xrightarrow[DMF]{e^-} PhCH{-}CH_2CO_2Me$$
$$\phantom{PhCH{=}CHCO_2Me + CO_2 \xrightarrow[DMF]{e^-} Ph}|\phantom{CH{-}CH_2CO_2Me}$$
$$\phantom{PhCH{=}CHCO_2Me + CO_2 \xrightarrow[DMF]{e^-} Ph}CO_2H$$

$$Me_2C{=}CHCO_2Et + H_2C{=}CH{-}CH{=}CH_2 \xrightarrow[DMF]{e^-} Me_2CH{-}CH_2CO_2Et$$
$$|$$
$$CH_2CH{=}CHCH_3$$

Scheme 21

The kinetics of hydrodimerization of 4-methylbenzalmalonodinitrile[175] in DMF indicate that the coupling step is a bimolecular radical–radical reaction, with $k = 5.8 \times 10^6$ dm^3 mol^{-1} s^{-1}. Attempts have been made to improve the economics of electro-organic reactions by pairing an anodic with a cathodic reaction; the cathodic reaction chosen has been a hydrodimerization process.[176]

Oxygen- (as Further Functional Groups), Sulphur-, and Selenium-containing Compounds.—The radical-anion of *cis*-thioindigo isomerizes to the *trans*-form and dimerizes much more rapidly than the *trans*-radical-anion. The rate of cross-coupling between these radical-anions and carbon dioxide, acrylonitrile, and cinnamonitrile has been measured.[177] Reduction of 2,6-diphenylpyrylium salts leads

[171] P. H. Zoutendam and P. T. Kissinger, *J. Org. Chem.*, 1979, **44**, 758.
[172] N. L. Askerov, S. I. Mekhtiev, V. M. Mamedova, and A. P. Tomilov, *Zh. Prikl. Khim.*, 1978, **51**, 1173; *Azerb. Khim. Zh.*, 1978, 121; T. Shono, H. Ohmizu, S. Kawakami, and H. Sugiyama, *Tetrahedron Lett.*, 1980, **21**, 5029.
[173] E. Lamy, L. Nadjo, and J. M. Saveant, *Nouv. J. Chim.*, 1979, **3**, 21; D. A. Tyssee, US P. 3 864 225, 1975 (*Chem. Abstr.*, 1975, **82**, 161 871).
[174] H. G. Thomas and F. Thönnessen, *Chem. Ber.*, 1979, **112**, 2786.
[175] L. Nadjo and J. M. Saveant, *J. Electroanal. Chem. Interfacial Electrochem.*, 1976, **73**, 163; 1978, **91**, 189.
[176] M. M. Baizer, R. C. Hallcher, and O. Deex, *Res. Discl.*, 1977, **159**, 5.
[177] L-S. Yeh and A. J. Bard, *J. Electroanal. Chem. Interfacial Electrochem.*, 1976, **70**, 157; 1977, **81**, 319.

to dimerization through the 4-position.[178]

The reduction of phenyl phosphates occurs with elimination of the phosphate group, probably by cleavage of the radical-ion; this procedure may be useful for the removal of hydroxyl substituents from a benzene ring (Scheme 22).[179] The radical-anions that are derived from benzyl esters also undergo cleavage of the carbon–oxygen bond to yield toluene. A study of this reaction showed no obvious correlation between the structure of the leaving group and the rate of cleavage for a series of 4-methoxybenzyl carboxylates.[180] A series of pyridylmethanols (19a) and (19b) undergo cleavage of the carbon–oxygen bond on reduction in aqueous acidic solution. Optically active 2-pyridyl derivatives are cleaved with overall retention of configuration, while the 4-pyridyl derivatives are almost racemized.[181] Phenyl-substituted epoxides are reductively cleaved at the benzyl ether function in DMF.[182] The radical-anion from the thioether (20) undergoes cleavage of the carbon–sulphur bond at a rate which is measurable by cyclic voltammetry.[183] The reduction of 1,2-diacetoxystilbene in anhydrous DMF affords diphenylacetylene, while acetoxystilbene is formed in moist DMF.[184]

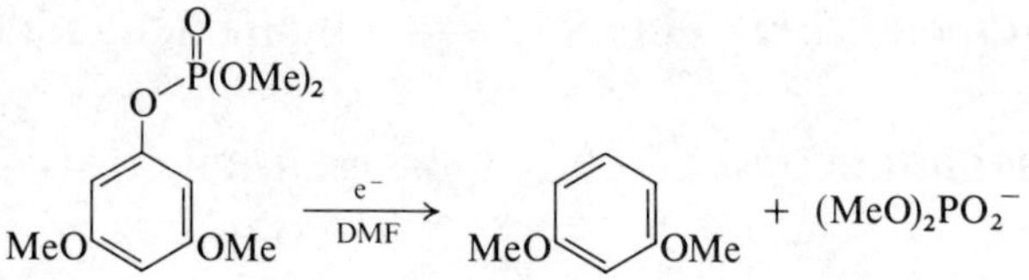

Scheme 22

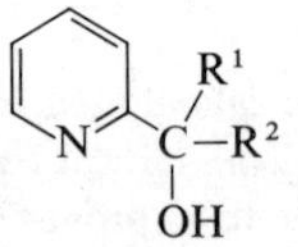

(19a)

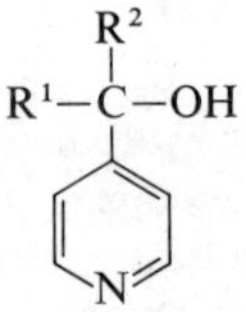

(19b)

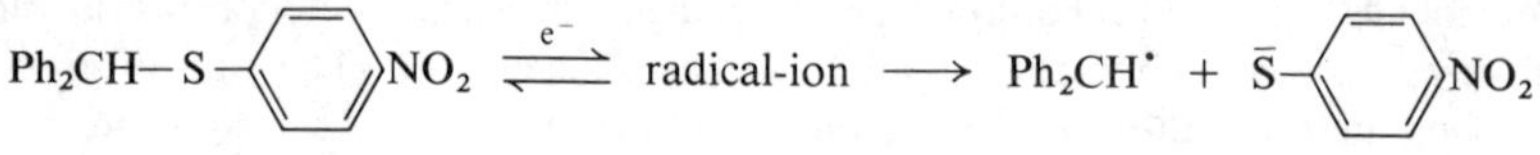

(20)

---

[178] F. Pragst and U. Seydewitz, J. Prakt. Chem., 1977, **319**, 952; F. Pragst, R. Ziebig, U. Seydewitz, and G. Driesel, Electrochim. Acta, 1980, **25**, 341.
[179] T. Shono, Y. Matsumura, K. Tsubata, and Y. Sugihara, J. Org. Chem., 1979, **44**, 4508.
[180] N. Berenjian and J. H. P. Utley, J. Chem. Soc., Chem. Commun., 1979, 550.
[181] T. Nonaka, T. Ota, and T. Fuchigami, Bull. Chem. Soc. Jpn., 1977, **50**, 2965.
[182] K. Boujlel and J. Simonet, Electrochim. Acta, 1979, **24**, 481.
[183] G. Farnia, M. G. Severin, G. Capobianco, and E. Vianello, J. Chem. Soc., Perkin Trans. 2, 1978, 1.
[184] C. Adams, N. M. Kambar, and J. H. P. Utley, J. Chem. Soc., Perkin Trans. 2, 1979, 1767.

Several reports indicate that an olefin is formed by the reductive 1,2-elimination of phenylthio- and either hydroxy- or acetoxy-groups (Scheme 23), and the reaction has been used by Shono as a key step in a novel synthesis of olefins.[185] A related 1,2-elimination of arylsulphinate and hydroxide ions is also observed (Scheme 23).[186] A 1,3-elimination of a phenylthio-group and an arylsulphonyl group (Scheme 23) has been used as the key step in a synthesis of cyclopropanes.[187] Alcohols can be smoothly transformed into the corresponding alkane by reduction of the alkyl methanesulphonate in DMF.[188] Reduction of dialkyl disulphides in DMF that is saturated with oxygen leads to the alkylsulphinate ion.[189]

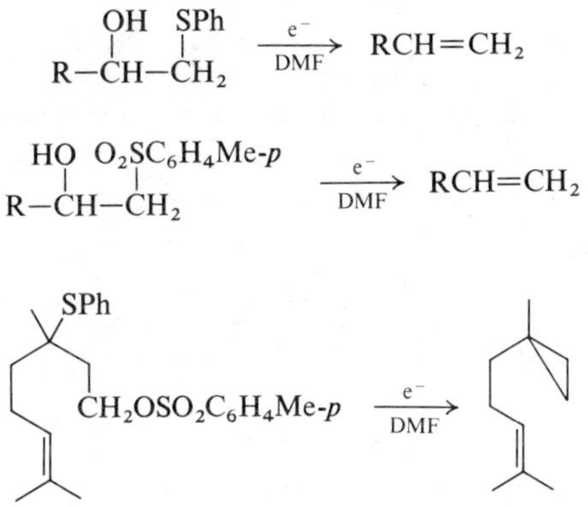

$$\underset{R-CH-CH_2}{\overset{OH\ \ SPh}{|\ \ \ \ \ |}} \xrightarrow[DMF]{e^-} RCH=CH_2$$

$$\underset{R-CH-CH_2}{\overset{HO\ \ O_2SC_6H_4Me\text{-}p}{|\ \ \ \ \ \ \ \ \ \ \ |}} \xrightarrow[DMF]{e^-} RCH=CH_2$$

Scheme 23

Sulphur dioxide gives the anion-radical on reduction in aprotic solvents. This reacts with alkylating agents to form the dialkyl sulphone.[190]

Alkyl aryl sulphones undergo cleavage of the carbon–sulphur bond on reduction in DMF. Further examples have been put forward to show that, where the aryl group is phenyl, exclusive cleavage of the alkyl–sulphur bond occurs.[191] Introduction of an *ortho*-substituent into the phenyl ring causes some cleavage of the aryl–sulphur bond, the extent of such cleavage increasing with increasing size of the *ortho*-group.[192] When the alkyl group is benzyl, cleavage of the alkyl–sulphur bond always predominates, even when an *ortho*-t-butyl substituent has been introduced onto the phenyl ring.[192] In contrast to methyl phenyl sulphone, methyl

---

[185] P. Martigny and J. Simonet, *J. Electroanal. Chem. Interfacial Electrochem.*, 1977, **81**, 407; T. Shono, Y. Matsumura, S. Kashimura, and H. Kyutoku, *Tetrahedron Lett.*, 1978, 2807; T. Shono, Y. Matsumura, and S. Kashimura, *ibid.*, 1980, **21**, 1545.
[186] T. Shono, Y. Matsumura, and S. Kashimura, *Chem. Lett.*, 1978, 69; S. Gambino, P. Martigny, G. Mousset, and J. Simonet, *J. Electroanal. Chem. Interfacial Electrochem.*, 1978, **90**, 105.
[187] T. Shono, Y. Matsumura, S. Kashimura, and H. Kyutoku, *Tetrahedron Lett.*, 1978, 1205.
[188] T. Shono, Y. Matsumura, K. Tsubata, and Y. Sugihara, *Tetrahedron Lett.*, 1979, 2157.
[189] C. Degrand and H. Lund, *Acta Chem. Scand.*, Ser. B, 1979, **33**, 512.
[190] B. Kastening and D. Knittel, Ger. Offen. 2 500 727 and 2 629 320 (*Chem. Abstr.*, 1976, **85**, 94 064 and 1978, **88**, 112 558); H. J. Wille, D. Knittel, B. Kastening, and J. Mergel, *J. Appl. Electrochem.*, 1980, **10**, 489.
[191] B. Lamm and K. Ankner, *Acta Chem. Scand.*, Ser. B, 1978, **32**, 31, 193.
[192] B. Lamm and K. Ankner, *Acta Chem. Scand.*, Ser. B, 1977, **31**, 375.

naphthyl sulphone undergoes cleavage of the aryl–sulphur bond, although the benzyl naphthyl sulphones are cleaved across the alkyl–sulphur bond, as expected.[193] Highly selective cleavage of the alkyl–sulphur bond in the disulphone (21) occurred only after indirect electrolysis in DMF, using anthracene as the electron-transfer agent.[194]

$$PhSO_2CH_2CH=CHCH_2CH_2SO_2Ph \xrightarrow[anthracene]{e^-,\ DMF} \begin{array}{c} MeCH=CHCH_2CH_2SO_2Ph \\ + \\ PhSO_2^- \end{array}$$
(21)

Diaryl sulphones are also known to undergo reductive cleavage of the carbon–sulphur bond in DMF. For one particular substrate, it has been possible to show that this reaction proceeds by cleavage of the radical-anion to give sulphinate ion and aryl radical by trapping the radical in an intramolecular reaction (Scheme 24).[195] Phenyl vinyl sulphone is cleaved at the vinyl–sulphur bond.[196] Reduction of methyl styryl sulphone and the corresponding sulphoxide in the presence of t-butyl chloride gave rise to t-butylated products.[197]

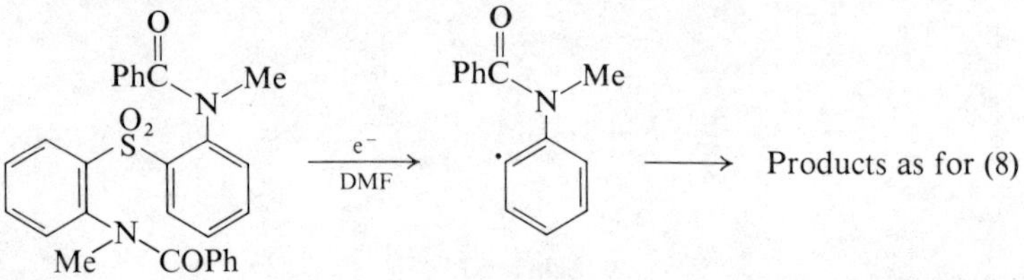

Scheme 24

Carbon disulphide and carbon diselenide[198] undergo reductive trimerization in DMF to form the related species (22) and (23). Alkyl aryl trithiocarbonates (24) are reduced in DMF to dimeric species, which are alkylated by an alkyl iodide to form the tetra(alkylthio)ethene.[199] Ethylene trithiocarbonate (26) undergoes a more

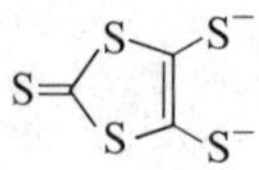

(22) 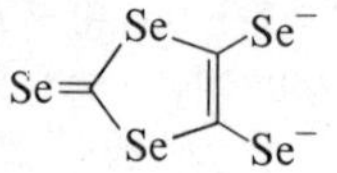   (23)

[193] B. Lamm and K. Ankner, *Acta Chem. Scand., Ser. B*, 1978, **32**, 264.
[194] J. Simonet and H. Lund, *Acta Chem. Scand., Ser. B*, 1977, **31**, 909.
[195] J. Grimshaw and J. Trocha-Grimshaw, *J. Chem. Soc., Perkin Trans. 1*, 1979, 779.
[196] K. Ankner, B. Lamm, and J. Simonet, *Acta Chem. Scand., Ser. B*, 1977, **31**, 742.
[197] H. Lund and C. Degrand, *C.R. Hebd. Seances Acad. Sci., Ser. C*, 1978, **287**, 535.
[198] W. P. Krug, A. N. Bloch, and D. O. Cowan, *J. Chem. Soc., Chem. Commun.*, 1977, 660; E. M. Engler, D. C. Green, and J. Q. Chambers, *ibid.*, 1976, 148.
[199] M. Falsig and H. Lund, *Acta Chem. Scand., Ser. B*, 1980, **34**, 545.
[200] F. J. Goodman and J. Q. Chambers, *J. Org. Chem.*, 1976, **41**, 626.
[201] M. Falsig and H. Lund, *Acta Chem. Scand., Ser. B*, 1980, **34**, 591.
[202] M. Falsig, H. Lund, L. Nadjo, and J. M. Saveant, *Nouv. J. Chim.*, 1980, **4**, 445.

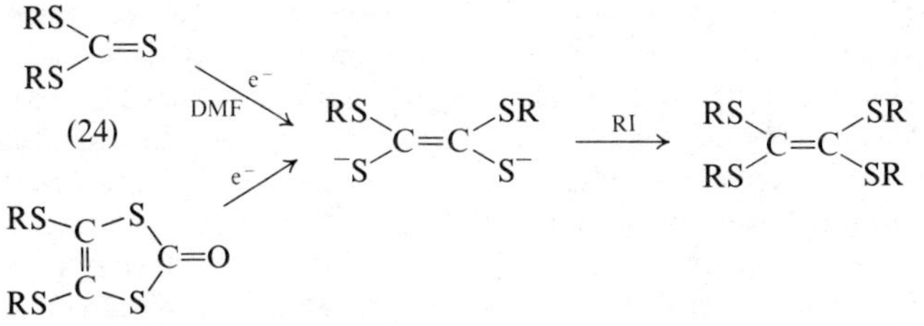

complex reaction, the first stage of which is the cleavage to ethylene and trithiocarbonate.[200] Further reactions occur between the trithiocarbonate ion and the starting radical-anion. The cyclic dithiocarbonates (25) are reduced with loss of carbon monoxide.[201] Kinetics of transfer of electrons and of the breaking of bonds in diphenyl trithiocarbonates have been measured.[202]

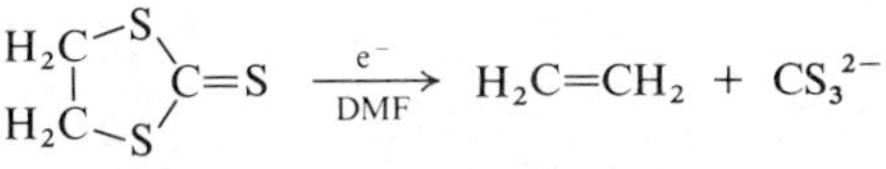

The dithioesters (27) undergo an interesting rearrangement on reduction in DMF. This is pictured as an intramolecular $S_{RN}1$ reaction (Scheme 25).[203]

Radical-anions that are derived from thio- and seleno-ketones have been characterized by e.s.r. spectroscopy.[204]

Nitro- and Nitroso-compounds.—Several examples[205,206] of the reduction of aliphatic nitro-compounds to hydroxylamines in acidic ethanol solution are

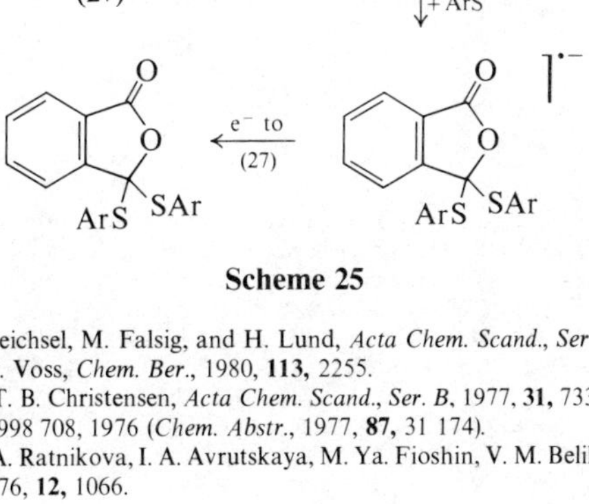

Scheme 25

[203] K. Praefcke, C. Weichsel, M. Falsig, and H. Lund, *Acta Chem. Scand.*, Ser. B, 1980, **34**, 403.
[204] C-P. Klages and J. Voss, *Chem. Ber.*, 1980, **113**, 2255.
[205] P. E. Iversen and T. B. Christensen, *Acta Chem. Scand.*, Ser. B, 1977, **31**, 733; P. Kabasakalian and S. Kalliney, US P. 3 998 708, 1976 (*Chem. Abstr.*, 1977, **87**, 31 174).
[206] V. T. Novikov, L. A. Ratnikova, I. A. Avrutskaya, M. Ya. Fioshin, V. M. Belikov, and K. K. Babievskii, *Elektrokhimiya*, 1976, **12**, 1066.

available. Further reduction under similar conditions affords the amine.[206,207] In anhydrous acetonitrile the intermediate nitroalkane anion-radical has a sufficiently long lifetime to undergo fragmentation reactions, and advantage has been taken of these processes to modify the structure of a natural sugar derivative (Scheme 26).[208] Aliphatic 1-nitro-2-acetoxy-compounds are reduced in DMF in a two-electron step to form the olefin with loss of nitrite and acetate ions in a 1,2-elimination.[209]

The counter-ion and the ionic strength are factors which influence the degree of ion-pairing for nitro-aryl radical-anions in aprotic solvents. Their effects on the half-wave potential and on the e.s.r. spectrum of these radical-anions have been examined in detail.[210] Acylation[50] and alkylation[211] of the nitrobenzene radical-anion in DMF proceeds to give the diacyl- or dialkyl-hydroxylamine.

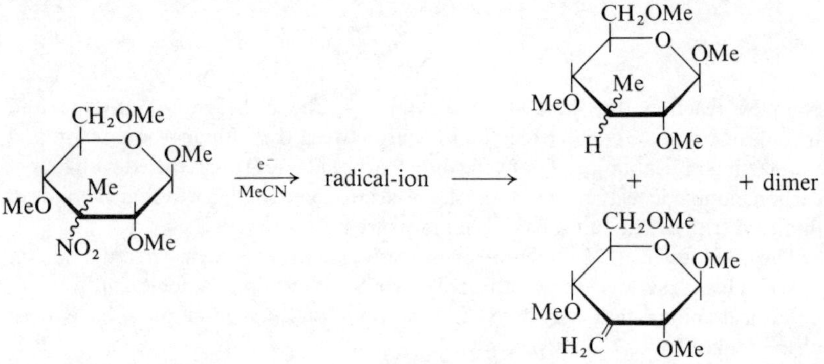

Scheme 26

An example of intramolecular abstraction of hydrogen by the nitro-aryl radical-anion is provided in the course of the reduction of (28), although the mechanism of the reaction has not been established in detail.[212] The hydroxylamine intermediate in the reduction of 4-isopropylnitrobenzene in concentrated sulphuric acid

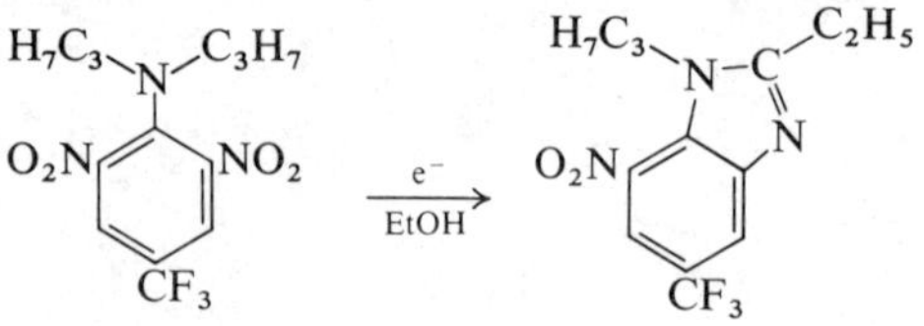

(28)

---

[207] V. T. Novikov, I. A. Avrutskaya, and M. Ya. Fioshin, *Elektrokhimiya*, 1976, **12**, 1486.
[208] A. K. Ganguly, P. Kabasakalian, J. Morton, O. Sarre, A. Wesertt, S. Kalliney, P. Mangiaracina, and A. Papaphilippou, *J. Chem. Soc., Chem. Commun.*, 1980, 56.
[209] A. Petsom and H. Lund, *Acta Chem. Scand., Ser. B*, 1980, **34**, 615.
[210] L. Lipsztajn, T. M. Krygowski, M. Smolarz, and W. Szelagowska, *Pol. J. Chem.*, 1978, **52**, 1069; W. R. Fawcett and A. Lasia, *J. Phys. Chem.*, 1978, **82**, 1114; B. G. Chaukan, W. R. Fawcett, and A. Lasia, *ibid.*, 1977, **81**, 1476.
[211] J. H. Wagenknecht, *J. Org. Chem.*, 1977, **42**, 1836.
[212] C. P. Keszthelyi, B. A. Kenney, P. J. Buras, L. M. Southwick, and G. H. Willis, *J. Electrochem. Soc.*, 1978, **125**, 714.

undergoes a rearrangement and dehydration at a rate that is competitive with that of its further reduction to 4-isopropylaniline (Scheme 27).[213] A wide range of examples have been given of the intramolecular condensation of the hydroxylamine that is obtained by reduction of a nitro-arene with an adjacent hydroxylamino-, carbonylamino-, or thiocyanato-group to form a heterocycle.[214]
Reduction of N-nitrodiphenylamine in alkaline aqueous methanol gives good yields of NN-diphenylhydroxylamine.[215]

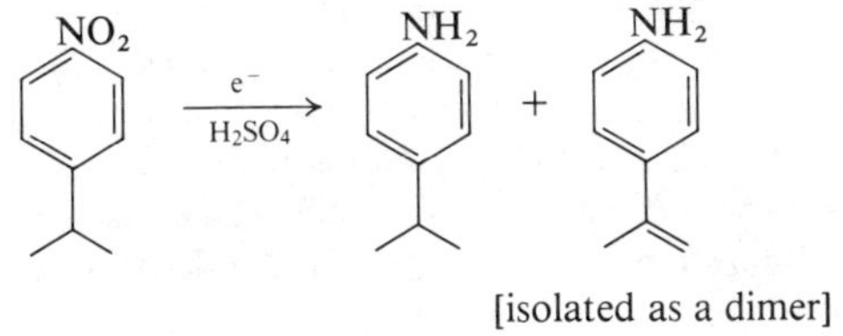

[isolated as a dimer]

Scheme 27

Other Nitrogen-containing Compounds.—The radical that is obtained from the reduction of some N-alkyl-pyridinium salts can be detected in aqueous solution. The 4-carbethoxy derivative shows reversible behaviour on cyclic voltammetry at a sweep rate of 2 V min^{-1} while the 4-carboxy-derivative is stable on the time-scale of cyclic voltammetry.[216] In the absence of a 4-substituent, the radicals rapidly dimerize. Radicals derived from 1-alkyl-pyridinium[217] and 1-benzyl-3-carbamoyl-pyridinium[218] salts have been shown to dimerize through the 4-position. Reduction of the pyridinium salt (29) gives first the 4,4'-dimer (by kinetic control) and ultimately the 4,2'-dimer (by thermodynamic control of the dimerization step).[219] Examples are available of the reduction of pyridines and pyridinium salts to the piperidine in acidic aqueous media.[220]

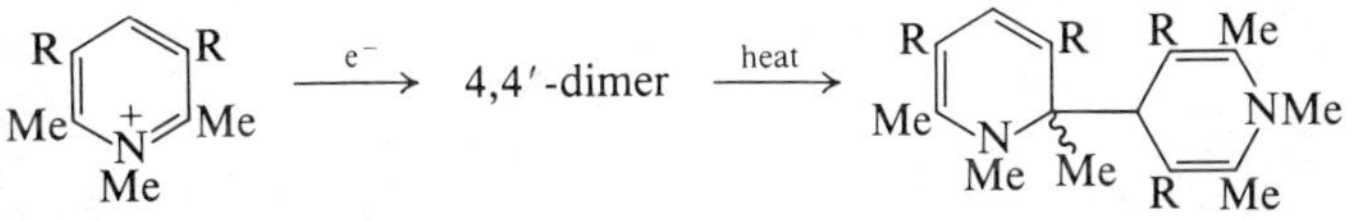

(29) R = CO$_2$Et

[213] A. Petsom and H. Lund, *Acta Chem. Scand.*, Ser. B, 1980, **34**, 693.
[214] J. Hlavaty, J. Volke, and O. Manousek, *Collect. Czech. Chem. Commun.*, 1975, **40**, 3751; Y. Mugnier and E. Laviron, *Bull. Soc. Chim. Fr.*, Part 2, 1978, 39; *Bull. Soc. Chim. Fr.*, 1976, 1496; D. Bernard, Y. Mugnier, G. Tainturier, and E. Laviron, *ibid.*, 1975, 2364; R. Hazard and A. Tallec, *ibid.*, 1976, 433.
[215] G. M. Sarsenbaeva, O. N. Novikova, I. A. Avrutskaya, and M. Ya. Fioshin, *Elektrokhimiya*, 1977, **13**, 581.
[216] V. Volkeova, J. Klima, and J. Volke, *Electrochim. Acta*, 1978, **23**, 1215; M. Mohamed, S. U. Sheikh, M. Iqbal, R. Ahmed, M. Razaq, and A. Y. Khan, *J. Electroanal. Chem. Interfacial Electrochem.*, 1978, **89**, 431.
[217] R. Raghavan and R. T. Iwamoto, *J. Electroanal. Chem. Interfacial Electrochem.*, 1978, **92**, 101.
[218] M. Micheletti, L. F. Franco, V. Carelli, A. Arnone, I. Carelli, and M. E. Cardinali, *J. Org. Chem.*, 1978, **43**, 3420.
[219] F. T. McNamara, J. W. Nieft, J. F. Ambrose, and E. S. Hugser, *J. Org. Chem.*, 1977, **42**, 988.
[220] M. Ferles, O. Kocian, M. Lebl, J. Lovy, S. Radl, A. Silhankova, and P. Stern, *Collect. Czech. Chem. Commun.*, 1976, **41**, 598; Yu. N. Forostyan, E. I. Lyushina, V. M. Artemova, and V. G. Govorukha, *Elektrokhimiya*, 1976, **12**, 73.

Reactions between hetrocyclic radical-anions and nucleophiles such as acetic anhydride and alkyl halides have been mentioned as general reactions of radical-anions. Further such reactions include the carboxylation of quinoline and of phenanthrene by reduction in DMF that is saturated with carbon dioxide[221] and the adamantylation of heterocycles.[222]

The reductions of Schiffs bases[232] as solutions in aqueous ethanol and of acyclic radical-anion, which decomposes by abstraction of a proton from unreacted substrate to form a radical, which then dimerizes.[223] An X-ray crystal structure analysis has been carried out on the dimer from 2-hydroxy-4,6-dimethylpyrimidine.[224] 4,6-Dimethyl-2-phenylpyrimidine is reduced in the primary step to a dihydro-derivative, which can undergo further reactions, with the eventual formation of a pyrrole (Scheme 28).[225] In the absence of a 2-phenyl substituent, the reaction proceeds only to the dihydropyrimidine stage. A number of condensed nitrogen heterocycles have been reduced to dihydro-derivatives in DMF.[226]

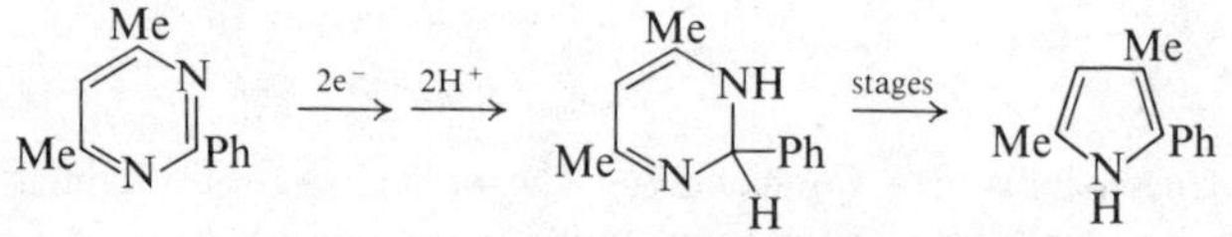

Scheme 28

A common feature of the reduction of aromatic triaza-systems is hydrogenation, with cleavage of a nitrogen–nitrogen bond. This is illustrated by the reduction of benzotriazole[227] and of benzo-1,2,3-triazin-4(3H)-ones (Scheme 29).[228] Reduction of 1,4-dimethyl-2,3-dioxo-1,2,3,4-tetrahydroquinoxaline to 1,4-dimethyl-2-oxo-1,2,3,4-tetrahydroquinoxaline in aqueous ethanol serves as a model for the reduction of related pteridines.[229]

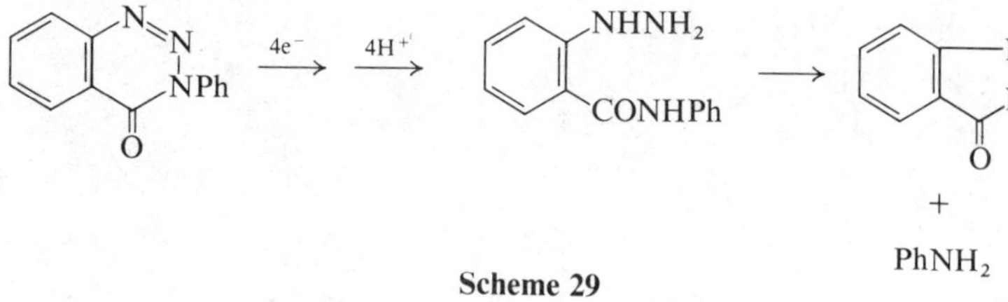

Scheme 29

---

[221] U. Hess, P. Fuchs, E. Jacob, and H. Lund, *Z. Chem.*, 1980, **20**, 64.
[222] U. Hess, D. Huhn, and H. Lund, *Acta Chem. Scand., Ser. B*, 1980, **34**, 413.
[223] T. Yao, T. Wasa, and S. Musha, *Nippon Kagaku Kaishi*, 1976, 704; B. Czochralska, H. Fritsche, and D. Shugar, *Z. Naturforsch., Teil. C*, 1977, **32**, 488; T. Wasa and P. J. Elving, *J. Electroanal. Chem. Interfacial Electrochem.*, 1978, **91**, 249; L. S. Tikhonova, Z. N. Timofeeva, K. L. Muravich-Aleksandr, and A. V. El'tsov, *Zh. Obshch. Khim.*, 1976, **46**, 881.
[224] B. Czochralska, D. Shugar, S. K. Arora, R. B. Bates, and R. S. Cutler, *J. Am. Chem. Soc.*, 1977, **99**, 2583.
[225] P. Martigny and H. Lund, *Acta Chem. Scand., Ser. B*, 1979, **33**, 575.
[226] J. Armand, K. Chekir, and J. Pinson, *C.R. Hebd. Seances Acad. Sci., Ser. C*, 1977, **284**, 391.
[227] M. Falsig and P. E. Iversen, *Acta Chem. Scand., Ser. B*, 1977, **31**, 15.
[228] H. H. Holst and H. Lund, *Acta Chem. Scand., Ser. B*, 1979, **33**, 233.
[229] R. Gottlieb and W. Pfleiderer, *Chem. Ber.*, 1978, **111**, 1753, 1763.

Triazenium ions (30) show reversible one-electron reduction in acetonitrile.[230] The analogous 5,7-diphenyl-2,3-dihydro-1,4-diazepinium ion (31) affords a much less stable radical on reduction in DMF, while the 6-phenyl compound (32) rapidly dimerizes on reduction in DMF and then undergoes further chemical reactions.[231]

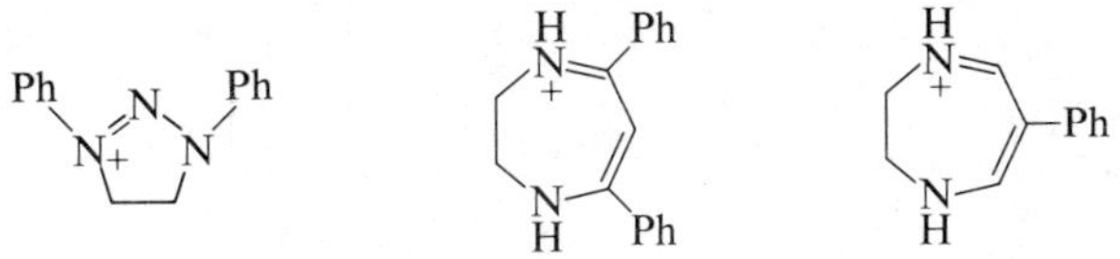

(30) (31) (32)

The reductions of Schiffs bases[232] as solutions in aqueous ethanol and of acyclic immonium salts[233] in DMF have been explored as methods for the synthesis of secondary and tertiary amines. Enantioselective electroreduction of Schiffs bases is achieved by using an optically active supporting electrolyte, but the optical yield is not high.[234] The reactive intermediates in the reduction of immonium salts have been trapped by inter- and intra-molecular reaction with alkylating agents. Examples[235] of the use of this reaction in the synthesis of alkaloid ring-systems are given in Scheme 30. Other examples of the same class of reaction are the reductions of benzalaniline[236] and of azobenzene[237] in the presence of α,ω-dialkylating agents

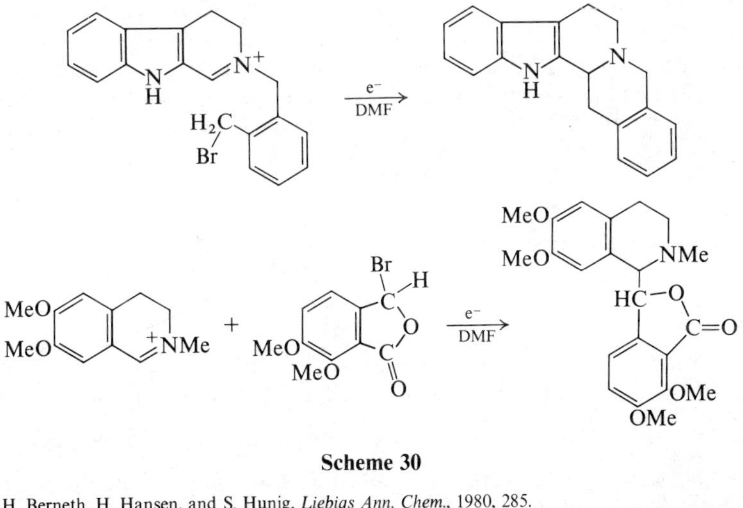

Scheme 30

[230] H. Berneth, H. Hansen, and S. Hunig, *Liebigs Ann. Chem.*, 1980, 285.
[231] D. Lloyd, C. A. Vincent, D. J. Walton, J. P. Declercq, G. Germain, and M. Van Meerssche, *J. Chem. Soc., Chem. Commun.*, 1978, 499; D. Lloyd, C. A. Vincent, and D. J. Walton, *J. Chem. Soc., Perkin Trans. 2*, 1980, 668.
[232] A. M. Osman, F. El-Cheikh, and Z. H. Khalill, *J. Appl. Chem. Biotechnol.*, 1976, **26**, 126.
[233] J. B. Kerr and P. E. Iversen, *Acta Chem. Scand., Ser. B*, 1976, **32**, 405.
[234] L. Horner and D. H. Skaletz, *Liebigs Ann. Chem.*, 1977, 1365.
[235] T. Shono, K. Yoshida, K. Ando, Y. Usui, and H. Hamaguchi, *Tetrahedron Lett.*, 1978, 4819; T. Shono, Y. Usui, T. Mizutani, and H. Hamaguchi, *ibid.*, 1980, **21**, 3073; T. Shono, Y. Usui, and H. Hamaguchi, *ibid.*, p. 1351.
[236] C. Degrand, C. Grosdemouge, and P. L. Compagnon, *Tetrahedron Lett.*, 1978, 3023.
[237] C. Degrand and D. Jacquin, *Tetrahedron Lett.*, 1978, 4955.

such as 1,3-dibromopropane and 4-bromobutanoyl chloride. Related also is the reductive cyclization of the di-imine (33).[238]

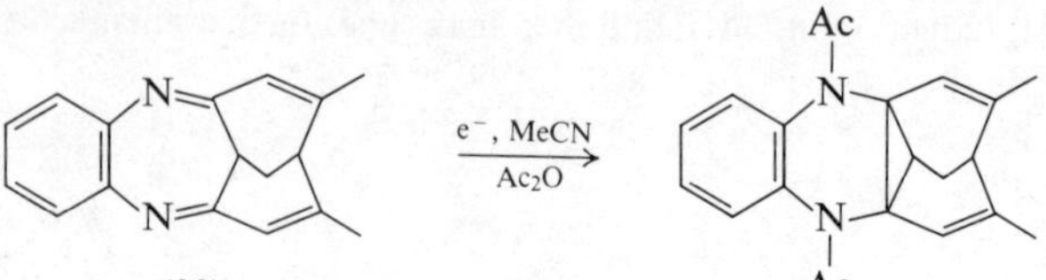

(33) → [e⁻, MeCN / Ac₂O] → Ac-substituted product

In aprotic solvents, the radical-anion of benzonitrile decomposes by a slow first-order process ($k = 5 \times 10^{-3}$ s^{-1}) to cyanide ion and a phenyl radical.[239] Examples of the satisfactory reduction of nitriles to primary amines in aqueous hydrochloric acid at a platinum-coated carbon electrode have been given.[240]

Reductive cleavage of the benzyl carbon–nitrogen bond occurs with 4-nitrobenzyltrimethylammonium salts. Benzyl radicals that are so formed dimerize to give 4,4′-dinitrobibenzyl.[241] The nitrogen–nitrogen bond in a wide range of hydrazine derivatives is reductively cleaved in methanol–DMF.[242] The nitrogen–oxygen bond in various hydroxylamine derivatives is also cleaved by reduction in methanol.[243] Hydroxylamine itself is reduced by titanium(III), which is generated electrochemically from a catalytic amount of titanium(IV) in the presence of phosphoric acid. The process generates $NH_2^{\cdot}$ radicals, which can be added to olefins. Thus maleic acid is converted into aspartic acid in one step.[244]

The sulphur–nitrogen bond in toluene-4-sulphonamides is cleaved on reduction in DMF, and this constitutes a method for removal of the sulphonylamine protecting-group.[245] Reduction of NN-dichloro-compounds generates a nitrene species which can be detected by its insertion reaction with dioxan[246] (Scheme 31).

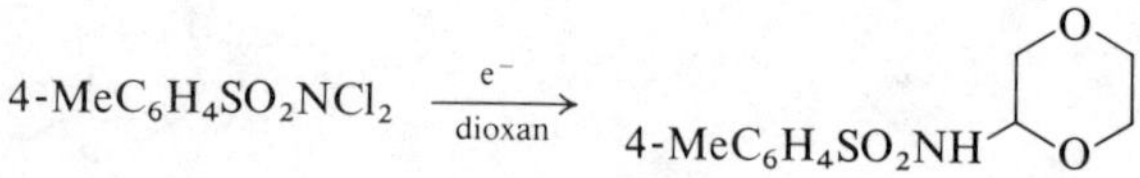

$$4\text{-MeC}_6\text{H}_4\text{SO}_2\text{NCl}_2 \xrightarrow[\text{dioxan}]{e^-} 4\text{-MeC}_6\text{H}_4\text{SO}_2\text{NH-(dioxanyl)}$$

Scheme 31

[238] J. M. Mellor, B. S. Pons, and J. H. A. Stibbard, *J. Chem. Soc., Chem. Commun.*, 1979, 761; *J. Chem. Soc., Perkin Trans. 1*, 1981, 3097.
[239] A. M. Romanin, A. Gennaro, and E. Vianello, *J. Electroanal. Chem. Interfacial Electrochem.*, 1978, **88**, 175.
[240] V. Krishnan, K. Raghupathy, and H. V. K. Udupa, *J. Electroanal. Chem. Interfacial Electrochem.*, 1978, **88**, 433; *J. Appl. Electrochem.*, 1978, **8**, 169.
[241] A. Desbene-Monvernay, J. E. Dubois, P. Clauce, and P. C. Lacaze, *J. Electroanal. Chem. Interfacial Electrochem.*, 1977, **81**, 377.
[242] L. Horner and M. Jordan, *Liebigs Ann. Chem.*, 1978, 1505.
[243] J. Volke, V. Volkeova, and H. Oelschläger, *Electrochim. Acta*, 1980, **25**, 1177; L. Horner and M. Jordan, *Liebigs Ann. Chem.*, 1978, 1518; I. G. Markova, M. K. Polievktov, and S. D. Sokolov, *Zh. Obshch. Khim.*, 1976, **46**, 398.
[244] G. Farnia, G. Sandona, and E. Vianello, *J. Electroanal. Chem. Interfacial Electrochem.*, 1978, **88**, 147.
[245] R. Kossai, G. Jeminet, and J. Simonet, *Electrochim. Acta*, 1977, **22**, 1395; S. P. Singer and K. B. Sharpless, *J. Org. Chem.*, 1978, **43**, 1448.
[246] T. Fuchigami and T. Nonaka, *Chem. Lett.*, 1977, 1087; T. Fuchigami, T. Nonaka, and K. Iwata, *J. Chem. Soc., Chem. Commun.*, 1976, 951.

Phosphorus- and Arsenic-containing Compounds.—A 1,2-disubstituted ethylene, together with 1,2- and 1,3,5-substituted derivatives of benzene where the substituent is the trialkyl- or triaryl-phosphonium ion, show reversible one-electron addition in aprotic solvents.[247] More commonly, phosphonium salts undergo cleavage of the carbon–phosphorus bond on reduction. The ratio of cleavage of alkyl–phosphorus to that of aryl–phosphorus bonds in alkyltriphenylphosphonium salts has been determined under a variety of experimental conditions, in methanol. The ratio depends on the electrode potential and on the temperature; added acetic acid favours the formation of methyltriphenylphosphine.[248] Benzyl–phosphorus and allyl–phosphorus bonds are preferentially cleaved, and the initial product is the benzyl or allyl radical. There is potential for synthesis in this process if conditions which promote the dimerization of radicals can be found. Yields of bibenzyl from this process are acceptable, and they are highest at high concentrations of the salt in DMF.[249] Dimerization of allyl radical is promoted at a high current density and by the use of an aluminium cathode.[250]

Reduction of the carbon–chlorine bond has been used to generate a phosphonate ylide which will undergo the Horner–Emmons reaction *in situ* with an added carbonyl compound (Scheme 32).[251] A three-membered-ring intermediate has been postulated in the conversion of α,α'-dibromophosphinates into olefins (Scheme 32).[252]

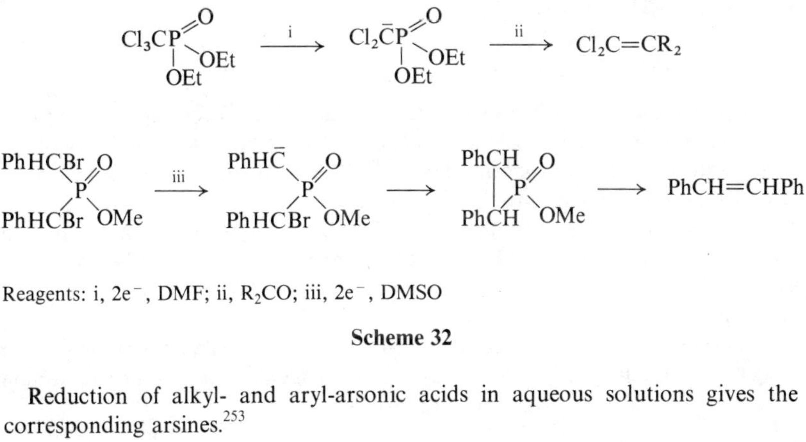

$$Cl_3CP\underset{OEt}{\overset{O}{\diagdown}}OEt \xrightarrow{i} Cl_2\bar{C}P\underset{OEt}{\overset{O}{\diagdown}}OEt \xrightarrow{ii} Cl_2C=CR_2$$

$$\underset{PhHCBr}{PhHCBr}\overset{O}{\diagdown}P\overset{O}{\diagdown}OMe \xrightarrow{iii} \underset{PhHCBr}{PhH\bar{C}}\overset{O}{\diagdown}P\overset{O}{\diagdown}OMe \longrightarrow \underset{PhCH}{PhCH}\overset{O}{\diagdown}P\overset{O}{\diagdown}OMe \longrightarrow PhCH=CHPh$$

Reagents: i, 2e⁻, DMF; ii, R_2CO; iii, 2e⁻, DMSO

Scheme 32

Reduction of alkyl- and aryl-arsonic acids in aqueous solutions gives the corresponding arsines.[253]

---

[247] E. A. Berdnikov, F. R. Tantasheva, V. I. Morozov, A. V. Il'yasov, and A. A. Vafina, *Izv. Akad. Nauk SSSR, Ser. Khim.*, 1977, 803; C. K. White and R. D. Riehe, *J. Org. Chem.*, 1978, **43**, 4638.
[248] L. Horner and J. Röder, *Liebigs Ann. Chem.*, 1977, 2067; *Phosphorus Relat. Group V Elem.*, 1976, **6**, 147.
[249] J. M. Saveant and S. K. Binh, *J. Electroanal. Chem. Interfacial Electrochem.*, 1978, **88**, 43.
[250] J. H. P. Utley and A. Webber, *J. Chem. Soc., Perkin Trans. 1*, 1980, 1154.
[251] K. Karrenbrock, H. J. Schafer, and I. Langer, *Tetrahedron Lett.*, 1979, 2915.
[252] A. J. Fry and L. L. Chung, *Tetrahedron Lett.*, 1976, 645.
[253] R. K. Elton and W. E. Geiger, *Anal. Chem.*, 1978, **50**, 712; A. Tzschach, H. Matschiner, and H. Biering, *Z. Anorg. Allg. Chem.*, 1977, **436**, 60.

PART II: Oxidations by *D. Pletcher*

1 Aliphatic Compounds

Carboxylic Acids.—Although the Kolbe reaction was the first organic electrode process to be described in the literature, the anodic oxidation of carboxylic acids continues to attract attention. The formation of radicals or carbenium ions by decarboxylation is recognized as having considerable utility in the development of improved syntheses of large organic molecules with biological activity or that are of physiological interest. Moreover, the mechanism of such reactions remains uncertain; there continues to be debate as to whether radical intermediates are adsorbed on the anode surface or are free in solution, and, under many conditions, on the relative importance of radical and cation intermediates.

Schäfer and co-workers[1—4] have re-investigated the oxidation, at a platinum anode, of a solution that contains two carboxylic acids to form the mixed Kolbe dimer; they have demonstrated the application of this oxidation in the synthesis of pheromones. They studied[1] reaction (1) in some detail, demonstrated that the

$$C_4H_9\text{-}CH=CH\text{-}CH_2CH_2COO^- + MeO_2C(CH_2)_3COO^- \xrightarrow[-2CO_2]{-2e^-} C_4H_9\text{-}CH=CH\text{-}CH_2(CH_2)_4CO_2Me \quad (1)$$

essential stereochemistry is unchanged by the electrode process, and showed that the yield of mixed dimer depends on current density, the % neutralization of the acids, and their relative concentrations. The yield can be as high as 55%. The product can be converted into the pheromone looplure by hydrolysis, reduction of the free acid to alcohol with lithium aluminium hydride, and acetylation; the overall yield of looplure from readily available compounds is 18%. Other reactions reported[2—4] include reactions (2)—(6), and the yields in alkaline methanol and at a platinum anode were between 42 and 66%.

Knolle and Schäfer[5] have also sought to optimize the yield of Kolbe dimer from the anodic oxidation of allylic carboxylic acids. They showed that the counter-ion

$$MeO_2CCH_2CH=CH\text{-}CH_2CH_2COO^- + Me(CH_2)_7COO^-$$

$$\downarrow -2e^- \quad\quad\quad\quad\quad\quad\quad\quad (2)$$

$$Me(CH_2)_8CH=CH\text{-}CH_2CH_2CO_2Me$$

[1] W. Seidel, J. Knolle, and H. J. Schäfer, *Chem. Ber.*, 1977, **110**, 3544.
[2] H. Klünenberg and H. J. Schäfer, *Angew. Chem., Int. Ed. Engl.*, 1978, **17**, 47.
[3] W. Seidel and H. J. Schäfer, *Chem. Ber.*, 1980, **113**, 451.
[4] W. Seidel and H. J. Schäfer, *Chem. Ber.*, 1980, **113**, 3898.
[5] J. Knolle and H. J. Schäfer, *Electrochim. Acta*, 1978, **23**, 5.

$$\begin{array}{c} Me(CH_2)_8\overset{\frown}{CH_2} \quad \overset{\frown}{CH_2}CH_2COO^- \\ + \\ Pr^i(CH_2)_2COO^- \end{array} \xrightarrow{-2e^-} Me(CH_2)_8\overset{\frown}{CH_2} \quad \overset{\frown}{CH_2}(CH_2)_3Pr^i \quad (3)$$

$$\begin{array}{c} C_{18}H_{37}\underset{Me}{\overset{|}{C}}HCH_2COO^- \\ + \\ MeO_2C(CH_2)_3COO^- \end{array} \xrightarrow{-2e^-} C_{18}H_{37}\underset{Me}{\overset{|}{C}}H(CH_2)_4CO_2Me \quad (4)$$

$$\begin{array}{c} C_{18}H_{37}\underset{Me}{\overset{|}{C}}H(CH_2)_4COO^- \\ + \\ MeCO\underset{Me}{\overset{|}{C}}H(CH_2)_3COO^- \end{array} \xrightarrow{-2e^-} C_{18}H_{37}\underset{Me}{\overset{|}{C}}H(CH_2)_7\underset{Me}{\overset{|}{C}}HCOMe \quad (5)$$

$$\begin{array}{c} Me(CH_2)_xC\equiv C(CH_2)_3COO^- \\ (x = 1\text{—}3) \\ + \\ MeO_2C(CH_2)_yCOO^- \\ (y = 2\text{—}6) \end{array} \xrightarrow{-2e^-} Me(CH_2)_xC\equiv C(CH_2)_{y+3}CO_2Me \quad (6)$$

and the current density were key parameters in preventing tarring of the electrode surface; using tri-n-butylamine to half-neutralize the acid, and a high current density, they could obtain 44% of hexa-1,5-diene from vinylacetic acid. Utley and Holman[6] have reported the addition of methyl and trifluoromethyl radicals, formed by the Kolbe reaction, to pyridine; nuclear substitution occurred in acid solution but the distribution of isomers depends on the electrolysis medium and the radical. With methyl radical, only α- and γ-picolines are formed, but α-, β-, and γ-substitution takes place with $CF_3^·$. The addition of trifluoromethyl radicals to alkenes, alkynes, and activated olefins has also been investigated,[7] but a mixture of a number of products is always obtained.

A synthetic application of the di-decarboxylation of a 1,2-dicarboxylic acid to form an olefin has been reported by Kopecky and Lau.[8] The yield is 45% for the formation of the olefin (1).

The above reactions are all one-electron oxidations, leading to an alkyl radical by decarboxylation. Depending on the structure of the carboxylic acid, and also on the electrolysis conditions, carboxylates can sometimes be oxidized in an overall two-electron process, and it is the corresponding carbenium ion that determines the

---

[6] J. H. P. Utley and R. J. Holman, *Electrochim. Acta*, 1976, **21**, 987.
[7] C. J. Brookes, P. L. Coe, A. E. Pedler, and J. C. Tatlow, *J. Chem. Soc., Perkin Trans. 1*, 1978, 202.
[8] K. R. Kopecky and M-P. Lau, *J. Org. Chem.*, 1978, **43**, 525.

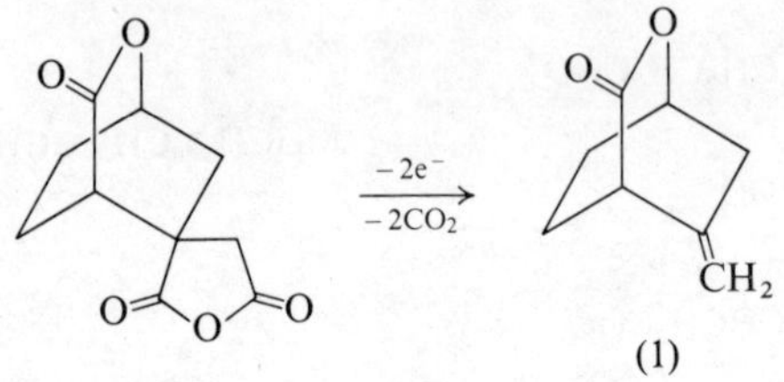

spectrum of products. Such reactions are also finding a number of applications in the synthesis of natural products.

The two-electron oxidation path provides a convenient way of replacing a carboxylic acid group by a methoxy- or ester group in high yield. Such reactions are well illustrated by the reactions shown in Scheme 1,[9–11] and several similar reactions have been reported.[12–15] The lifetime of the cationic intermediate can be sufficient for skeletal rearrangement to occur prior to trapping; depending upon structure and the reaction conditions, the rearranged product can be a major or a minor component. The acetate or methyl ether that is formed in the electrolysis is not always stable to hydrolysis, and this provides a route to cyclic compounds. For example, such reactions provide a route for the conversion of aspartic acids into

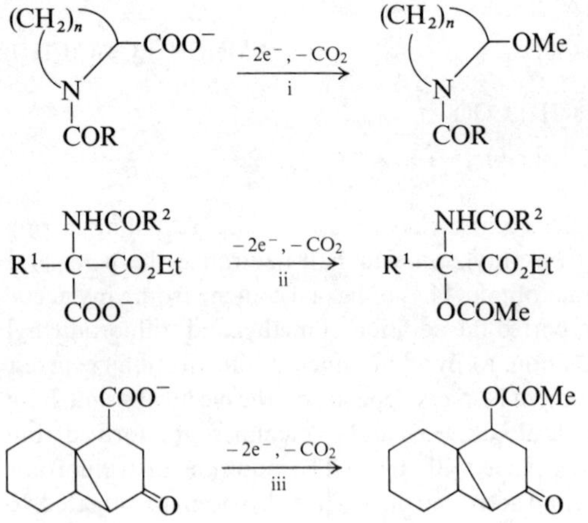

Conditions: i, MeOH, MeO$^-$, graphite electrodes; ii, HOAc, NaOAc, graphite electrodes; iii, HOAc, Bu^{t}OH, Et$_3$N

Scheme 1

[9] T. Iwasaki, H. Horikawa, K. Matsumoto, and M. Miyoshi, *J. Org. Chem.*, 1979, **44**, 1552.
[10] T. Iwasaki, H. Horikawa, K. Matsumoto, and M. Miyoshi, *J. Org. Chem.*, 1977, **42**, 2419.
[11] S. Torii, T. Okamoto, G. Tanida, H. Hino, and Y. Kitsuya, *J. Org. Chem.*, 1976, **41**, 166.
[12] S. Torii, T. Inokuchi, K. Mizuguchi, and M. Yamazaki, *J. Org. Chem.*, 1979, **44**, 2303.
[13] F. M. Banda and R. Brettle, *J. Chem. Soc., Perkin Trans. 1*, 1977, 1773.
[14] D. Lelandais, C. Bacquet, and J. Einhorn, *J. Chem. Soc., Chem. Commun.*, 1978, 194.
[15] T. Akiyama, T. Fujii, H. Ishiwara, T. Imagawa, and M. Kawanisi, *Tetrahedron Lett.*, 1978, 2165.

uracil[16] and the final step in the sequence for the synthesis of cis-jasmone[17] shown in Scheme 2.

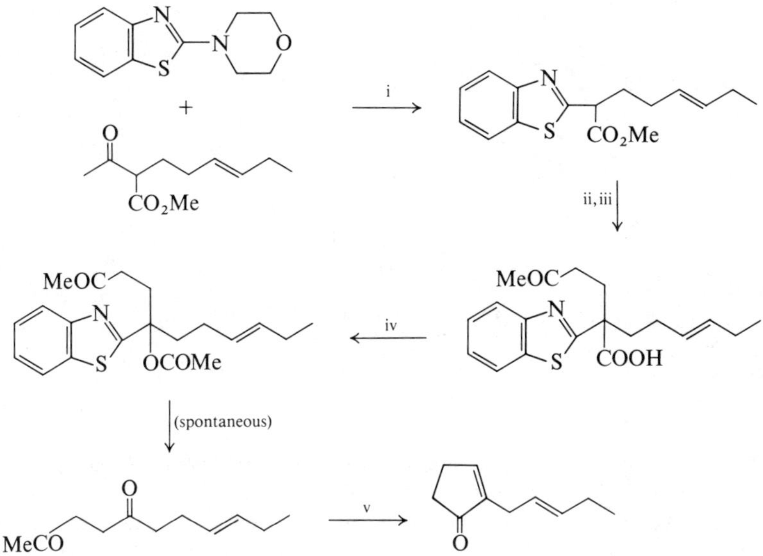

Reagents: i, MeOH, reflux; ii, MeCOCH=CH$_2$, acetone, K$_2$CO$_3$, reflux for 12 hours; iii, K$_2$CO$_3$, MeOH, H$_2$O, at 45 °C, for 14 hours; iv, electrolysis, HOAc, Bu^{t}OH, Et$_3$N; v, aq. KOH

Scheme 2

The two-electron oxidation of carboxylic acids also offers a route to the introduction of double-bonds into molecules by loss of a proton from the carbenium ion. Straightforward examples of this reaction are the decarboxylation of dihydroaromatic acids to aromatic hydrocarbons[18] and the final step in the synthesis of some substituted pyrroles.[19] The yields of the latter reactions were 90—95%, but only when the medium was THF–water. Otherwise the further oxidation of the product markedly decreased the yield. Bobbitt and co-workers[20,21] have used the reaction to introduce unsaturation into substituted heterocycles and discussed both the site of the initial loss of an electron and the relationship of the mechanism of this anode reaction to that for similar reactions in Nature. α-Methylene-γ-lactones, e.g. (2), this being a structure which features in several biologically active molecules, have also been synthesized by a decarboxylation reaction.[22] The yield for this reaction is, however, only 30—35%, but if the substrate

---

[16] T. Iwasaki, H. Horikawa, K. Matsumoto, and M. Miyoshi, *Tetrahedron Lett.*, 1978, 4799.
[17] S. Torii, H. Tanaka, Y. Kobayasi, J. Nokami, and M. Kawata, *Bull. Chem. Soc. Jpn.*, 1979, **52**, 1553.
[18] J. Slobbe, *J. Chem. Soc., Chem. Commun.*, 1977, 82.
[19] H. Horikawa, T. Iwasaki, K. Matsumoto, and M. Miyoshi, *J. Org. Chem.*, 1978, **43**, 335.
[20] J. M. Bobbitt and T. Y. Cheng, *J. Org. Chem.*, 1976, **41**, 443.
[21] J. M. Bobbitt and J. P. Willis, *J. Org. Chem.*, 1980, **43**, 1978.
[22] S. Torii, T. Okamoto, and T. Oida, *J. Org. Chem.*, 1976, **43**, 2294.

is a related molecule (3), containing a sulphur atom to provide a site for facile initial loss of an electron, the yield rises substantially to 75—90 % (note, however, that the current yield is very low).

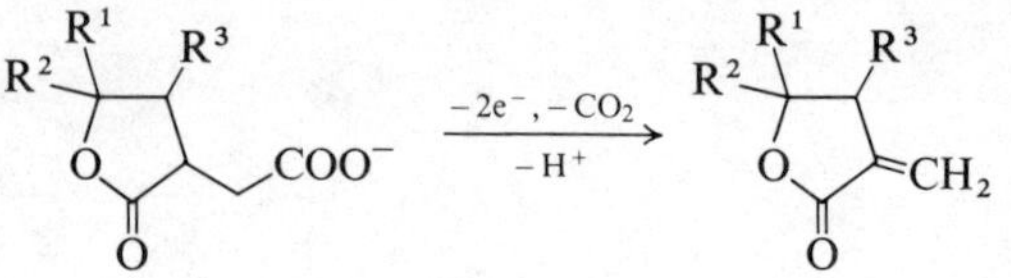

$$\text{(2)}$$

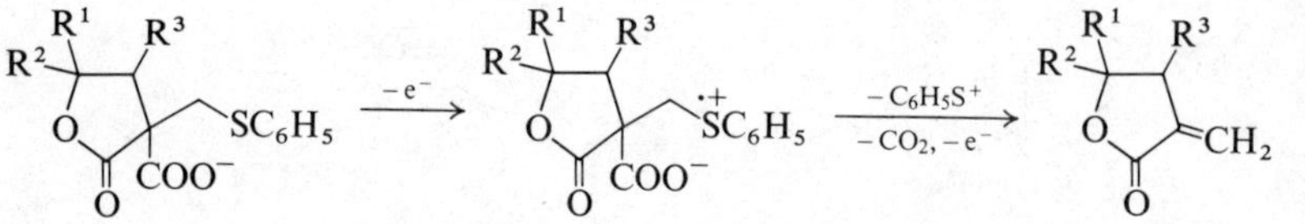

$$\text{(3)}$$

These electrolyses were also interesting because they were carried out with continuous extraction, by using an emulsion of aqueous pyridine and benzene–ether in the cell. Again, since carbenium ions are intermediates, skeletal rearrangements can occur before protons are lost. Hence the oxidation of β-hydroxycarboxylic acids occurs by the pathway[23] shown in Scheme 3 at carbon in acetonitrile–triethylamine. It was shown that the migrationary aptitude is similar for all simple alkyl groups, but benzylic, cyclopropyl, and (particularly) olefinic groups migrate preferentially. This latter observation was used in the synthesis of a large ring ketone, dl-muscone; the rearrangement of the cation is, in this case, a ring-expansion. Proton is not the only cationic leaving group in the formation of an olefin from the oxidation of carboxylates, and Shono et al.[24] have reported that the oxidation of acids with a trimethylsilyl group in the β-position leads to a good yield of terminal olefin.

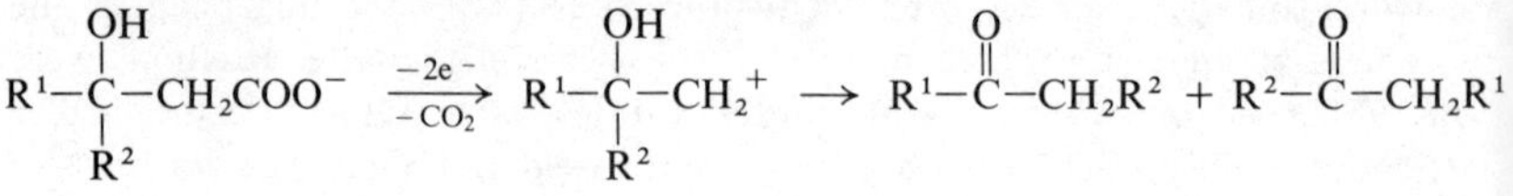

$$R^1-\underset{R^2}{\underset{|}{\overset{OH}{\overset{|}{C}}}}-CH_2COO^- \xrightarrow[-CO_2]{-2e^-} R^1-\underset{R^2}{\underset{|}{\overset{OH}{\overset{|}{C}}}}-CH_2^+ \rightarrow R^1-\overset{O}{\overset{\|}{C}}-CH_2R^2 + R^2-\overset{O}{\overset{\|}{C}}-CH_2R^1$$

Scheme 3

Nokami et al.[25] have reported the anodic di-decarboxylation of derivatives of malonic acid, e.g. (4), in a reaction which gives yields in the range 55—63 % and which provides a route to complex ketones. The optimum conditions for electrolysis varied with the structure of the substrate, but continuous extraction in the cell could again be used to advantage.

---

[23] T. Shono, J. Hayashi, H. Omoto, and Y. Matsumura, *Tetrahedron Lett.*, 1977, 2667.
[24] T. Shono, H. Omizu, and N. Kise, *Chem. Lett.*, 1980, 1517.
[25] J. Nokami, T. Yamamoto, M. Kawada, M. Izumi, N. Ochi, and R. Okawara, *Tetrahedron Lett.*, 1979, 1047.

$$\underset{\underset{(4)}{\text{MeCOCH}_2\text{CH}_2}}{\overset{n\text{-H}_{13}\text{C}_6}{\diagdown}}\!\!\!\underset{\diagup}{\overset{\diagup}{\text{C}}}\!\!\!\overset{\text{COO}^-}{\diagdown}\quad\xrightarrow[-2\text{CO}_2]{-4e^-,\,+\text{H}_2\text{O}}\quad\underset{\text{MeCOCH}_2\text{CH}_2}{\overset{n\text{-H}_{13}\text{C}_6}{\diagdown}}\!\!\!\text{C}=\text{O}$$

Russel[26] has reported the chronopotentiometry of a wide range of carboxylates as their Bu_4N^+ salts in acetonitrile. All give diffusion-controlled waves and dicarboxylic acids show two oxidation waves. The values of $E_{\frac{1}{4}}$ show a strong dependence on the structure of the carboxylate and, for example, trifluoroacetate, acetate, and malonate are oxidized at $+2.14$, $+1.23$, and $+0.82$ V vs Ag^+/Ag, respectively. Laurent and co-workers[27–29] have also studied oxidations in acetonitrile. They have used deuterium labelling to investigate the oxidation of n-butyrate[27,28] and also studied the oxidation of methacrylates[29] and of endo- and exo-trinorborn-5-ene-2-carboxylates.[30] The exo-compound gives both trinorborn-5-ene and trinortricycl-3-yl products, with a similar distribution to those found in homogeneous reactions, but the endo-compound gives only trinortricycl-3-yl products, and the authors suggest that adsorption of intermediates may lead to this difference.

As stated above, there is evidence both for and against adsorption of radical intermediate is put most strongly by Bewick and Brown.[34] They have used specular Nyberg, and Servin[31] have reported that complete racemization occurs during the mixed coupling of D-(−)-2-methyloctadecanoate and monomethylmalonate and compare this result with the expectation that adsorption of the radical would ensure at least partial retention of configuration. The conclusion that the radical is not adsorbed is also drawn by Utley et al.,[32] from a stereochemical argument. They examined the oxidation of cis- and of trans-4-substituted-cyclohexanecarboxylic acids and determined the stereochemistry of the dimers that were formed along with the alkanes, alkenes, and ethers. The ratio of stereoisomers was completely statistical. This was not the case with the corresponding cyclohexene compounds, and hence it was felt that adsorption of the cyclohexenyl radicals may occur.[33] With these acids, the spectrum of products shows a strong dependence on the concentration of the starting material. When the concentration of the acid was in the range 0.4—1.0 mol dm^{-3}, the major products were the methyl ethers (formed by two-electron oxidation), but the dimers (from one-electron oxidation) were the major products at higher concentrations. The case for adsorption of the radical intermediate is put most strongly by Bewick and Brown.[34] They have used specular reflectance spectroscopy to examine both the oxidation of acetate and of monoethylmalonate and to produce evidence for ranges of potential where a platinum anode is covered by oxide and by oxide + a monolayer of radicals, and,

---

[26] C. D. Russel, J. Electroanal. Chem. Interfacial Electrochem., 1976, **71**, 81.
[27] E. Laurent and M. Thomella, Tetrahedron Lett., 1976, 4727.
[28] E. Laurent, M. Thomella, B. Marquet, and U. Berger, J. Org. Chem., 1980, **45**, 4193.
[29] E. Laurent and M. Thomella, C.R. Hebd. Seances Acad. Sci., Ser. C, 1978, **282**, 441.
[30] E. Laurent and M. Thomella, Bull. Soc. Chim. Fr., 1977, 834.
[31] L. Eberson, K. Nyberg, and R. Servin, Acta Chem. Scand., Ser. B, 1978, **30**, 906.
[32] G. E. Hawkes, J. H. P. Utley, and G. B. Yates, J. Chem. Soc., Perkin Trans. 2, 1976, 1709.
[33] J. H. P. Utley and G. B. Yates, J. Chem. Soc., Perkin Trans. 2, 1978, 395.
[34] A. Bewick and D. J. Brown, Electrochim. Acta, 1976, **21**, 979.

where the Kolbe reaction takes place, by oxide + two layers of radicals. They also report spectra for the carbanion $\overline{C}H_2CO_2Et$ during pulse electrolysis.

Finally, Kraeutler and Bard[35] have shown that the oxidation of acetate in acetonitrile may be photo-assisted. Using an n-type titanium dioxide, illuminated by a 450 W xenon lamp, the oxidation was found to commence at -1.6 V vs Ag/Ag^+, which is 2.4 V negative to the potential that is essential at a platinum electrode. Indeed, the oxidation of acetate ion occurs so readily under illumination that a platinized titanium dioxide power will, under irradiation, catalyse the decomposition of an aqueous solution of acetate.[36] Surprisingly, however, the products are methane and hydrogen ($[CH_4]/[C_2H_6] = 11$), rather than ethane and hydrogen, although ethane is certainly the product from the photo-assisted electrolysis. The photochemical catalytic decomposition of aqueous carboxylates $RCOO^-$ at platinized TiO_2 to give RH as the major organic product has been generalized to longer-chain analogues.[37]

Alcohols and Aldehydes.—A procedure for the oxidation of primary alcohols to aldehydes which is general, selective, rapid and uses only cheap and non-toxic reagents remains a high priority for synthesis. Moreover, the oxidations of secondary alcohols to ketones and of primary alcohols to acids are not always as straightforward as the textbooks would have the reader believe. Hence a number of groups have investigated the electrochemical oxidation of alcohols.

Scholl, Lentsch, and Van de Mark[38] have surveyed a range of electrolysis conditions for the oxidation of n-butanol and found that the highest yield of n-butanol (77%) was obtained when the medium was the alcohol itself, containing $LiBF_4$. Neat alcohol was also found[39] to be the optimum medium for the conversion of secondary alcohols into ketones, although in this case $LiNO_3$ was considered to be a better electrolyte; it led to a higher yield of ketone and also to a lower oxidation potential (indicating an indirect oxidation mechanism via $NO_3^{\cdot}$). The conductivity of long-chain alcohols was, however, insufficient, but 9:1 acetonitrile:water was then a useful electrolysis medium. Using either the alcohols themselves or acetonitrile–water as the solvent, the yields of ketones were good except when the alcohol contained one or, worse, several tertiary C–H bonds. Such structures gave a mixture of products. White and Coleman[40] have examined the oxidation of alcohols that contain anhydrous hydrogen chloride. The sequence of reactions shown in Scheme 4 occurs with a good current efficiency and excellent selectivity under conditions where chlorine is generated at the anode. Methylene chloride can also be used as a solvent for the reaction without adverse effects on the performance.

Shono and co-workers[41—43] have investigated several indirect oxidations of alcohols. Using an aqueous KI medium, they demonstrated the conversion (in high yield) of primary alcohols into esters and of secondary alcohols into ketones; the

---

[35] B. Kraeutler and A. J. Bard, *J. Am. Chem. Soc.*, 1977, **99**, 7729.
[36] B. Kraeutler and A. J. Bard, *J. Am. Chem. Soc.*, 1978, **100**, 2239.
[37] B. Kraeutler and A. J. Bard, *J. Am. Chem. Soc.*, 1978, **100**, 5985.
[38] P. C. Scholl, S. E. Lentsch, and M. R. Van de Mark, *Tetrahedron*, 1976, **32**, 303.
[39] J. E. Leonard, P. C. Scholl, T. P. Steckel, S. E. Lentsch, and M. R. Van de Mark, *Tetrahedron Lett.*, 1980, **21**, 4695.
[40] D. A. White and J. P. Coleman, *J. Electrochem. Soc.*, 1978, **125**, 1401.
[41] T. Shono, Y. Matsumura, J. Hayashi, and M. Mizoguchi, *Tetrahedron Lett.*, 1979, 165.
[42] T. Shono, Y. Matsumura, M. Mizoguchi, and J. Hayashi, *Tetrahedron Lett.*, 1979, 3861.
[43] T. Shono, Y. Matsumura, J. Hayashi, and M. Mizoguchi, *Tetrahedron Lett.*, 1980, **21**, 1867.

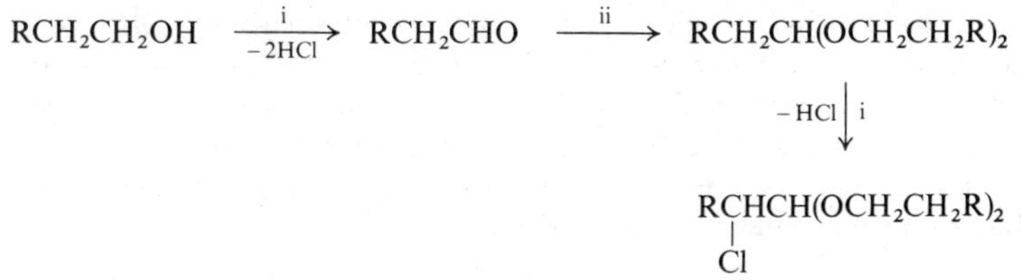

$$\text{RCH}_2\text{CH}_2\text{OH} \xrightarrow[-2\text{HCl}]{\text{i}} \text{RCH}_2\text{CHO} \xrightarrow{\text{ii}} \text{RCH}_2\text{CH}(\text{OCH}_2\text{CH}_2\text{R})_2$$

$$-\text{HCl} \downarrow \text{i}$$

$$\underset{\underset{\text{Cl}}{|}}{\text{RCHCH}(\text{OCH}_2\text{CH}_2\text{R})_2}$$

Reagents: i, Cl$^-$, platinum electrode; ii, RCH$_2$CH$_2$OH, H$^+$

Scheme 4

active intermediate was thought to be I^{+}.[41] The other papers[42,43] describe the oxidation of secondary alcohols to ketones, using benzonitrile as the solvent. The catalysts were methyl phenyl sulphide and n-octyl methyl sulphide + bromide, respectively. The latter system was considered better since it required a lower oxidation potential and the catalyst turnover was higher. While the organic yields were good in all three procedures, the current efficiencies were very poor because the electrolyses were continued to 5—15 F mol^{-1}.

Workers at Southampton[44] have described an indirect method which uses an emulsion of aqueous sodium bromide and the alcohol in amyl acetate, and which makes use of phase-transfer catalysis. The anode reaction is the generation of hypobromite, which, after transfer into the organic medium (by Bu$_4$N$^+$), reacts with the alcohol, the bromide that is formed returning to the aqueous phase and therefore playing the role of catalyst. The procedure is good for the conversion of benzyl alcohols into the corresponding aldehydes; the organic yields are always high, but the current yields varied (20—70%) with the substituent on the ring (lowest for electron-withdrawing groups) because, when the oxidation of the alcohol is slow, the hypobromite disproportionates into inactive bromate and bromide. A second paper[45] discusses cell design for such two-phase oxidations.

Two papers[46,47] have sought to extend the synthetic utility of nickel anodes in alkaline solution for the oxidation of alcohols. A mixture of t-butyl alcohol and water has been shown to be suitable for the oxidation of compounds which are insoluble in water. Amjad, Pletcher, and Smith[46] have reported the oxidation of 3-phenylpropanol, 2-phenylethanol, and substituted benzyl alcohols while Schäfer and Kaulen[47] have described the oxidation of long-chain aliphatic alcohols, including those with olefinic and acetylenic bonds. With the exception of the benzyl alcohols, the products are always carboxylic acids, in very good yield, and this electrolysis should be considered an excellent synthetic procedure, despite the limitation of a low current density because of kinetic control of the current. With benzyl alcohols, both aldehyde and acid are formed, the ratio depending on the electrolysis medium and the structure of the alcohol; low concentrations of base and electron-donating groups in the ring favour the formation of aldehyde. Robertson[48] has commented further on the mechanism of the oxidation of alcohols at nickel

---

[44] D. Pletcher and N. R. Tomov, *J. Appl. Electrochem.*, 1977, **7**, 501.
[45] R. E. W. Jansson and N. R. Tomov, *J. Appl. Electrochem.*, 1980, **10**, 583.
[46] M. Amjad, D. Pletcher, and C. Z. Smith, *J. Electrochem. Soc.*, 1977, **124**, 203.
[47] H. J. Schäfer and J. Kaulen, *Synthesis*, 1979, 513.
[48] P. M. Robertson, *J. Electroanal. Chem. Interfacial Electrochem.*, 1980, **111**, 97.

anodes in basic media, a topic of some controversy in recent years. He reports the variation of the I vs E curves with pH, concentration of substrate, and rate of rotation of a disc electrode, and he suggests some modification of the accepted mechanism, involving a rate-determining chemical step between alcohol substrate and a surface layer of a higher nickel oxide. He also indicates the electrolysis conditions for maximum current density.

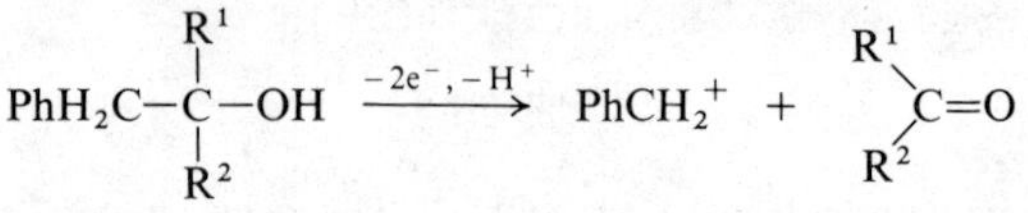

$$PhH_2C-\underset{R^2}{\overset{R^1}{\underset{|}{\overset{|}{C}}}}-OH \xrightarrow{-2e^-,-H^+} PhCH_2^+ + \underset{R^2}{\overset{R^1}{\diagdown}}C=O$$

(5)

The oxidation of a tertiary alcohol that includes a benzyl substituent, e.g. (5), in acetonitrile leads to a ketone by the cleavage of the C–C bond.[49] The benzyl cation is isolated as either benzylacetamide or benzaldehyde, depending on whether it is quenched by acetonitrile or water. Shono et al.[50] have shown that a similar cleavage reaction, that of glycol derivatives, also leads to ketones (see Scheme 5), and they claim that the reaction routes are competitive procedures for the synthesis of symmetrical and unsymmetrical ketones. The same group have reported the oxidation of some enol acetates according to reaction (7).[51]

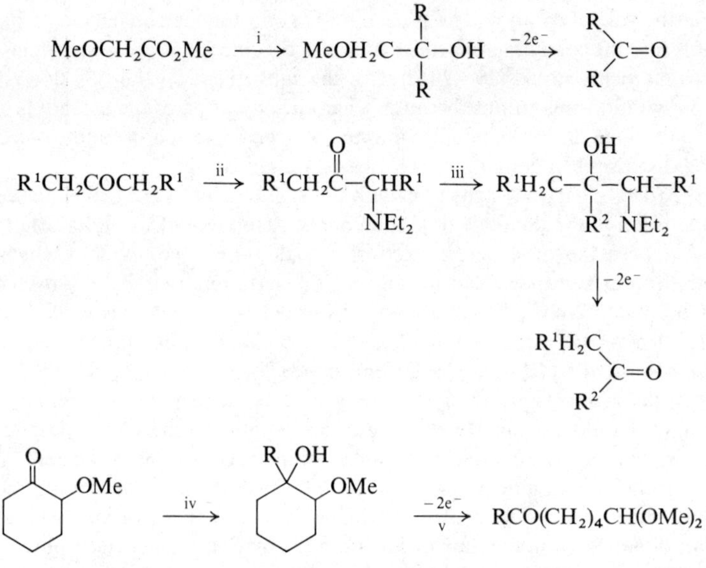

$$MeOCH_2CO_2Me \xrightarrow{i} MeOH_2C-\underset{R}{\overset{R}{\underset{|}{\overset{|}{C}}}}-OH \xrightarrow{-2e^-} \underset{R}{\overset{R}{\diagdown}}C=O$$

$$R^1CH_2COCH_2R^1 \xrightarrow{ii} R^1CH_2\overset{O}{\overset{\|}{C}}-\underset{NEt_2}{\overset{|}{C}}HR^1 \xrightarrow{iii} R^1H_2C-\underset{R^2}{\overset{OH}{\underset{|}{\overset{|}{C}}}}-\underset{NEt_2}{\overset{|}{C}}H-R^1$$

$$\downarrow -2e^-$$

$$\underset{R^2}{\overset{R^1H_2C}{\diagdown}}C=O$$

Reagents: i, 2RMgX; ii, Et$_2$NH; iii, R^{2}MgBr; iv, RMgX; v, MeOH

Scheme 5

[49] M. H. Khalifa and A. Rieker, J. Chem. Res. (S), 1977, 316.
[50] T. Shono, H. Hamaguchi, Y. Matsumura, and K. Yoshida, Tetrahedron Lett., 1977, 3625.
[51] T. Shono, I. Nishiguchi, S. Kashimura, and M. Okawa, Bull. Chem. Soc. Jpn., 1978, **51**, 2181.

$$R^1-\underset{R^2}{\underset{|}{C}}=\overset{OAc}{\underset{|}{C}}-(CH_2)_3CH=CH_2 \xrightarrow{-2e^-} \text{[cyclohexene with } R^2, COR^1 \text{ substituents]} \quad (7)$$

The oxidation of alcohols to aldehydes in aqueous solutions of potassium cyanide and ammonium carbonate has been reported to give amino-acids in moderate yields;[52] this electrolysis can be considered as an analogue of the normal Stecker reaction.

The oxidation of aldehydes at several metal anodes in basic solution has been studied.[53–55] The oxidation occurs several volts less positive than at platinum in neutral acetonitrile because the electro-active species is the anion $RCH(O^-)OH$, formed by addition of hydroxide ion. At nickel, copper, platinum, and mercury the oxidation occurs by reaction between this anion and an anodically formed film on the electrode surface, but on gold and silver the reaction occurs on a film-free metal surface. Even so, the current is kinetically controlled. Silver appears to be the anode of choice for the conversion of aldehydes into carboxylic acids without side-reactions. Becker and Fritz[56] have reported the oxidation of acetaldehyde in methanol that contains ammonia to give a low yield (20 %) of 1,3,5-trimethyltriazine and also an unusual reaction of isobutyraldehyde (6) in basic methanol.[57] The reaction has been reported to give a yield of 42 % for the product (7). A similar

$$Me_2CHCHO \xrightarrow[Base]{MeOH} Me_2CHCH(OMe)_2 \xrightarrow[-H^+]{base} Me_2\bar{C}CH(OMe)_2$$
(6)

$$\downarrow MeOH \mid -2e^-$$

$$\underset{(7)}{Me_2\overset{OMe}{\underset{|}{C}}CH(OMe)_2}$$

reaction is the methoxylation of acetals to give orthoesters. Reactions of the type shown in reaction (8) have been reported;[58] the yields of (8) are variable (7—64 %, depending on the structure of the acetal).

$$\begin{array}{c}\text{[1,3-dioxolane with R, H]} \xrightarrow[base, MeOH]{-2e^-} \text{[1,3-dioxolane with R, OMe]}\end{array} \quad (8)$$
(8)

---

[52] E. P. Krysin, V. V. Trodikov, and V. A. Grinberg, *Sov. Electrochem. (Engl. Transl.)*, 1976, **12**, 1449.
[53] S. Sibille, J. Moiroux, J-C. Marot, and S. Deycard, *J. Electroanal. Chem. Interfacial Electrochem.*, 1978, **88**, 105.
[54] M. B. Fleury, S. Letellier, J. C. Dufresne, and J. Moiroux, *J. Electroanal. Chem. Interfacial Electrochem.*, 1978, **88**, 123.
[55] R. M. Van Effen and D. H. Evans, *J. Electroanal. Chem. Interfacial Electrochem.*, 1979, **103**, 383.
[56] B. F. Becker and H. P. Fritz, *Chem. Ber.*, 1976, **109**, 1346.
[57] B. F. Becker and H. P. Fritz, *Z. Naturforsch., Teil. B*, 1976, **31**, 175.
[58] J. W. Scheeren, H. J. M. Goossens, and A. W. H. Top, *Synthesis*, 1978, 283.

Nitrogen-containing Compounds.—Lindsay Smith and Masheder[59] have reported the effect of molecular structure on the oxidation potential of a series of tertiary amines at glassy carbon in basic methanol–water. They have concluded that the rate-determining step is transfer of an electron to form an aminium ion and that loss of a proton from the α-carbon atoms in this species is an unselective process. This investigation was extended to some hydroxylamines and hydrazines[60] and the oxidation of hydrazines in acetonitrile has also been studied.[61,62] Chabaud[61] has reported a detailed electroanalytical study of the oxidation of a substituted hydrazine while Kinlen, Evans, and Nelson[62] have investigated three fully substituted compounds. These latter molecules are oxidized in two one-electron steps and the cation radical is stable; the kinetics of the neutral molecule/cation radical couple are, however, very dependent on the electrode material. Nelson et al.[63,64] have made the interesting observation that, by cyclic voltammetry at low temperatures, the oxidation behaviour of the different conformers of hydrazines may be defined separately.

The cyanation of tertiary amines by oxidation at a platinum anode in aqueous methanol that contains NaCN has been reported.[65] The major products are cyanated at the carbon atoms α to the nitrogen, and the yields are particularly good (60—70%) for substitution in the ring in N-alkylated piperidine and pyrrolidine. Okita et al.[66] have reported the hydroxylation of five-, six-, and seven-membered-ring lactams in wet acetonitrile. The five- and six-membered-ring compounds, e.g. (9), are substituted in the ring (yield ≃60%) with some product by further oxidation.

$$\underset{R}{\underset{(9)}{O=\!\!\!\diagdown\!\!\!N\!\!\!\diagdown}} \xrightarrow[H_2O]{-2e^-, -H^+} \underset{R}{O=\!\!\!\diagdown\!\!\!N\!\!\!\diagdown\!\!\!OH}$$

4 PhCH$_2$N⟨
(10)
(monomer, M)

⟶   PhH$_2$C—N⟨⟩N—CH$_2$Ph with CH$_2$Ph groups on the other two nitrogens
(11)
(tetramer, T)

[59] J. R. Lindsay Smith and D. Masheder, *J. Chem. Soc., Perkin Trans. 2*, 1976, 47.
[60] J. R. Lindsay Smith and D. Masheder, *J. Chem. Soc., Perkin Trans. 2*, 1977, 1732.
[61] B. Chabaud, *Electrochim. Acta*, 1978, **23**, 757.
[62] P. J. Kinlen, D. H. Evans, and S. F. Nelson, *J. Electroanal. Chem. Interfacial Electrochem.*, 1979, **97**, 265.
[63] S. F. Nelson, L. Echegoyen, E. L. Clennan, D. H. Evans, and D. A. Corrigan, *J. Am. Chem. Soc.*, 1977, **99**, 1130.
[64] S. F. Nelson, E. L. Clennan, and D. H. Evans, *J. Am. Chem. Soc.*, 1978, **100**, 4012.
[65] T. Chiba and Y. Tanaka, *J. Org. Chem.*, 1977, **42**, 2973.
[66] M. Okita, T. Wakamatsu, and Y. Ban, *J. Chem. Soc., Chem. Commun.*, 1979, 749.

The seven-membered-ring compounds are oxidized less selectively, but all the products arise by initial loss of proton from the exocyclic α-carbon atom. Simonet et al.[67] have reported the oxidation of the three-membered-ring compound benzylaziridine (10) in methanol, methylene chloride, and acetonitrile. The product of the oxidation is a macrocycle, i.e. (11). The tetramerization is not a net oxidation reaction and, indeed, it requires only a catalytic quantity of charge. On the other hand the monomer is quite stable in the electrolysis media, and the mechanism postulated is:

$$M \rightarrow M^{+\cdot} + e^-$$

$$M^{+\cdot} + 3M \rightarrow T^{+\cdot}$$

$$T^{+\cdot} + M \rightarrow T + M^{+\cdot}$$

The reaction goes particularly well in acetonitrile in the presence of an electron-transfer mediator, e.g. thianthrene or tris-p-bromophenylamine, when the yield is 80% at the expense of only 0.14 F mol^{-1}.

Hydrazines are more readily oxidized than amines, and this has always prevented the direct oxidative dimerization of amines. The amine anions that are formed by treatment of the amines with lithium metal are, however, very readily oxidized, and this has been shown to be a route to hydrazines in moderate yield.[68] For example, the anion from di-n-butylamine gives tetrabutylhydrazine in 45% yield. The same reaction with the anions from primary amines leads to azoalkanes because the dialkylhydrazine that is formed in the first stage can be deprotonated by the substrate and is then oxidized further.

Enamines are useful substrates for anodic synthesis. Shono et al.[69] have examined the methoxylation of the enamine of cyclohexanone and morpholine and shown that it is a route to a substituted cyclohexanone (12) (yield 80%).

Chiba et al.,[70] in a more extensive investigation, have examined the oxidation of the enamines of cyclohexanone and cyclopentanone with morpholine, piperidine, and pyrrolidine in methanol–methoxide solution that contains the carbanions of acetylacetone, methyl acetylacetate, or diethyl malonate. This is a moderate-yield approach to the synthesis of substituted cyclohexanones, cyclopentanones, or

---

[67] R. Kossai, J. Simonet, and G. Dauphin, *Tetrahedron Lett.*, 1980, **21**, 3575.
[68] R. Bauer and H. Wendt, *Angew. Chem., Int. Ed. Engl.*, 1978, **17**, 202.
[69] T. Shono, Y. Matsumura, H. Hamaguchi, T. Imanishi, and K. Yoshida, *Bull. Chem. Soc. Jpn.*, 1978, **51**, 2179.
[70] T. Chiba, M. Okimoto, H. Nagai, and Y. Takata, *J. Org. Chem.*, 1979, **44**, 3519.

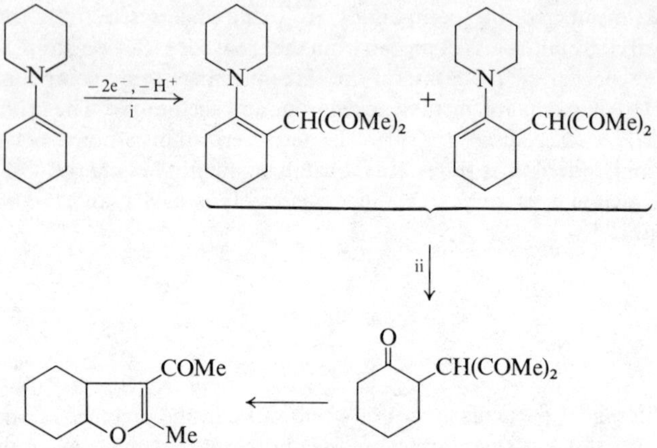

Reagents: i, C̄H(COMe)₂; ii, dilute HCl in CHCl₃; iii, 20% HCl

Scheme 6

substituted furans. A typical reaction route is shown in Scheme 6. The best yields (≃60%) are obtained for readily oxidized enamines (the carbanions are also easily oxidized) in the presence of excess carbanion (the methoxide is also a nucleophile; see above).

Finkelstein et al.[71] have examined the poly-methoxylation and -ethoxylation of dimethylformamide and discuss the distribution of the several isomers formed.

Hydrocarbons and Alkyl Chains.—While the period of this review has seen a large volume of papers on the anodic oxidation of aliphatic hydrocarbons, both saturated and unsaturated, the control of the conditions of electrolysis to obtain yields high enough to be synthetically useful remains a problem.

Fritz and Wuerminghausen[72,73] have continued their studies of the oxidation of alkanes in methylene chloride, trifluoroacetic acid, and their mixtures. With hydrocarbons of favourable structure, e.g. cyclohexane, the yield of trifluoroacetate ester can be very high, typically >90%. The oxidation potential of alkanes has been shown to depend on the concentration of trifluoroacetic acid, the oxidation becoming easier as the concentration of acid is increased. The authors have also studied the isomer distribution for the products from the oxidation of n-decane, and they drew the interesting conclusion that the ratio of secondary esters is non-statistical (i.e. some secondary C–H bonds in the cation radical are weaker than others). This work was extended to some highly branched hydrocarbons.

The oxidation of adamantanes has received attention both because the presence of tertiary C–H bonds leads to high yields of a single isomer and because the products are interesting intermediates for synthesis. Bewick, Mellor, and co-

---

[71] M. Finkelstein, K. Nyberg, R. D. Rossard, and R. Servin, *Acta Chem. Scand.*, Ser. B, 1978, **32**, 182.
[72] H. P. Fritz and T. Wuerminghausen, *J. Chem. Soc., Perkin Trans. 1*, 1976, 610.
[73] H. P. Fritz and T. Wuerminghausen, *Z. Naturforsch.*, Teil. B, 1977, **32**, 241.

workers[74,75] have noted that adamantane (as, for example, do cyclohexane and decalin) gives I vs E curves which show two oxidation processes, and they have demonstrated that electrolysis at potentials of the second process leads to disubstituted products via cationic intermediates. Thus, by selection of the potential and the electrolysis medium, the products (13)—(16) may be isolated after an aqueous work-up.

Solvents are: (A) MeCN; (B) $CF_3CO_2H + (CF_3CO)_2O$

The investigation has been extended to the further remote substitution of adamantyl alcohols, esters, and acetamides. 1-Alkyl-adamantanes have also been studied.[76] The methyl, ethyl, and isopropyl derivatives undergo substitution at the 3-position, but with 1-t-butyladamantane the two-electron oxidation leads to cleavage of the C–C bond. Vincent et al.[77] have investigated the oxidation of 2-adamantyl halides in acetonitrile. With 2-iodoadamantane the oxidation results in cleavage of the C–I bond and the formation of the corresponding acetamide, but with the other halides the major reaction is substitution at the bridgehead C–H bond to give the 1-acetamido-2-halogeno-adamantane. With the bromo- and fluoro-compounds, cleavage of the C–X bond is a minor competing process. Vincent et al.[78] have also studied the oxidation of other bicyclic hydrocarbons in acetonitrile and found good yields of the substituted acetamides that arise by oxidation of a tertiary C–H bond. The oxidation potentials for some strained polycyclic aliphatic

---

[74] A. Bewick, G. J. Edwards, S. R. Jones, and J. M. Mellor, Tetrahedron Lett., 1976, 631.
[75] A. Bewick, G. J. Edwards, S. R. Jones, and J. M. Mellor, J. Chem. Soc., Perkin Trans. 1, 1977, 1831.
[76] G. J. Edwards, S. R. Jones, and J. M. Mellor, J. Chem. Soc., Perkin Trans. 2, 1977, 505.
[77] F. Vincent, R. Tardivel, and P. Mison, Tetrahedron, 1976, 32, 1681.
[78] F. Vincent, R. Tardivel, P. Mison, and P. von R. Schleyer, Tetrahedron, 1977, 33, 325.

hydrocarbons in acetonitrile have been measured and related to molecular structure.[79]

Fluorosulphonic acid has retained some interest as a solvent for the anodic oxidation of alkanes. Pletcher and Smith[80] have reported the electrosynthesis of the powerful alkylating agents methyl and ethyl fluorosulphonate from methane and ethane, respectively, in good yield; these products arise by chemical reaction between the hydrocarbons and the peroxide $(FSO_3)_2$, formed by anodic oxidation of fluorosulphonate ion. The same mechanism has been proposed[81] for the oxidation of the fluorinated alkane $C_5F_{11}CF_2H$, which also yields the fluorosulphonate on electrolysis in FSO_3H–FSO_3K. Pitti et al.[82] have commented on the mechanism for the direct oxidation of alkanes in fluorosulphonic acid; they have studied the cyclic voltammetry of cyclopentane at $-60\,°C$ and concluded that oxidation occurs in two one-electron steps via a radical to a carbenium ion. This conclusion contradicts those of other studies and seems to be based on an over-interpretation of the data.

Tomat and Rigo[83] have demonstrated that electrochemically generated Fenton's reagent may be used to oxidize cyclohexane to cyclohexanone (maximum yield of 48 %). The use of a chloride medium is essential to obtain more than nominal yields, and the Fenton's reagent is formed by co-reduction of Fe^{III} and oxygen; the solvent employed was t-butyl alcohol–water.

Medium- and long-chain aliphatic compounds in which the functional group is electro-inactive can often be oxidized as if they were alkanes. Thus Campbell and Pletcher[84] have discussed the remote substitution of aliphatic ketones in trifluoroacetic acid. The ketones are oxidized directly in a two-electron process and in some cases, for example pentan-2-one, hexan-2-one, and 5-methylhexan-2-one, substitution occurs at the $(\omega-1)$ position to give the keto-ester in moderate yield. On work-up with water, the final product that is isolated may be an olefin, ester, or alcohol. Breslow and Goodin[85] have reported the remote chlorination of a steroid (17) at a platinum or carbon anode, in acetonitrile that contains chloride ion, in up

(17)　　　　　　　　　　　　　　　[ClSI]

[79] P. G. Gassman and R. Yamaguchi, J. Am. Chem. Soc., 1979, 101, 1308.
[80] D. Pletcher and C. Z. Smith, Chem. Ind. (London), 1976, 371.
[81] A. Germain and A. Commeyras, J. Chem. Soc., Chem. Commun., 1978, 118.
[82] S. Pitti, M. Herlem, and J. Jordan, Tetrahedron Lett., 1976, 3221.
[83] R. Tomat and A. Rigo, J. Appl. Electrochem., 1980, 10, 549.
[84] C. B. Campbell and D. Pletcher, Electrochim. Acta, 1978, 23, 923.
[85] R. Breslow and R. Goodin, Tetrahedron Lett., 1976, 2675.

to 70% yield. The reaction only occurs at potentials where the steroid (SI) is oxidized, and it is tempting to propose a simple nucleophilic mechanism. The authors, however, reject this mechanism because it seems certain that initial oxidation will occur at the iodine atom, and therefore it is not clear why substitution takes place at the 9-position. Moreover, if there are radical traps in solution they inhibit the synthesis. Instead the authors have proposed a mechanism involving oxidation of both the steroid and chloride ion, *i.e.*

$$SI \xrightarrow{-e^-} SI^{+\cdot} \xrightarrow{Cl^-} SICl\cdot \xrightarrow{-HCl} \cdot SI \xrightarrow{Cl_2} ClSI + Cl\cdot$$

in which the oxidation of the steroid is only necessary to start a chain reaction that is initiated by chlorine atoms. The specificity in the oxidation site arises by a through-space interaction between the iodine atom and the C–H bond at the 9-position.

Remote oxidations by indirect anode reactions have also been reported. In acetonitrile, the remote oxidation of esters is catalysed by Cl_2, ClO_2, and ClO_4^-; the oxidation of each leads to a radical species, at a potential less positive than the oxidation of methyl hexanoate, that is capable of causing abstraction of $(\omega - 1)$- and $(\omega - 2)$-H and hence eventually to acetamidation at these sites, in 19—28 % yield.[86] The remote oxidation of carboxylic acids that have an alkane chain of medium length, *e.g.* (18), in fluorosulphonic acid *via* $(FSO_3)_2$ leads to lactones (19),[87] *via* the sequence of reactions shown in Scheme 7, and the yield can be good if oxidation occurs at a carbon site which permits the formation of a five- or six-membered ring. A double-pulse method has confirmed that the rate-determining step is abstraction of a hydrogen atom.[88]

Synthetically, the reports of the direct anodic oxidation of unsaturated hydrocarbons have been surprisingly disappointing. The oxidations of olefins,

$$2FSO_3^- \longrightarrow (FSO_3)_2 + 2e^-$$

$$R^1R^2CH(CH_2)_nCOOH + HSO_3F \longrightarrow R^1R^2CH(CH_2)_nCO^+ + H_2O + FSO_3^-$$

(18)

$$(FSO_3)_2 + R^1R^2CH(CH_2)_nCO^+ \longrightarrow HSO_3F + R^1R^2\overset{\overset{\displaystyle OSO_2F}{|}}{C}(CH_2)_nCO^+$$

$$\Big\downarrow H_2O$$

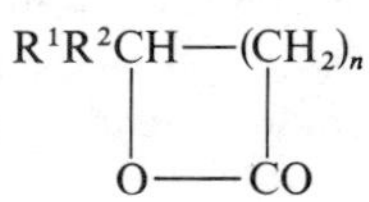

(R^1, R^2 = H or alkyl, n = 2 or 3)

$$\begin{array}{c} R^1R^2CH\!-\!(CH_2)_n \\ | \quad\quad\quad\quad | \\ O\!-\!-\!-\!CO \end{array}$$

(19)

Scheme 7

---

[86] L. L. Miller and M. Katz, *J. Electroanal. Chem. Interfacial Electrochem.*, 1976, **72**, 329.
[87] C. J. Myall, D. Pletcher, and C. Z. Smith, *J. Chem. Soc., Perkin Trans. 1*, 1976, 2035.
[88] C. J. Myall and D. Pletcher, *J. Electroanal. Chem. Interfacial Electrochem.*, 1977, **85**, 371.

acetylenes, dienes, diynes, and allenes in methanol[89,90] have each been studied and shown to lead to complex mixtures of products, while attempts to cyanate cyclohexene in methanol–NaCN led to methoxylated compounds and a low percentage of isocyanides, but only a trace of the cyanide.[91] On the other hand, some substituted olefins can be oxidized to give more acceptable yields. For example, the acetoxylation of methyl oleate[92] is a sequential reaction, and it is possible to obtain a 60% yield of allylic acetate or an 85% yield of saturated diacetates. The product is, in each case, a mixture of isomers. Utley and co-workers[93] have also demonstrated that the oxidation of some unsaturated carboxylic acids occurs at the olefin centre, and, in aqueous acetonitrile, the oxidation of the cyclohexene derivative (20) leads to lactones. For steric reasons, the product from the cyclopentene derivative (21) is quite different.

The anodic bromination of cyclohexene has been shown to give products which vary with the solvent.[94] In acetonitrile and dimethylformamide the major product is 1,2-dibromocyclohexane, with mixed addition across the double-bond as the minor reaction. In methanol, however, the major product is 1-methoxy-2-bromocyclohexane.

Bauer and Wendt[95] have shown that the coupling of acetaldehyde ions to butadiynes is a possible reaction in a carefully designed flow cell. The reaction in THF–LiClO$_4$ is complicated by the addition of the starting anion to the product to form polymer, but this can be avoided by forcing the electrolyte to flow through a cell which allows a high conversion per pass and then immediately quenching the solution with acetic acid. The oxidation of solid polyacetylene has been studied.[96] This is an interesting reaction because while polyacetylene is a rather poor semiconductor, the products, i.e. $(CHI_{0.07})_x$ from oxidation in aqueous KI and $[CH(ClO_4)_{0.06}]_x$ from oxidation in CH_2Cl_2–Bu_4NClO_4, are metallic conductors.

Some indirect oxidations of olefins have attracted some commercial interest.

---

[89] M. Katz and H. Wendt, *Electrochim. Acta*, 1976, **21**, 215.
[90] B. Zinger and J. Y. Becker, *Electrochim. Acta*, 1980, **25**, 791.
[91] K. Yoshida, T. Kanbe, and T. Fueno, *J. Org. Chem.*, 1977, **42**, 2313.
[92] J. H. P. Utley, C. Adams, and E. N. Frankel, *J. Chem. Soc., Perkin Trans. 1*, 1979, 353.
[93] C. Adams, N. Jacobsen, and J. H. P. Utley, *J. Chem. Soc., Perkin Trans. 2*, 1978, 1071.
[94] P. Pouillen, R. Minko, M. Verniette, and P. Martinet, *Electrochim. Acta*, 1979, **24**, 1184.
[95] R. Bauer and H. Wendt, *J. Electroanal. Chem. Interfacial Electrochem.*, 1977, **80**, 395.
[96] P. J. Nigrey, A. G. MacDiarmid, and A. J. Heager, *J. Chem. Soc., Chem. Commun.*, 1979, 594.

Organic Electrochemistry – Synthetic Aspects: Oxidations 221

Drakesmith and Hughes[97] have examined the perfluorination of propylene, using a medium of HF that contains 6% of NaF and a nickel foam anode. They assert that reproducible results may be obtained by careful control of the electrolysis conditions and by conditioning the anode by applying +6 V for 30 minutes prior to the introduction of the organic substrate. The parameters that were found to be important were the anode potential (too positive a value favours fragmentation), the flow rate, and the composition of the gas feed (ratio of N_2 to $H_2C{=}CHMe$). Under optimum conditions, the current efficiency for fluorination is 94%, but six fluorinated products were identified; the major product is octafluoropropane (yield 38%), with some heptafluoro- (10%) and hexafluoro-propane (9%). The same authors[98] have studied the perfluorination of octanoyl chloride, and by a similar careful optimization they were able to isolate perfluoro-octanoyl fluoride (47% yield). This reaction was also run on a pilot-scale plant.

It has also been demonstrated that epoxidation of olefins can, with advantage, be carried out electrolytically. The favoured route is *via* hypobromite, in an undivided cell, so that the cathode reaction can be used to form the base that is necessary for ring-closure, *i.e.* as shown in Scheme 8.

$$Br^- + H_2O \longrightarrow HOBr + H^+ + 2e^-$$

$$2H_2O + 2e^- \longrightarrow H_2 + 2OH^-$$

$$HOBr + RCH{=}CH_2 \longrightarrow \underset{\underset{OH\ \ Br}{|\ \ \ \ |}}{RCH{-}CH_2} \xrightarrow{OH^-} RCH\underset{O}{\overset{}{-}}CH_2 + H_2O + Br^-$$

Scheme 8

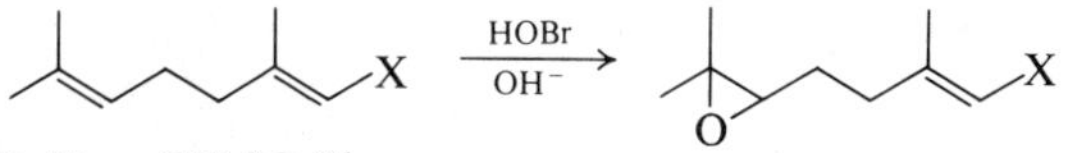

(22) X = CH$_2$SO$_2$Ph, CH$_2$OAc, or Ac

Jansson and co-workers[99] have reported the epoxidation of propylene by this pathway in aqueous sodium bromide, while Torii *et al.*[100] have used NaBr in acetonitrile–THF–water to epoxidize a series of substituted isoprenes (22) to give intermediates for the synthesis of terpenoid structures. Both types of epoxidation give close to quantitative yields.

Schäfer and co-workers[101–104] have published a series of papers on the anodic functionalization of conjugated dienes. A series of dienes each undergo a two-

---

[97] F. G. Drakesmith and D. A. Hughes, *J. Appl. Electrochem.*, 1976, **6**, 23.
[98] F. G. Drakesmith and D. A. Hughes, *J. Appl. Electrochem.*, 1979, **9**, 685.
[99] J. Ghoroghchian, R. E. W. Jansson, and D. Jones, *J. Appl. Electrochem.*, 1977, **7**, 437.
[100] S. Torii, K. Uneyama, M. Ono, H. Tazawa, and S. Matsumura, *Tetrahedron Lett.*, 1979, 4661.
[101] H. Baltes, E. Steckhan, and H. J. Schäfer, *Chem. Ber.*, 1978, **111**, 1294.
[102] H. Baltes, L. Stork, and H. J. Schäfer, *Chem. Ber.*, 1979, **112**, 807.
[103] H. Baltes, L. Stork, and H. J. Schäfer, *Angew. Chem., Int. Ed. Engl.*, 1977, **16**, 413.
[104] H. Baltes, H. J. Schäfer, and L. Stork, *Liebigs Ann. Chem.*, 1979, 318.

electron oxidation in methanol[101] or acetonitrile–water,[102] but a complex spectrum of products, including disubstituted 3-enes and disubstituted dimeric dienes, is always isolated; a typical reaction is that of (23), other isomers being formed in addition to those shown. The oxidation of dienes in the presence of substituted ureas can be a more rewarding reaction,[103,104] and imidazolidin-2-ones can be formed in moderate yield [44% of (24) for the reaction shown]. Schmidt and Steckhan[105] have sought to optimize the yield of dibromo-dimers (25)—(27) from the bromination of butadiene. In order to maximize the amounts of dimers (which are synthons for the synthesis of natural products), an electron-transfer mediator, tri-(p-bromophenyl)-amine, was employed, in CH_2Cl_2, at -60 °C, so that the bromine atoms are formed away from the electrode surface and do not form Br_2.

The hydrocarbon 1,3,5-tri-t-butylpentalene has been studied. It is readily oxidized ($E = +0.75$ V vs the S.C.E.), and in CH_2Cl_2, at -77 °C, it forms a stable cation radical.[106]

Miscellaneous Aliphatic Compounds.—Torii et al.[107] have described the oxidation of enol acetates and ethers and of silyl enol ethers in acetonitrile–water that contains ammonium bromide or iodide, and have shown it to be an effective route to halogeno-ketones, e.g. (28).

[105] W. Schmidt and E. Steckhan, J. Electroanal. Chem. Interfacial Electrochem., 1979, **101**, 123.
[106] R. W. Johnson, J. Am. Chem. Soc., 1977, **99**, 1461.
[107] S. Torii, T. Inokuchi, S. Misima, and T. Kobayashi, J. Org. Chem., 1980, **43**, 2731.

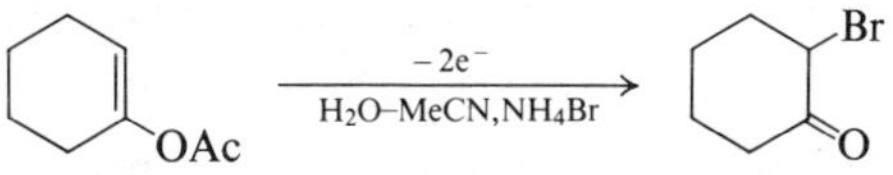

(28)

[95%]

In a series of papers, Becker and co-workers[108–111] have discussed the oxidation of alkyl bromides in acetonitrile. The electrode reaction always leads to cleavage of the C–Br bond, the formation of a carbenium ion, and usually to acetamides as final products. The effect of the electrolysis parameters (*e.g.* concentration of substrate, temperature, electrolyte, and charge) on the overall yield and on the product distribution has been examined. The yield can be high, *e.g.* 83% for t-butylacetamide from the oxidation of t-butyl bromide, but, in general, n-alkyl bromides give lower yields (40—50%) and a mixture of (secondary alkyl)acetamides. Later papers examine the effect of the length of the carbon chain in n-alkyl bromides and of the size of the ring in cycloalkyl bromides on current efficiency and product distribution. The oxidation of alkyl bromides at a platinum anode, in liquid hydrogen fluoride, has also been studied;[112] well-shaped but irreversible oxidation peaks are seen on cyclic voltammograms. Electrolysis of methyl and ethyl bromides and of 1,2-dibromoethane leads to very selective replacement of one bromine by a fluorine atom (presumably by cleavage of the C–Br bond to give a carbenium ion), but the larger analogues give increasingly complex product mixtures.

Torii et al.[113] have described the synthesis of bis-dialkylthiocarbamoyl disulphides (which are used for vulcanization and as fungicides) by the anodic coupling of carbon disulphide and secondary amines, in acetonitrile, or in dimethylformamide, or in water–methylene chloride emulsions. Using a 2:1 ratio of amine:carbon disulphide, the product is formed with a yield in the range 87—100% by the route shown in Scheme 9. Further oxidation of a mixture of the product and a secondary amine is a novel route to forming a S–N bond.[114] The mechanism is

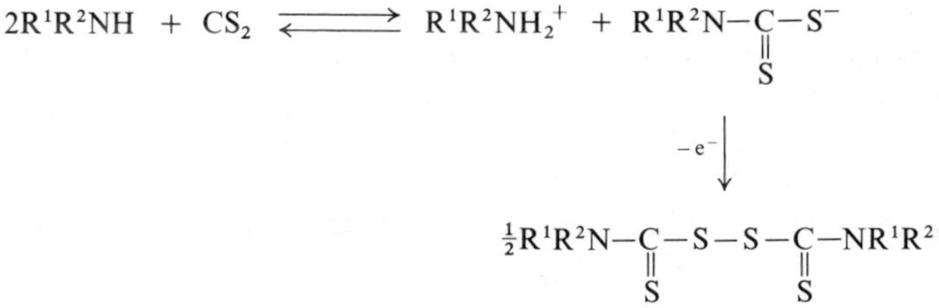

$$2R^1R^2NH + CS_2 \rightleftharpoons R^1R^2NH_2^+ + R^1R^2N-\underset{\underset{S}{\|}}{C}-S^-$$

$$\downarrow -e^-$$

$$\tfrac{1}{2} R^1R^2N-\underset{\underset{S}{\|}}{C}-S-S-\underset{\underset{S}{\|}}{C}-NR^1R^2$$

Scheme 9

[108] J. Y. Becker and M. Munster, *Tetrahedron Lett.*, 1977, 455.
[109] J. Y. Becker, *J. Org. Chem.*, 1977, **42**, 3997.
[110] J. Y. Becker, *Tetrahedron Lett.*, 1978, 1331.
[111] J. Y. Becker and D. Zemach, *J. Chem. Soc., Perkin Trans. 2*, 1979, 914.
[112] J. Badoz-Lambling, A. Thiebault, and P. Oliva, *Electrochim. Acta*, 1979, **24**, 1029.
[113] S. Torii, H. Tanaka, and K. Mishima, *Bull. Chem. Soc. Jpn.*, 1978, **51**, 1575.
[114] S. Torii, H. Tanaka, and M. Ukida, *J. Org. Chem.*, 1978, **43**, 3223.

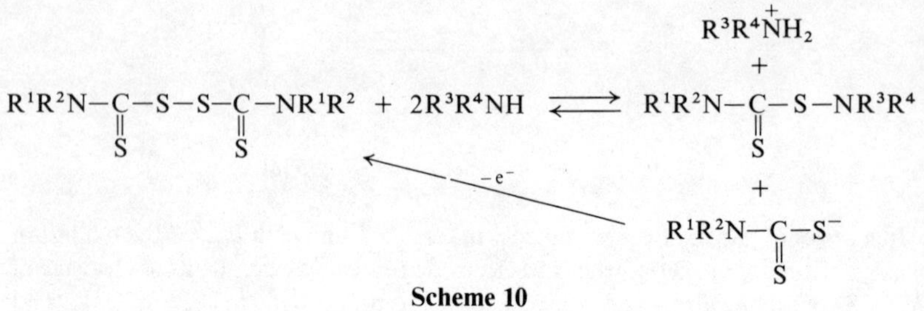

Scheme 10

$$R^1R^2N-\underset{S}{\underset{\|}{C}}-S-S-\underset{S}{\underset{\|}{C}}-NR^1R^2 + 2R^3R^4NH \rightleftharpoons R^1R^2N-\underset{S}{\underset{\|}{C}}-S-NR^3R^4$$

$$+ R^3R^4\overset{+}{N}H_2$$

$$+ R^1R^2N-\underset{S}{\underset{\|}{C}}-S^-$$

probably that shown in Scheme 10, but certainly the yields of sulphenamides that are reported are excellent. Torii et al.[115] have also reported the coupling of phthalimide or succinimide with dicyclohexyl disulphide, but recommend the presence of Br^- as a catalyst; the S–N-bonded products are again formed in good yield.

Bewick et al.[116] have reported the anodic cleavage of S–S bonds in the presence of olefins, in acetonitrile. The products are acetamido-sulphides, in yields between 43 and 84%. A typical reaction is the formation of (29), as shown in Scheme 11.

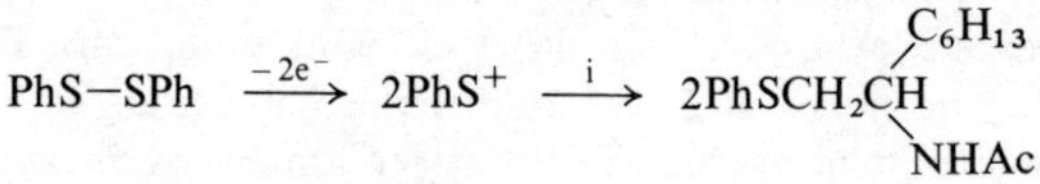

$$PhS-SPh \xrightarrow{-2e^-} 2PhS^+ \xrightarrow{i} 2PhSCH_2\underset{NHAc}{\overset{C_6H_{13}}{CH}}$$

(29)

Reagents: i, $2H_2C=CHC_6H_{13}$, $MeCN-H_2O$

Scheme 11

Nokami et al.[117] have reported the acetoxylation of alkyl sulphides at the carbon atom α to the sulphur atom. The yield is 50—80%, and pyrolysis of the acetates leads to unsaturated sulphides. The same group have also described the anodic oxidation of α-(phenylthio)carboxylic acids (30),[118] while Wilson and co-workers[119] have reported the oxidation of some mesocyclic dithioethers in acetonitrile. Porter and Utley[120] have reported the cleavage of C–S bonds and described procedures for the anodic deprotection of ketones that are protected by a 1,3-dithian group, e.g. (31), using aqueous acetonitrile for simple ketones and acetonitrile–THF–water for steroidal ketones.

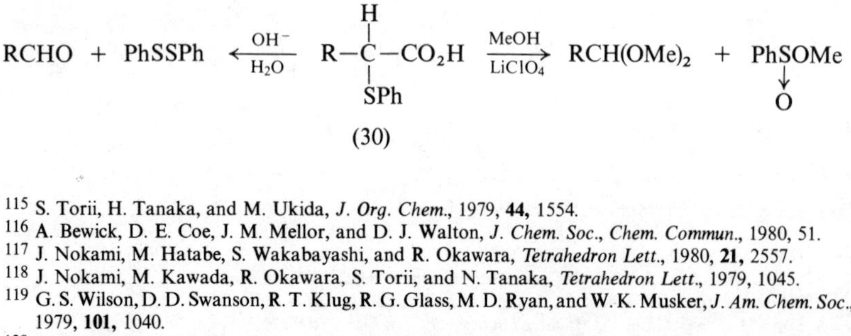

$$RCHO + PhSSPh \xleftarrow[H_2O]{OH^-} R-\underset{SPh}{\overset{H}{\underset{|}{\overset{|}{C}}}}-CO_2H \xrightarrow[LiClO_4]{MeOH} RCH(OMe)_2 + PhSOMe$$

(30)

---

[115] S. Torii, H. Tanaka, and M. Ukida, J. Org. Chem., 1979, **44**, 1554.
[116] A. Bewick, D. E. Coe, J. M. Mellor, and D. J. Walton, J. Chem. Soc., Chem. Commun., 1980, 51.
[117] J. Nokami, M. Hatabe, S. Wakabayashi, and R. Okawara, Tetrahedron Lett., 1980, **21**, 2557.
[118] J. Nokami, M. Kawada, R. Okawara, S. Torii, and N. Tanaka, Tetrahedron Lett., 1979, 1045.
[119] G. S. Wilson, D. D. Swanson, R. T. Klug, R. G. Glass, M. D. Ryan, and W. K. Musker, J. Am. Chem. Soc., 1979, **101**, 1040.
[120] Q. N. Porter and J. H. P. Utley, J. Chem. Soc., Chem. Commun., 1978, 255.

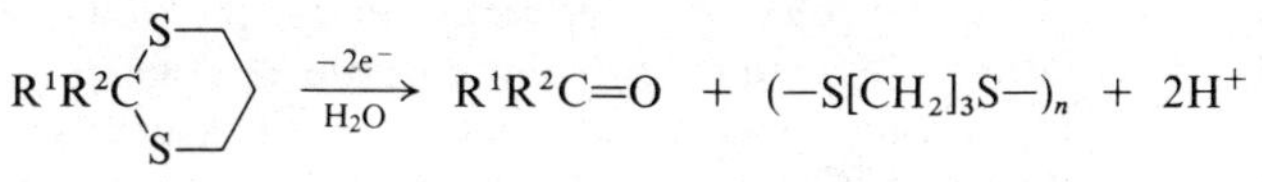

$$R^1R^2C\underset{S}{\overset{S}{\diagdown\diagup}}\xrightarrow[H_2O]{-2e^-} R^1R^2C{=}O + (-S[CH_2]_3S-)_n + 2H^+$$

(31)

The anodic oxidation of some phosphorus-containing molecules has been reported. Gara and Roberts[121] have studied the oxidation of trialkylphosphines in acetonitrile and used e.s.r. spectroscopy to identify the dimeric intermediates $[R_3P-PR_3]^{+\cdot}$ while two papers have reported the behaviour of alkyl phosphates in the same solvent. The behaviour is complex in solutions that contain a trialkyl phosphate alone,[122] but coupled products (32) are obtained (yield 34—66%) in the presence of an aromatic hydrocarbon, although after a work-up with a reducing agent (e.g. NaI) the phosphonium ion is converted into an aryl phosphate. With alkyl-benzenes, a mixture of isomers is formed. The coupling reaction to form (33)[123]

$$(RO)_3P \xrightarrow{-e^-} (RO)_3P^{\cdot+} \xrightarrow[-e^-,-H^+]{PhH} (RO)_3\overset{+}{P}Ph$$

(32)

$$(R^1O)_2\overset{O}{\overset{\|}{P}}H + HNR^2_2 \xrightarrow[-2H^+]{-2e^-} (R^1O)_2P{-}NR^2_2$$

(33)

gives almost quantitative yields, but only if the oxidation is carried out indirectly, *via* iodine, by using an electrolyte of sodium iodide in acetonitrile. Wagenknecht[124] has used the well-known anodic cleavage of a C–N bond in tertiary amines to construct a synthesis of *N*-substituted iminomethylenephosphonic acids (34) by electrolysis in aqueous hydrochloric acid. The disubstituted substrate is readily prepared, but the monosubstituted compound is only formed in poor yield by direct synthesis.

$$RN(CH_2PO_3H_2)_2 \xrightarrow[H_2O]{-2e^-} RNHCH_2PO_3H_2 + \underset{CHO}{\overset{|}{P}O_3H_2}$$

(34)

Knunyants and co-workers[125] have described the synthesis of some fluorosilane compounds (35) by anodic displacement of alkyl groups by fluoride. The electrolyte is acetonitrile–Et$_4$NF·3HF; the yields varied in the range 15—80%.

$$R^2_3SiR^1 + F^- \xrightarrow{-e^-} R^2_3SiF + R^{1\cdot}$$

(35)

---

[121] W. B. Gara and B. P. Roberts, *J. Chem. Soc., Perkin Trans. 2*, 1978, 150.
[122] H. Ohmori, S. Nakai, and M. Masui, *J. Chem. Soc., Perkin Trans. 1*, 1979, 2023.
[123] S. Torii, N. Sayo, and H. Tanaka, *Tetrahedron Lett.*, 1979, 4471.
[124] J. H. Wagenknecht, *J. Electrochem. Soc.*, 1976, **123**, 620.
[125] I. Ya. Alyev, I. N. Rozhkov, and I. L. Knunyants, *Tetrahedron Lett.*, 1976, 2469.

In a series of papers, Mengoli et al.[126–129] have discussed the synthesis of tin and lead alkyls by oxidation of a zinc alkyl at the appropriate metal anode in an aprotic solvent. The zinc alkyl is always formed at a zinc cathode by reduction of an alkyl bromide, alkyl iodide, or dialkyl sulphate. It was found that a low concentration of alkyl iodide was an effective catalyst for the reduction of alkyl bromides. In another paper,[130] the coupling of acetyl or benzoyl halides with alkyl halides to give unsymmetrical ketones is reported; the reaction is accompanied by the dissolution of the cadmium anode and presumably proceeds through a cadmium alkyl intermediate.

Several papers have discussed the anodic coupling of diethyl malonate [reaction (9)] both by direct anodic oxidation and via a halogen. The direct reaction has been attempted in both acetonitrile that contains a base[131] and in an aqueous organic

$$2\,\bar{C}H(CO_2Et)_2 \xrightarrow{-2e^-} (EtO_2C)_2-CH-CH-(CO_2Et)_2 \qquad (9)$$

emulsion,[132] but the yield is poor. The reaction is much more satisfactory in the presence of halide ion.[132,133] It is, however, necessary for the diethyl malonate to be present as its carbanion, and this is possible either by working in a basic solution or by generating the carbanion at the cathode in an aprotic solvent. In the latter system the yield of dimer can be as high as 98 %. Jansson and Tomov[134] have discussed cell design for the paired reaction (i.e. electrolysis employing both the anode and the cathode reaction) and showed the effects of the mixing pattern of the anode and the cathode products on the yield. Torii et al.[135] have shown that, by using a high current density and a CH_2Cl_2–MeONa–Et_4NBr medium, it is possible to isolate α-brominated intermediates from the oxidation of methylated active-methylene groups, e.g. (37) from (36).

$$\underset{(36)}{\text{MeCOCHCO}_2\text{Me}} \overset{\text{Me}}{|} \xrightarrow[-H^+]{Br^-,\,-2e^-} \underset{\underset{Br}{|}}{\text{MeCOCCO}_2\text{Me}} \overset{\text{Me}}{|}$$

(37)
[84%]

Alkyl carbonates have been synthesized from carbon monoxide and an alcohol that contains a bromide salt.[136] The reactions are:

$$2Br^- \xrightarrow{-2e^-} Br_2 \xrightarrow{CO} COBr_2 \xrightarrow{2ROH} (RO)_2C{=}O + 2HBr$$

---

[126] G. Mengoli and F. Furlanetto, *J. Electroanal. Chem. Interfacial Electrochem.*, 1976, **73**, 119.
[127] G. Mengoli and S. Daolio, *Electrochim. Acta*, 1976, **21**, 889.
[128] G. Mengoli and S. Daolio, *J. Appl. Electrochem.*, 1976, **6**, 521.
[129] G. Mengoli and S. Daolio, *J. Chem. Soc., Chem. Commun.*, 1976, 96.
[130] J. J. Habeeb and D. G. Tuck, *J. Chem. Soc., Chem. Commun.*, 1976, 696.
[131] H. G. Thomas, M. Streukins, and R. Peek, *Tetrahedron Lett.*, 1978, 45.
[132] T. C. Franklin and T. Honda, *Electrochim. Acta*, 1978, **23**, 439.
[133] D. A. White, *J. Electrochem. Soc.*, 1977, **124**, 1177.
[134] R. E. W. Jansson and N. R. Tomov, *Electrochim. Acta*, 1980, **25**, 497.
[135] S. Torii, K. Uneyama, and N. Yamasaki, *Bull. Chem. Soc. Jpn.*, 1980, **53**, 819.
[136] D. Cipris and I. L. Mador, *J. Electrochem. Soc.*, 1978, **125**, 1954.

Using a current density of 120 mA cm^{-2} and a pressure of 100 atmospheres, the yield of dimethyl carbonate from methanol is 70% and of ethylene carbonate from ethylene glycol is 45%. Breslow, Klubtz, and Khanna[137] have reported the oxidation of a peroxide (38) in the presence of tosylamide; the interest in the reaction

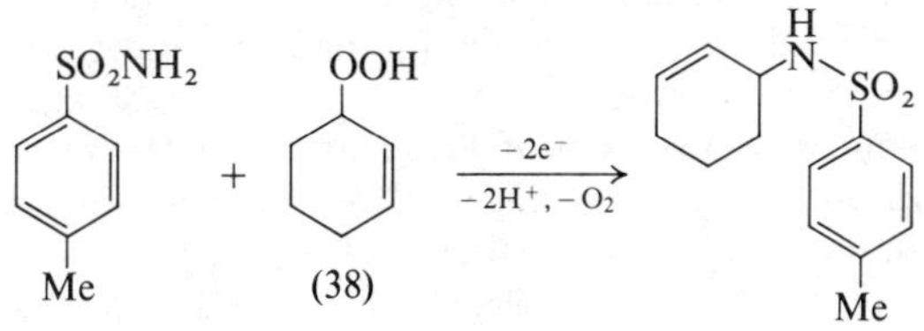

$$\text{SO}_2\text{NH}_2\text{-}C_6H_4\text{-Me} + (38) \xrightarrow[-2H^+, -O_2^-]{-2e^-} \text{product}$$

is that it is specific to a vanadium anode. Torii et al.[138] have reported a reaction [reaction (10)] which requires the presence of a ferric salt. In methanol, at -10 °C, and for $n = 4$ or 5, the yield is 70—90%.

$$\text{(CH}_2)_n\text{-cyclopropane-Cl, Cl, OSiMe}_3 \longrightarrow \text{(CH}_2)_n\text{-C(=O)-CO}_2\text{Me} \qquad (10)$$

2 Aromatic Compounds

Hydrocarbons: Substitution in the Ring and in the Side-chain.—Electroanalytical techniques continue to be used to define conditions where the cation radicals and dications of aromatic hydrocarbons are stable. Media considered during the period of this review include liquid sulphur dioxide at -40 °C,[139] the low-temperature molten salt mixture aluminium trichloride–N-butylpyridinium choride at 40 °C,[140] trifluoroacetic acid,[141] and propylene carbonate.[142] There have also been several similar studies of cyclophanes[143,144] and of metacyclophanes.[145,146] Fritz et al.[147,148] have reported the formation of a dimer cation radical from the oxidation of naphthalene in methylene chloride; when the electrolysis is carried out at -45 °C, the salt $[(C_{10}H_8)_2]^+ \text{PF}_6^-$ precipitates as red-violet crystals; an X-ray structure shows the crystals to be layered, with a short distance between parallel cation radical and neutral naphthalene molecules. Two papers[149,150] have reported the oxidation of polyvinylferrocene in aprotic solvents. The interest in such molecules is

---

[137] R. Breslow, R. Q. Klubtz, and P. L. Khanna, *Tetrahedron Lett.*, 1979, 3273.
[138] S. Torii, T. Okamoto, and N. Ueno, *J. Chem. Soc., Chem. Commun.*, 1978, 293.
[139] L. A. Tinker and A. J. Bard, *J. Am. Chem. Soc.*, 1979, **101**, 2316.
[140] J. Robinson and R. A. Osteryoung, *J. Am. Chem. Soc.*, 1979, **101**, 323.
[141] H. P. Fritz and H. Gebauer, *Z. Naturforsch.*, Teil. B, 1978, **33**, 702.
[142] C. Madec and J. Courtot-Coupez, *J. Electroanal. Chem. Interfacial Electrochem.*, 1977, **84**, 169.
[143] J. Y. Becker, L. L. Miller, V. Boekelheide, and T. Morgan, *Tetrahedron Lett.*, 1976, 2939.
[144] T. Sato and K. Torizuka, *J. Chem. Soc., Perkin Trans. 2*, 1978, 1199.
[145] T. Sato and M. Kamada, *J. Chem. Soc., Perkin Trans. 2*, 1977, 384.
[146] T. Sato, K. Torizuka, R. Komaki, and H. Atobe, *J. Chem. Soc., Perkin Trans. 2*, 1980, 561.
[147] H. P. Fritz, H. Gebauer, P. Friedrich, R. Artes, and U. Schubert, *Z. Naturforsch.*, Teil. B, 1978, **33**, 498.
[148] H. P. Fritz, H. Gebauer, P. Friedrich, and U. Schubert, *Angew. Chem., Int. Ed. Engl.*, 1978, **17**, 275.
[149] T. W. Smith, J. E. Kuder, and D. Wychick, *J. Polym. Sci., Polym. Chem. Ed.*, 1976, **14**, 2433.
[150] J. B. Flanagan, S. Margel, A. J. Bard, and F. C. Anson, *J. Am. Chem. Soc.*, 1978, **100**, 4248.

that the ferrocenes seem to behave as non-interacting centres for electron transfer, and hence the I vs E curves recorded for their solutions are similar to those for a reversible one-electron process; the later paper considers the thermodynamics of multi-electron transfer to or from non-interacting centres in a molecule. Phelps and Bard[151] have discussed a similar phenomenon in the oxidation of substituted ethylenes and, in particular, the structural features which determine whether the oxidation occurs in two one-electron steps or in a single two-electron step.

Recent developments in instrumental methods and in the interpretation of the resulting experimental data are making possible a far greater understanding of the detailed mechanism of the anodic substitution of aromatic hydrocarbons. Bewick et al.[152,153] have reported very detailed studies of the anodic oxidation of methylbenzenes in acetonitrile. They confirm the mechanism to be:

$$PhCH_3 \rightleftharpoons PhCH_3^{+} + e^{-}$$

$$PhCH_3^{+} \underset{k_2}{\overset{k_1}{\rightleftharpoons}} PhCH_2\cdot + H^{+}$$

$$PhCH_2\cdot + PhCH_3^{+} \underset{k_4}{\overset{k_3}{\rightleftharpoons}} PhCH_2^{+} + PhCH_3$$

$$PhCH_2^{+} + CH_3CN \underset{k_6}{\overset{k_5}{\rightleftharpoons}} PhCH_2-\overset{+}{N}=C-CH_3$$

using product analysis, electroanalytical experiments, and modulated specular reflectance spectroscopy (u.v.—visible). The latter technique is particularly powerful since it is possible to identify the spectroscopic characteristics of the cation radical, radical, and carbenium ion during the anodic oxidation, and hence to monitor the concentration–time behaviour of each intermediate, both during anodic generation and during open-circuit decay. The slowest steps in the reactions are the deprotonation of the cation radicals; for pentamethylbenzene the rate-constants reported are $k_1 = 700$ s^{-1}, $k_3 = 7.5 \times 10^9$ dm^3 mol^{-1} s^{-1}, $k_4 \simeq 1$ dm^3 mol s^{-1}, $k_5 = 1 \times 10^5$ s^{-1}, and $k_6 = 1.5 \times 10^3$ s^{-1}. Parker and co-workers[154,155] have also studied the oxidation of hexamethylbenzene and of durene in acetonitrile and in CH_2Cl_2–CF_3CO_2H. While they agree with Bewick et al. concerning the overall reaction pathway, they conclude that the loss of proton from the cation radical is rapid and that the electron exchange between the cation radical and the radical is the slow step; this is clearly not compatible with the data quoted above.

Parker has stressed the importance of deducing reaction mechanisms from the determination of reaction orders, and hence a knowledge of the detailed rate equation (in contrast to the common electrochemical approach of estimating apparent rate constants), and with his co-workers[156—160] applied the method to the

---

[151] J. Phelps and A. J. Bard, *J. Electroanal. Chem. Interfacial Electrochem.*, 1976, **68**, 313.
[152] A. Bewick, G. J. Edwards, J. M. Mellor, and B. S. Pons, *J. Chem. Soc., Perkin Trans. 2*, 1977, 1952.
[153] A. Bewick, J. M. Mellor, and B. S. Pons, *Electrochim. Acta*, 1980, **25**, 931.
[154] R. S. Baumberger and V. D. Parker, *Acta Chem. Scand., Ser. B*, 1980, **34**, 537.
[155] J. Barck, E. Ahlberg, and V. D. Parker, *Acta Chem. Scand., Ser. B*, 1980, **34**, 85.
[156] U. Svanholm and V. D. Parker, *J. Chem. Soc., Perkin Trans. 2*, 1976, 1567.
[157] U. Svanholm and V. D. Parker, *J. Am. Chem. Soc.*, 1976, **98**, 2942.
[158] U. Svanholm and V. D. Parker, *J. Am. Chem. Soc.*, 1976, **98**, 997.
[159] U. Svanholm and V. D. Parker, *Acta Chem. Scand., Ser. B*, 1980, **34**, 5.
[160] E. Ahlberg and V. D. Parker, *Acta Chem. Scand., Ser. B*, 1980, **34**, 97.

study of the reactions of cation radicals ($Ar^{\ddot{+}}$) with nucleophiles. In several systems (*e.g.* thianthrene cation radical + anisole, 9,10-diphenylanthracene cation radical + pyridine) the reaction is found to be, at least in some conditions, second-order in cation radical, first-order in nucleophile, and inverse first-order in neutral hydrocarbon. Hence a mechanism involving an initial formation of a complex between the cation radical and nucleophile is proposed, *i.e.*

$$Ar^{\ddot{+}} + Nu \rightleftharpoons [ArNu]^{\ddot{+}}$$

$$[ArNu]^{\ddot{+}} + Ar^{\ddot{+}} \rightleftharpoons Ar + [ArNu]^{2+}$$

$$[ArNu]^{2+} \rightarrow products \quad (r.d.s.)$$

Steckhan[161,162] has used transmission spectroscopy through an optically transparent electrode to deduce the mechanism for the dimerization of the cation radicals from highly substituted ethylenes. In these systems the simple disproportionation mechanism predominates, and the products arise by attack of the dication on the neutral olefin.

Eberson and his group have continued to use product analysis after the low-conversion oxidation of selected substrates to probe two important features of anodic substitution, *i.e.* the role of adsorption of the substrate and the factors which determine the ratio of nuclear to side-chain substitution. Thus two studies, of the oxidation of two methoxylated indenes[163] and of the acetoxylation of anisole in the presence of other compounds that are likely to be adsorbed on the anode surface,[164] have sought (without success) to demonstrate the need for adsorption of the substrate hydrocarbon. A further paper[165] discusses the role of structural factors on the ratio of nuclear to side-chain substitution; it was shown that with both triptycene (where loss of a proton from the benzylic position would lead to a highly strained radical and cation) and fluorene (where the benzyl cation is anti-aromatic), substitution in the ring is highly preferred.

The mechanism of the anodic oxidation of benzyl ethers has been the subject of dispute.[166,167] Two quite different mechanisms have been proposed, one an overall one-electron oxidation, the other a two-electron process, as shown in Scheme 12.

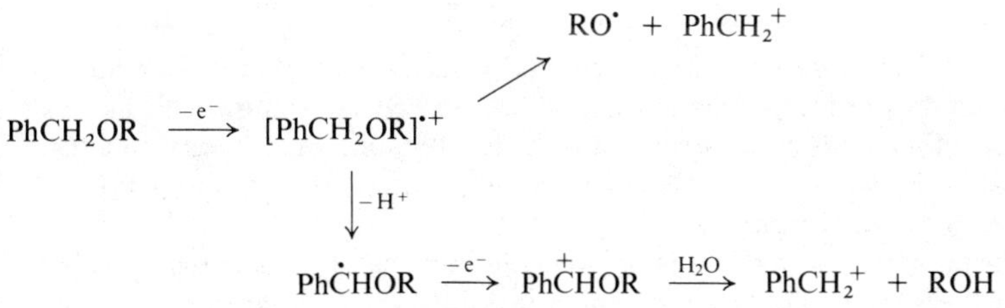

$$PhCH_2OR \xrightarrow{-e^-} [PhCH_2OR]^{\cdot +} \nearrow RO^{\cdot} + PhCH_2^+$$
$$\downarrow -H^+$$
$$Ph\dot{C}HOR \xrightarrow{-e^-} Ph\overset{+}{C}HOR \xrightarrow{H_2O} PhCH_2^+ + ROH$$

Scheme 12

---

[161] E. Steckhan, *Electrochim. Acta*, 1977, **22**, 395.
[162] E. Steckhan, *J. Am. Chem. Soc.*, 1978, **100**, 3526.
[163] L. Cedheim and L. Eberson, *Acta Chem. Scand.*, Ser. B, 1976, **30**, 527.
[164] W. J. M. Van Tilborg, J. J. Scheele, and L. Eberson, *Acta Chem. Scand.*, Ser. B, 1978, **32**, 36.
[165] Z. Blum, L. Cedheim, and L. Eberson, *Acta Chem. Scand.*, Ser. B, 1977, **31**, 662.
[166] R. Lines and J. H. P. Utley, *J. Chem. Soc., Perkin Trans. 2*, 1977, 803.
[167] J. W. Boyd, P. W. Schalzl, and L. L. Miller, *J. Am. Chem. Soc.*, 1980, **102**, 3856.

The evidence formulated to support each mechanism is strong, and there also appear to be some experimental discrepancies, so a final decision as to the mechanism must await further studies by other groups.

Bewick, Mellor, and Pons[168] have described the use of a sulphonic acid cation-exchange resin, *in situ* in the anolyte, for trapping positively charged intermediates from the oxidation of alkyl-benzenes. This experimental procedure leads to an increased yield of the amide after an aqueous work-up; for example, the yield of amide from the oxidation of *p*-xylene and toluene was 27% and 17% whereas in the absence of resin it was less than 1%. Bewick *et al.*[169,170] have discussed the disubstitution of hexamethylbenzene (39) and durene in acetonitrile that can be achieved by oxidation at the potential of the second wave. For the former, the yields are reasonable.

$$\text{Me}_3\text{C}_6\text{H}_2\text{-CH}_2\text{NHCOMe} \xleftarrow[\text{[+1.25 V]}]{-2e^-} \text{Me}_4\text{C}_6\text{H}_2 \xrightarrow[\text{[+1.70 V]}]{-4e^-} \text{Me}_2\text{C}_6\text{H}_2(\text{CH}_2\text{NHCOMe})_2$$

(39)

The oxidation of isodurene (40) in acetic-acid-based media has also been reported.[171] The conversion can be carried out in good yield whether the electrolyte is an acetate, a fluoroborate, or a nitrate, although in the latter case the mechanism is probably different, *i.e.* indirect (*via* NO_3^-) rather than direct. The anode reaction in nitrate medium has also been compared with oxidation by ceric ammonium nitrate; a notable difference is the lower nitrate/acetate ratio in the electrolysis product.

$$\text{Me}_4\text{C}_6\text{H}_2 \xrightarrow[\text{MeCO}_2\text{H}, -2\text{H}^+]{-2e^-} \text{Me}_3\text{C}_6\text{H}_2\text{-CH}_2\text{OCOMe}$$

(40)

Koch, Miller, and Osteryoung[172] have published a detailed description of the electrochemical oxidation of hexamethylbenzene in the room-temperature melt aluminium trichloride–pyridinium chloride and in a 50:50 mixture of the melt and benzene. Electrolysis in the second medium is interesting since it leads to a high-yield transalkylation process; the products are pentamethylbenzene and diphenylmethane (from the benzene).

An application to synthesis of the oxidation of a substituted toluene has been reported. The anodic conversion of 4-acetoxytoluene to 4-acetoxybenzyl acetate could be achieved in 69% yield[173] and was used as one step in a new route to vanillin.

---

[168] A. Bewick, J. M. Mellor, and B. S. Pons, *J. Chem. Soc., Chem. Commun.*, 1978, 738.
[169] A. Bewick, G. J. Edwards, and J. M. Mellor, *Electrochim. Acta*, 1976, **21**, 1101.
[170] A. Bewick, G. J. Edwards, and J. M. Mellor, *Liebigs Ann. Chem.*, 1978, 41.
[171] L. Eberson and E. Oberrauch, *Acta Chem. Scand., Ser. B*, 1979, **33**, 343.
[172] V. R. Koch, L. L. Miller, and R. A. Osteryoung, *J. Am. Chem. Soc.*, 1976, **98**, 5277.
[173] S. Torii, H. Tanaka, T. Siroi, and M. Akada, *J. Org. Chem.*, 1979, **44**, 3305.

The electrolysis step, however, requires carefully selected conditions in order to achieve a good yield without parallel nuclear substitution; the recommended conditions were 9:1 acetic acid:t-butyl alcohol, containing tetraethylammonium tosylate and copper acetate, and a carbon anode. Eberson et al.[174] have reported the application of their large tubular capillary-gap cell to the acetoxylation of 4-halogeno-toluenes (at a rate of 1 mole per hour) in acetic acid–tetrafluoroborate medium (yields 20—40%). They also report the difunctionalization of p-xylene [reaction (11)] to be a selective process.

$$\text{Me-C}_6\text{H}_4\text{-Me} \xrightarrow[2\ \text{MeCO}_2\text{H},\ -4\text{H}^+]{-4e^-} \text{MeCOOCH}_2\text{-C}_6\text{H}_4\text{-CH}_2\text{OCOMe} \quad (11)$$

Helgee and Eberson have continued their studies of two-phase substitution. The electrolyses employ a methylene chloride–water emulsion and a phase-transfer agent is used to supply the nucleophile from a sodium salt in the aqueous layer. The reactions reported include the cyanation of stilbenes,[175] the carboxylation of substrates such as naphthalene, anisole, and durene[176] and the anodic displacement of an alkoxy-group by cyanide,[177] i.e. reaction (12). In general, however, the yields are very moderate. The exception is reaction (12); the material yield is above 70% when R is octyl, and the product is of interest as a liquid crystal.

$$\text{RO-C}_6\text{H}_4\text{-C}_6\text{H}_4\text{-OR} \xrightarrow[\text{CN}^-]{-2e^-} \text{RO-C}_6\text{H}_4\text{-C}_6\text{H}_4\text{-CN} \quad (12)$$

Further studies of anodic cyanation have also been reported. Yoshida and Nagase[178] have discussed the cyanation of methyl- and dimethyl-naphthalenes and show that the major products result from substitution at the para-position (yields of 50—70%). All of the products could be rationalized on the basis of a reaction between cation radical and cyanide ion. There have been two interesting synthetic applications of anodic cyanation. The first[179] concerns the substitution of a steroid (41) to give three major products (42)—(44), and the yield of each can be synthetically useful. For example, conditions are described for the isolation of the cyanodienone (42) in a 60% material yield (although the current yield is very poor). The second paper[180] discusses the cyanation of zinc octaethylporphyrin in DMF–NaCN; the substitution occurs stepwise and provides a route to a mono-, di-, tri-, or tetra-cyanoporphyrin, depending on the electrode potential and the ratio of porphyrin to CN^- ion.

---

[174] L. Eberson, J. Hlavaty, L. Jonsson, K. Nyberg, R. Servin, H. Sternerup, and G. L. Wistrand, *Acta Chem. Scand., Ser. B*, 1979, **33**, 113.
[175] L. Eberson and B. Helgee, *Acta Chem. Scand., Ser. B*, 1978, **32**, 313.
[176] L. Eberson and B. Helgee, *Acta Chem. Scand., Ser. B*, 1978, **32**, 157.
[177] L. Eberson and B. Helgee, *Acta Chem. Scand., Ser. B*, 1977, **31**, 813.
[178] K. Yoshida and S. Nagase, *J. Am. Chem. Soc.*, 1979, **101**, 4268.
[179] K. Ponsold and H. Kasch, *Tetrahedron Lett.*, 1979, 4465.
[180] H. J. Callot, A. Louati, and M. Gross, *Tetrahedron Lett.*, 1980, **21**, 3281.

Several groups have re-examined the methoxylation of benzene, naphthalene, and their derivatives,[181—188] and a procedure for the synthesis of a bis-ketal by the oxidation of 1,4-dimethoxybenzene in methanol has been detailed in *Organic Syntheses*.[189] Swenton and co-workers[181—185] have re-affirmed that the oxidation of 1,4-dimethoxy-compounds in alkaline methanol is a good route to bis-ketals and, following controlled hydrolysis, to mono-ketals. Furthermore, they have used the procedure in a synthesis of anthracycline antibiotics; the oxidation of 2-bromo-1,4-dimethoxybenzene is followed by metalation with butyl-lithium to give a quinone anion which is used in later molecule-building steps. Two surprising reactions have been reported; the oxidation of naphthalene in methanol that is saturated with calcium oxide, at platinum and at 0—5 °C, in darkness, is said to give the blue biradical (45),[187] while the oxidation of toluene in acidic methanol–methylene chloride has been reported to give a 20% yield of *p*-methoxybenzaldehyde (46) by the sequence shown.[188]

Ponsold and Kasch[190] have reported the benzylic methoxylation of the steroid (41) in CH_2Cl_2–MeOH–$NaClO_4$ that contains 2,6-lutidine, in almost quantitative yield [the ratio of the two isomers (47) and (48) is 60:40].

[181] M. J. Manning, P. W. Raynolds, and J. S. Swenton, *J. Am. Chem. Soc.*, 1976, **98**, 5008.
[182] M. J. Manning, D. R. Henton, and J. S. Swenton, *Tetrahedron Lett.*, 1977, 1679.
[183] M. G. Dolson, D. K. Jackson, and J. S. Swenton, *J. Chem. Soc., Chem. Commun.*, 1979, 327.
[184] D. R. Henton, B. L. Chenard, and J. S. Swenton, *J. Chem. Soc., Chem. Commun.*, 1979, 326.
[185] D. R. Henton, R. L. McCreery, and J. S. Swenton, *J. Org. Chem.*, 1980, **45**, 369.
[186] G. Bockmair and H. P. Fritz, *Electrochim. Acta*, 1976, **21**, 1099.
[187] F. Barba, A. Soler, and J. Varea, *Tetrahedron Lett.*, 1976, 557.
[188] M. Rakoutz, D. Michelet, B. Brossard, and J. Varagnat, *Tetrahedron Lett.*, 1978, 3723.
[189] P. Margaretha and P. Tissot, *Org. Synth.*, 1977, **57**, 92.

Me ~~-2e⁻→~~ Me/OMe ~~-2e⁻→~~ CH$_2$OMe/OMe ~~-2e⁻→~~ CHO/OMe
(46)

(41) ~~MeOH, -2H⁺→~~ (47) + (48)

The selective halogenation of aromatic compounds remains a useful goal. Anodic nucleophilic substitution, however, is generally unsuccessful; with iodide, bromide, and chloride the problem is normally transfer of electrons between the cation radical and the halide ion. Hence most investigators have concentrated on alternative electrolytic routes. Casalbore et al.[191–193] have studied the bromination of simple aromatic hydrocarbons such as benzene, toluene, and xylene in acetic acid or acetonitrile, but at potentials well positive to that for the conversion of Br⁻ into Br$_2$, and where further oxidation peaks are observed on cyclic voltammograms. Good yields of brominated products are obtained, although with the methylbenzenes, both side-chain and nuclear substitution occurs. The mechanism is, however, less certain, and it is not known whether the key step is the oxidation of Br$_2$ to Br⁺ or of a π-complex, ArH:Br$_2$. The same group[194] have extended these studies to chlorination while Simonet and co-workers[195] have also reported the chlorination of simple aromatic compounds in acetonitrile at more positive potentials, and have emphasized the advantages of the anolyte containing a Lewis acid. In these conditions the yields are impressive (e.g. 80% of chlorobenzene and 95% of chlorotoluenes from benzene and toluene) and influence the para to ortho ratio of chlorotoluenes. This ratio is determined by the concentration of toluene and Lewis acid, the choice of Lewis acid, and the current density, and it reaches 1.80, compared to 0.64 in the absence of a Lewis acid.

The cation I⁺, even if it exists as a solvent-complexed species (e.g. MeC⁺=N—I in acetonitrile), is a better defined intermediate than the other X⁺ cations. Hence there is no doubt concerning the mechanism of the iodination that occurs when substrate

[190] K. Ponsold and H. Kasch, *Tetrahedron Lett.*, 1979, 4463.
[191] G. Casalbore, M. Mastragostini, and S. Valcher, *J. Electroanal. Chem. Interfacial Electrochem.*, 1976, **68**, 123.
[192] G. Casalbore, M. Mastragostini, and S. Valcher, *J. Electroanal. Chem. Interfacial Electrochem.*, 1977, **77**, 373.
[193] G. Casalbore, M. Mastragostini, and S. Valcher, *J. Electroanal. Chem. Interfacial Electrochem.*, 1978, **87**, 411.
[194] M. Mastragostini, G. Casalbore, S. Valcher, and L. Pastorelli, *J. Electroanal. Chem. Interfacial Electrochem.*, 1978, **90**, 439.
[195] J. Gourcy, J. Simonet, and M. Lascaud, *Electrochim. Acta*, 1979, **24**, 1039.

and iodine are electrolysed. Miller and Watkins[196] have reported such substitution to occur in excellent yields for a wide range of substituted benzenes and have also used the determination of relative rate constants, from competitive reactions and isotope effects, to confirm the mechanism. They found that *para*-iodination was always favoured although the *para* to *ortho* ratio depended on the substituent in the substrate. The reaction was not possible in neutral acetonitrile or methylene chloride when the substrate was deactivated by an electron-withdrawing group. Lines and Parker,[197] however, demonstrated that even nitrobenzene and benzaldehyde could be iodinated if the electrolysis medium was trifluoroacetic acid and a solvent. The reactivity of the I$^+$ varied with the percentage of TFA and the co-solvent, but 3-iodonitrobenzene was formed in 78% yield in $ClCH_2CH_2Cl$–TFA.

Matsue *et al.*[198] have reported an interesting observation about the *ortho* to *para* ratio of the products from the anodic chlorination of anisole in aqueous sodium chloride. With a graphite anode that was covered by cyclodextrin (either chemically bonded to or adsorbed on the surface), the *para*/*ortho* ratio was 17 or 25, respectively, compared to 7 at an untreated graphite surface. The reaction is an example of indirect chlorination, by reaction between hypochlorite and anisole. Since it is known that cyclodextrin complexes anisole such that the anisole is within the tub structure of the cyclodextrin, it is proposed that it is the formation of an adduct which directs the *para*-substitution at the anode surface.

With anodic nucleophilic fluorination, the problem is not the transfer of an electron between the cation radical and fluoride ion but rather the low solubility of fluorides in aprotic media and the low nucleophilicity of fluoride ion. The solvate $Et_4NF \cdot 3HF$, in acetonitrile, has been shown to give a medium where some fluorination is possible. Both the formation of 9,10-difluoro-9,10-diphenyl-anthracene from diphenylanthracene[199] and the monofluorination of hexamethylbenzene and *p*-xylene[200] have been reported.

Many of the studies of acetoxylation were initiated in order to find new routes to phenols. It is now clear, however, that there are more promising routes. Thus trifluoroacetoxylation tends to give much higher yields because the products are stable to further oxidation if the potential is controlled, and the esters are more readily hydrolysed. Indeed, hydrolysis to phenols generally occurs during extraction of the trifluoroacetate esters. So and Miller[201] have described a very straightforward procedure, using 2:1 nitromethane:trifluoroacetic acid in an undivided cell, which with substrates such as benzene, chlorobenzene, acetophenone, and ethyl benzoate leads (on aqueous work-up) to phenols, with yields in the range 60—80%. Bockmair, Fritz, and Gebauer[202] have reported a detailed study of the oxidation of chlorobenzene in trifluoroacetic acid and have shown that the total yield of isomeric monotrifluoroacetates can be 94%; on continued constant-current electrolysis it was also possible to obtain diesters

---

[196] L. L. Miller and B. Watkins, *J. Am. Chem. Soc.*, 1976, **98**, 1515.
[197] R. Lines and V. D. Parker, *Acta Chem. Scand., Ser. B*, 1980, **34**, 47.
[198] T. Matsue, M. Fujihira, and T. Osa, *J. Electrochem. Soc.*, 1979, **126**, 500.
[199] I. N. Rozhkov, N. P. Gambaryan, and E. G. Galpern, *Tetrahedron Lett.*, 1976, 4819.
[200] A. Bensadat, G. Bodennec, E. Laurent, and R. Tardivel, *Tetrahedron Lett.*, 1977, 3799.
[201] Y. H. So and L. L. Miller, *Synthesis*, 1976, 468.
[202] G. Bockmair, H. P. Fritz, and H. Gebauer, *Electrochim. Acta*, 1978, **23**, 21.

(maximum yield 70%) and triesters (maximum yield 25%). Fritz and Kremer[203] have examined the trifluoroacetoxylation of benzene and alkyl-benzenes and found it advantageous to have a catalyst (Tl^I or Ce^{III}) in the electrolyte. The synthesis of phenols from benzene and hydrogen peroxide in the presence of Fe^{II} (Fenton's reagent) has also been shown to give reasonable yields if there is a large excess of both Fe^{III} and benzene in the system but the concentration of hydrogen peroxide does not greatly exceed that of Fe^{II}.[204] The conditions are readily obtained by the dropwise addition of hydrogen peroxide and by controlling the concentration of Fe^{II} electrolytically. The yield in such conditions is 64% (based on H_2O_2). A later paper[205] extends the procedure to the synthesis of phenols from chlorobenzene, fluorobenzene, and benzonitrile (yields are 20—80%). Tomat and Rigo[206] have studied oxidation of the side-chain in a similar system. They generated the hydrogen peroxide by reduction of oxygen and showed that, particularly in the presence of chloride ion, the co-reduction of $Fe^{III} + O_2$ leads to a powerful oxidizing agent. For example, methylbenzaldehydes could be produced from the corresponding xylenes in yields in excess of 80%.

Clarke et al.[207] have studied the mechanism for the oxidation of benzene, toluene, and anisole at a PbO_2 anode in sulphuric acid; they confirmed that quinone can be formed from benzene with a yield between 60 and 100%, depending on the potential. Moreover, Oloman[208] has described an attempt to scale up the oxidation of benzene in a packed lead-bed-anode reactor and achieved a reasonable performance. A study of the hydroxylation of polychlorinated biphenyls in acetonitrile–water has also been published;[209] its objectives were not synthetic but to gain an insight into the mechanism of oxidative degradation.

Nyberg and co-workers have reported the anodic displacement of fluoride by carboxylate groups.[210,211] The oxidation of 4-fluoroanisole in an acetic acid medium was found to lead to 4-acetoxyanisole rather than to the expected substitution products, and this principle was then used to synthesize polyfluoroquinones from hexafluorobenzene and octafluoronaphthalene (yields were 75% and 60%, respectively) by oxidation in trifluoroacetic acid. Bellamy[212] has reported the nitromethylation of benzene, using electrogenerated manganese(III) acetate in acetic acid–lithium tetrafluoroborate.

Further studies of aryl olefins have been described. Schäfer and co-workers[213] have investigated the dimerization of four styrenes and of indene under various electrolysis conditions, but the yields of products were only moderate. Verniette[214] et al. have reported the oxidation of olefins and acetylenes in dimethylformamide and in acetonitrile that contains chloride ion. The products arise by the reactions

[203] H. P. Fritz and H. J. Kremer, Z. Naturforsch., Teil. B, 1976, 31, 1565.
[204] E. Steckhan and J. Wellman, Angew. Chem., Int. Ed. Engl., 1976, 15, 294.
[205] E. Steckhan and J. Wellman, Chem. Ber., 1977, 110, 3561.
[206] R. Tomat and A. Rigo, J. Appl. Electrochem., 1979, 9, 301.
[207] J. S. Clarke, R. Ehigamaso, and A. T. Kuhn, J. Electroanal. Chem. Interfacial Electrochem., 1976, 70, 333.
[208] C. Oloman, J. Appl. Electrochem., 1980, 10, 553.
[209] R. J. Fenn, K. W. Krantz, and J. D. Stuart, J. Electrochem. Soc., 1976, 123, 1643.
[210] K. Nyberg and L. G. Wistrand, J. Chem. Soc., Chem. Commun., 1976, 898.
[211] Z. Blum and K. Nyberg, Acta Chem. Scand., Ser. B, 1979, 33, 73.
[212] A. J. Bellamy, Acta Chem. Scand., Ser. B, 1979, 33, 208.
[213] R. Engels, H. J. Schäfer, and E. Steckhan, Liebigs Ann. Chem., 1977, 204.
[214] M. Verniette, C. Daremon, and J. Simonet, Electrochim. Acta, 1978, 23, 929.

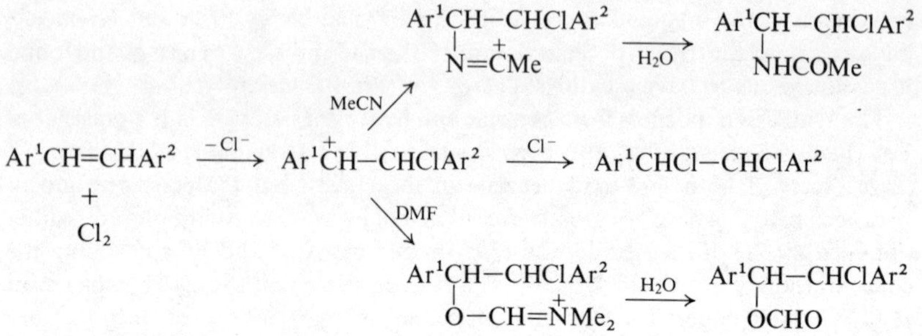

Scheme 13

$$Ar^1CH=CHAr^2 + Cl_2 \xrightarrow{-Cl^-} Ar^1\overset{+}{C}H-CHClAr^2$$

Routes from $Ar^1\overset{+}{C}H-CHClAr^2$:
- MeCN → $Ar^1CH-CHClAr^2$ with $\overset{+}{N}=CMe$ group $\xrightarrow{H_2O}$ $Ar^1CH-CHClAr^2$ with NHCOMe group
- Cl^- → $Ar^1CHCl-CHClAr^2$
- DMF → $Ar^1CH-CHClAr^2$ with $O-CH=\overset{+}{N}Me_2$ group $\xrightarrow{H_2O}$ $Ar^1CH-CHClAr^2$ with OCHO group

shown in Scheme 13 and the mixed addition products can be formed in reasonable yields. With the acetylenes, products from two-electron or four-electron oxidation can be isolated, and in DMF the products from the four-electron oxidation are chloro-ketones. Similar reactions in acetonitrile that contains $Et_4NF \cdot 3HF$ also give the mixed product, e.g. (49),[200,215] although the mechanism is presumably different; the initial step is almost certainly the removal of an electron from the olefin.

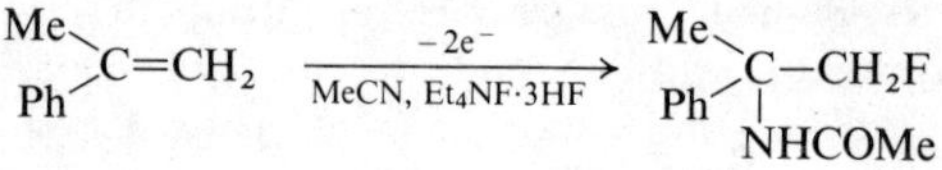

$$\underset{Ph}{\overset{Me}{>}}C=CH_2 \xrightarrow[MeCN, Et_4NF \cdot 3HF]{-2e^-} \underset{Ph}{\overset{Me}{>}}\underset{NHCOMe}{\overset{|}{C}}-CH_2F$$

(49)

The effect of substituents on the oxidation potentials of both triphenyl-methanes[216] and the corresponding anions[217] has been discussed.

Phenols.—Hammerich et al.[218] have shown that it is possible to measure formal potentials for the one-electron oxidation of phenols under conditions where the loss of proton from the cation radical is slow, i.e. $CH_2Cl_2-10\%$ HSO_3F at $-50\ °C$. The phenols show reversible one-electron oxidation processes on cyclic voltammograms and the values of the formal potentials confirm that, under most conditions, subsequent chemical reactions cause a large negative shift in the oxidation potential. The authors also compare the formal potentials with those for the corresponding methyl ethers and show that, as expected, the phenols are most difficult to oxidize.

The oxidation of a wide range of phenol structures in methanol has been investigated by Ronlan and co-workers.[219] Methoxylation and dimerization at the *ortho*- and *para*-positions can occur as well as further oxidation of the primary products. Hence with phenol itself, and lightly substituted compounds, the product spectrum is generally complex and very dependent on the experimental conditions (e.g. potential, concentrations, stirring) although it is sometimes possible to select the parameters so that one product is obtained in reasonable yield. With increasing

---

[215] I. N. Rozhkov, I. Ya. Aliev, and I. L. Knunyants, *Izv. Akad. Nauk SSSR, Ser. Khim.*, 1976, 1418.
[216] J. E. Kuder, W. W. Limburg, M. Stolka, and S. R. Turner, *J. Org. Chem.*, 1979, **44**, 761.
[217] S. Bank, G. L. Ehrlich, and J. A. Zubieta, *J. Org. Chem.*, 1979, **44**, 1454.
[218] O. Hammerich, V. D. Parker, and A. Ronlan, *Acta Chem. Scand., Ser. B*, 1976, **30**, 89.
[219] A. Nilsson, U. Palmquist, T. Pettersson, and A. Ronlan, *J. Chem. Soc., Perkin Trans. 1*, 1978, 696.

substitution, the products become simpler, and this and other papers[220–222] report studies of hindered phenols. Thus phenols with two t-butyl groups in the *ortho*-position and an alkyl group in the *para*-position, *e.g.* (50), give single major products in high yield (60—95%, for example, where R is Me, Et, Pr^i, or Bu^t). The behaviour of the phenolate anions has also been studied.[222] In acetonitrile that contains a base, the oxidation of the anions to the radicals is reversible on the time-scale of a cyclic voltammogram, but on a longer time-scale the radicals disproportionate to give a quinone methide. For example, for (50; R = Pr^i), the reaction proceeds as shown in Scheme 14. In dry acetonitrile, the quinone methide (51) isomerizes to the vinylphenol and is oxidized further, but in acetonitrile–10% water the yield is 95%. The quinone methide from the phenol (50; R = Et) is quite stable, and is formed almost quantitatively. The oxidation of the polyfunctional compound 1,6-dimethoxy-4-allylphenol in methanol has been described.[223] The spectrum of products is always complex. In basic solution, two major products (52) and (53) are

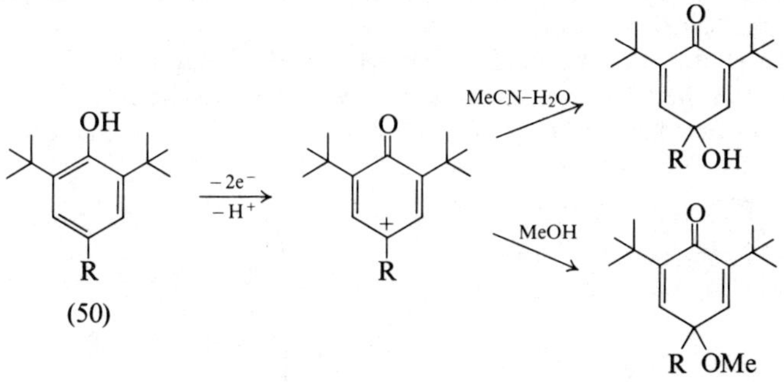

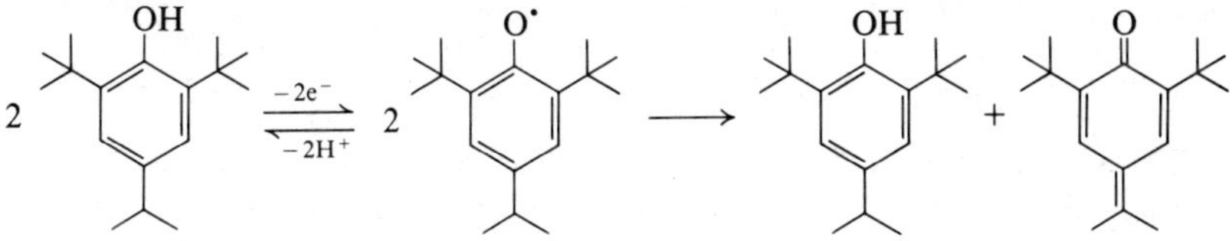

Scheme 14

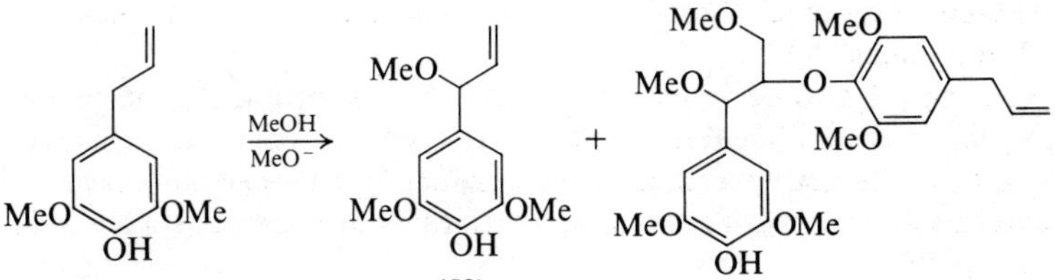

[220] B. Speiser and A. Rieker, *J. Chem. Res. (S)*, 1977, 314.
[221] A. Rieker, E-L. Dreher, H. Geisel, and M. H. Khalifa, *Synthesis*, 1978, 851.
[222] J. A. Richards and D. H. Evans, *J. Electroanal. Chem. Interfacial Electrochem.*, 1977, **81**, 171.
[223] M. Iguchi, A. Nishiyama, Y. Terada, and S. Yamamura, *Tetrahedron Lett.*, 1977, 4511.

formed, in 30% and 21% yield, while a very complex mixture results in neutral solution. One of those products is, however, asatone; although in very low yield, it is a one-step synthesis.

Ohmori et al.[224] have described the oxidation of 2-hydroxy-3-methoxy-5-methylbenzaldehyde and its Schiffs bases in acetonitrile that contains pyridine. Pyridination occurs at the methyl group of the parent compound but in the ring for most of the Schiffs bases.

Ronlan et al.[225] have examined the oxidation of a series of substituted diarylalkanes. Only the 1,3-diaryl-propanes (54) give non-polymeric products, but these compounds gave high yields of spiro-dienones (55) by intramolecular coupling. In CH_2Cl_2–CF_3SO_3H the products are different, arising from an acid-catalysed rearrangement of the spiro-dienones.

The electrolysis of phenol in acidic aqueous sodium bromide, using a platinum or a platinized titanium anode, leads to p-bromophenol (yield of 70%) or 2,4-dibromophenol (maximum yield of 80%), depending on the duration of the electrolysis.[226] The reaction does not involve initial transfer of an electron from the phenol, but the products are not the same as those from the homogeneous bromination of phenol. Therefore it is proposed that the active species is an adsorbed bromine atom.

Eggins and Harwood[227] have discussed the mechanism of oxidation of hydroquinone esters while Ryan et al.[228] have used electroanalytic techniques to investigate the chemistry of orthoquinones resulting from the oxidation of catechols in aqueous solution. Two papers have described electrochemical studies of large

---

[224] H. Ohmori, A. Matsumoto, and M. Masui, J. Chem. Soc., Perkin Trans. 2, 1980, 347.
[225] U. Palmquist, A. Nilsson, V. D. Parker, and A. Ronlan, J. Am. Chem. Soc., 1976, **98**, 2571.
[226] T. Bejeramo and E. Gileadi, Electrochim. Acta, 1976, **21**, 231.
[227] B. R. Eggins and R. Harwood, J. Electroanal. Chem. Interfacial Electrochem., 1977, **83**, 347.
[228] M. D. Ryan, A. Yueh, and W. Y. Chen, J. Electrochem. Soc., 1980, **127**, 1489.

Organic Electrochemistry – Synthetic Aspects: Oxidations

phenol derivatives that occur in Nature; one study concerns phenols that are related to lignin[229] and the other the halomicine group of antibiotics.[230]

Amines.—The cation radicals of many triarylamines are known to be stable in aprotic media. Steckhan and co-workers[231–235] have investigated the redox chemistry of a series of brominated triarylamines and emphasized the suitability of such amines as redox catalysts for synthetic reactions in aprotic solvents. The reactions reported include the decarboxylation of aliphatic carboxylates,[232] the cleavage of benzyl groups that are protecting alcohol groups,[233–235] and the oxidative removal of 1,3-dithian protecting groups from ketones.[236]

Serve has reported a study of the oxidation of the triarylamine o-$NH_2C_6H_4NPh_2$ in acetonitrile, both unbuffered and in the presence of acid and of base.[237] Similar investigations have been reported of diarylamines with various degrees of substitution;[238,239] the stabilities of the intermediates depend strongly on the structure of the parent amine and on the presence of free acid or base. Moreover, the decay pathways, and hence the products that are isolated, vary with the presence of such bases as are able to remove protons from reaction intermediates. For example, three different products may be obtained from di-p-anisylamine (56) in acetonitrile, depending on the nature of the base that is present in solution.

$$(p\text{-MeOC}_6H_4)_2NH \quad \begin{array}{l} \xrightarrow[-e^-]{\text{unbuffered MeCN}} (p\text{-MeOC}_6H_4)_2NH^{\cdot+} \\ \xrightarrow[-e^-, -H^+]{\text{MeCN, 2,6-lutidine}} \tfrac{1}{2} \text{(phenazine structure with } C_6H_4OMe\text{-}p \text{ groups)} \\ \xrightarrow[-e^-, -H^+]{\text{MeCN, CN}^-} \tfrac{1}{2}(p\text{-MeOC}_6H_4)_2N-N(C_6H_4OMe\text{-}p)_2 \end{array}$$

An attempt to methoxylate diphenylamine by oxidation in basic methanol was totally unsuccessful;[240] the yield of monomethoxylated product was only 7.6%.

The anodic behaviour of acridan, 9- and 10-substituted acridans, and 10,10'-dimethyl-9,9'-spirobiacridan in acetonitrile has been described.[241] Cyclic

---

[229] H. Chabaud, F. Sundholm, and G. Sundholm, *Electrochim. Acta*, 1978, **23**, 659.
[230] A. K. Ganguly, P. Kabasakalian, S. Kalliney, O. Sarre, S. Szmulewicz, and A. Westcott, *J. Org. Chem.*, 1976, **41**, 1258.
[231] W. Schmidt and E. Steckhan, *Chem. Ber.*, 1980, **113**, 577.
[232] W. Schmidt and E. Steckhan, *J. Electroanal. Chem. Interfacial Electrochem.*, 1978, **89**, 215.
[233] W. Schmidt and E. Steckhan, *Angew. Chem., Int. Ed. Engl.*, 1978, **17**, 673.
[234] W. Schmidt and E. Steckhan, *Angew. Chem., Int. Ed. Engl.*, 1979, **18**, 801.
[235] W. Schmidt and E. Steckhan, *Angew. Chem., Int. Ed. Engl.*, 1979, **18**, 802.
[236] M. Platen and E. Steckhan, *Tetrahedron Lett.*, 1980, **21**, 511.
[237] D. Serve, *Bull. Soc. Chim. Fr.*, 1976, 1567.
[238] C. Cauquis, H. Delhomme, and D. Serve, *Electrochim. Acta*, 1976, **21**, 557.
[239] D. Serve, *Electrochim. Acta*, 1976, **21**, 1171.
[240] K. Yoshida and T. Fueno, *J. Org. Chem.*, 1976, **41**, 731.
[241] E. Sturm, H. Kiesele, and E. Daltrozzo, *Chem. Ber.*, 1978, **111**, 227.

voltammetry at high rates of potential scan shows that they each initially undergo a reversible one-electron oxidation, but, on a longer time-scale the compounds that are fully alkylated at the 9-position couple at the 2- and 2'-positions and are oxidized further, while acridans that are unsubstituted at the 9-position give acridinium ions; examples are shown in Scheme 15.

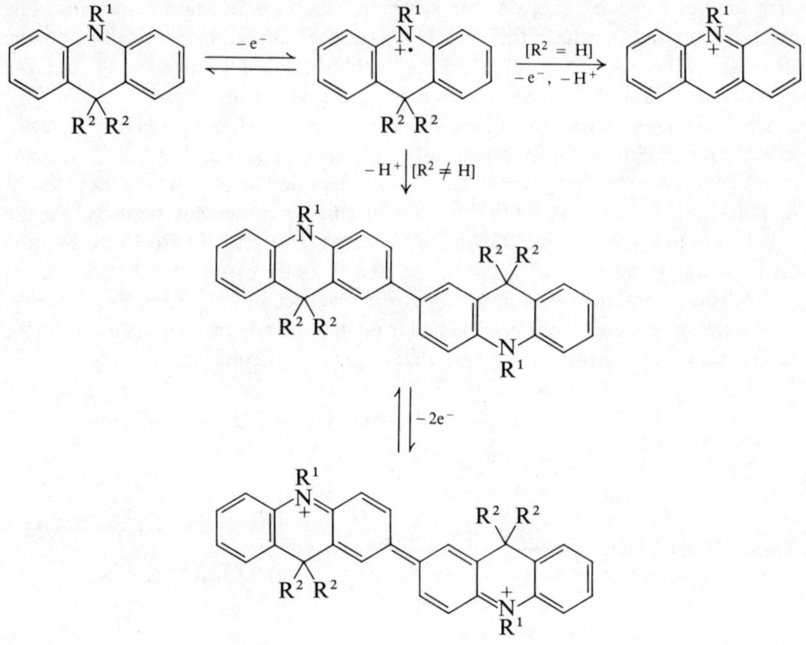

Scheme 15

The oxidation of 4-halogeno-NN-dimethylaniline in acetonitrile has been re-examined.[242] While it was confirmed that the 4,4'-dimer is formed with loss of halogen, a different mechanism has been proposed. It is now proposed that the cation radical dimerizes reversibly, with loss of halogen as the rate-determining step. The oxidation of N-methyl-N-benzylanisylamine in wet acetonitrile leads to benzoquinone, methanol, and N-methylbenzylamine.[243]

Hand and Nelson[244] have continued their work on the oxidation of anilines in aqueous solution with a study of *ortho*- and *meta*-substituted anilines at carbon electrodes in 6M-H_2SO_4. Cyclic voltammetry is well suited to the elucidation of the mechanism and of the kinetics of the complex coupled reactions which result from both tail-to-tail and head-to-tail coupling of the cation radicals and further oxidation and hydrolysis of the dimeric products. In some cases, reasonable yields of substituted *p*-benzoquinone can be isolated. A similar study, but using thin-layer

---

[242] E. Ahlberg, B. Helgee, and V. D. Parker, *Acta Chem. Scand.*, Ser. B, 1980, **34**, 187.
[243] M. Masui, H. Ohmori, H. Sayo, A. Ueda, and C. Ueda, *J. Chem. Soc., Perkin Trans. 2*, 1976, 1180.
[244] R. L. Hand and R. F. Nelson, *J. Electrochem. Soc.*, 1978, **125**, 1059.

cells, has been made of NN-dimethyl-p-phenylenediamines in neutral and weakly acidic aqueous solutions.[245,246] Nelson et al.[247] have also used cyclic voltammetry in benzonitrile to examine the effect of structure on the ease of removal of electrons from o-phenylenediamines. Where necessary, to obtain stable cation radicals and dications, the temperature was reduced to $-60\,°C$.

Miscellaneous Aromatic Compounds.—A number of papers have reported the oxidation of aromatic compounds by superoxide, itself formed by the cathodic reduction of oxygen in aprotic solvents.[248–252] Sagae et al.[248,249] have described the reaction of superoxide with nitrotoluenes. With the *ortho*- and *para*-isomers the products are the nitrobenzoic acid, in reasonable yield (40—55%), but *m*-nitrotoluene does not react; p-ethylnitrobenzene does react, but both the α- and β-carbons are attacked. The reaction is thought to proceed by abstraction of a hydrogen atom by the superoxide ion. The same group have studied the oxidation of picolines.[250] The rate of oxidation again depends on the position of the methyl group in the ring (the major products are the corresponding picolinic acid), and it was also found that the N-oxides react more rapidly than the picolines themselves. The reaction of superoxide with o- and p-halogeno-nitrobenzenes[251] is by nucleophilic attack of the O_2^- on the C–X bond to give the nitrophenol (yield is 39—52%) as the major product. The superoxide ion also reacts with α-diketones, with cleavage of the C–C bond and the formation of carboxylic acids.[252] Typical reactions that have recently been reported include those of (57) and (58). A preliminary note[253] records the reactivity of o-phenylenediamine towards superoxide ion.

Moyer and his co-workers have described some other, very interesting, indirect oxidations where the redox catalysts are Ru^{II}/Ru^{IV} couples. A number of related ruthenium complexes have been used, but the most active and stable is the mixed

---

[245] D. Lelievre, A. Henriet, and V. Plichon, *J. Electroanal. Chem. Interfacial Electrochem.*, 1977, **78**, 281.
[246] D. Lelievre, V. Plichon, and M. A. Dosal, *J. Electroanal. Chem. Interfacial Electrochem.*, 1977, **78**, 301.
[247] S. F. Nelson, E. L. Clennan, L. Echegoyan, and L. A. Grezzo, *J. Org. Chem.*, 1978, **43**, 2621.
[248] H. Sagae, M. Fujihira, T. Osa, and H. Lund, *Chem. Lett.*, 1977, 793.
[249] H. Sagae, M. Fujihira, H. Lund, and T. Osa, *Bull. Chem. Soc. Jpn.*, 1980, **53**, 1537.
[250] H. Sagae, M. Fujihira, H. Lund, and T. Osa, *Heterocycles*, 1979, **13**, 321.
[251] H. Sagae, M. Fujihira, K. Komozawa, H. Lund, and T. Osa, *Bull. Chem. Soc. Jpn.*, 1980, **53**, 2188.
[252] K. Boujlet and J. Simonet, *Tetrahedron Lett.*, 1979, 1063.
[253] C. L. Hussey, T. M. Laker, and J. M. Accord, *J. Electrochem. Soc.*, 1980, **127**, 1865.

complex with terpyridyl, bipyridyl, and water as ligands. The potential of the oxidation $Ru^{II} \rightarrow Ru^{IV}$ is then only +0.61 V vs the S.C.E. in a phosphate or borate buffer, but the presence of a catalytic quantity of the Ru^{II} complex allows anodic oxidation of, for example, p-xylene to terephthalic acid (the oxidation of alcohols to aldehydes or ketones and of olefins was also reported).[254] Similar redox couples have been employed as catalysts for the oxidation of triphenylphosphine to triphenylphosphine oxide.[255,256]

Other reactions involving phosphorus compounds have been discussed. Thus Ohmori et al.[257] have investigated the oxidation of triphenylphosphine in acetonitrile. In the dry solvent the behaviour is complex, but in acetonitrile–10% water the triphenylphosphine oxide may be isolated in almost quantitative yield. Moreover, if the oxidation is carried out in the presence of a primary amine, the coupled product (59) is formed, in a yield of 50—70%. The formation of

$$Ph_3P \xrightarrow{-e^-} Ph_3P^{\cdot +} \xrightarrow{RNH_2} [Ph_3P-NH_2R]^{\cdot +} \xrightarrow[-H^+]{-e^-} Ph_3\overset{+}{P}-NHR$$
(59)

phosphorus–sulphur bonds can be achieved by the co-electrolysis of a diaryl (or dialkyl) disulphide with a diaryl (or dialkyl) phosphate.[258] The yield is, however, much better if the oxidation is carried out in the presence of bromide ion. The mechanism is probably as shown in Scheme 16, although initial attack of the bromine on the disulphide is also possible. With bromide, the yield is >90%. Woerner and Gulick[259] have sought to define the mechanism of the oxidation of triphenylphosphine phenylimine in acetonitrile. The primary radical cation can be observed by e.s.r. spectroscopy but the final product mixture is complicated, containing triphenylphosphine oxide (from hydrolysis of P=N bond) and much unchanged starting material.

$$2Br^- \longrightarrow Br_2 + 2e^-$$

$$(RO)_2\overset{O}{\overset{\|}{P}}H + Br_2 \longrightarrow (RO)_2\overset{O}{\overset{\|}{P}}Br + HBr$$

$$(RO)_2\overset{O}{\overset{\|}{P}}Br + PhSSPh \longrightarrow (RO)_2\overset{O}{\overset{\|}{P}}-SPh + PhSBr$$

$$PhSBr + (RO)_2\overset{O}{\overset{\|}{P}}H \longrightarrow (RO)_2\overset{O}{\overset{\|}{P}}SPh + HBr$$

Scheme 16

---

[254] B. A. Moyer, M. S. Thompson, and T. J. Meyer, J. Am. Chem. Soc., 1980, **102**, 2310.
[255] F. R. Keene, D. T. Salmon, and T. J. Meyer, J. Am. Chem. Soc., 1977, **99**, 4821.
[256] B. A. Moyer and T. J. Meyer, J. Am. Chem. Soc., 1978, **100**, 3601.
[257] H. Ohmori, S. Nakai, and H. Masui, J. Chem. Soc., Perkin Trans. 1, 1978, 1333.
[258] S. Torii, H. Tanaka, and N. Sayo, J. Org. Chem., 1979, **44**, 2938.
[259] C. J. Woerner and W. M. Gulick, Electrochim. Acta, 1977, **22**, 445.

$$4X^- \longrightarrow 2X_2 + 4e^-$$

$$2X_2 + \text{PhSeSePh} \longrightarrow 2\text{PhSeX} + 2X^-$$

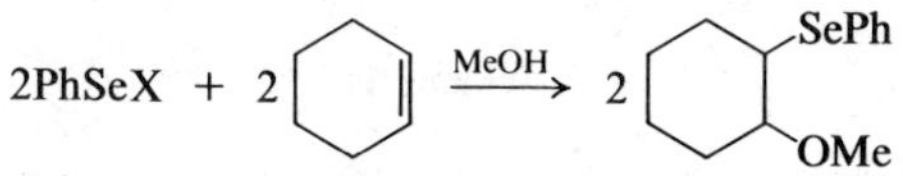

2PhSeX + 2 [cyclohexene] $\xrightarrow{\text{MeOH}}$ 2 [cyclohexane with SePh and OMe]

Scheme 17

2 [ketone CH₂R] + PhSe—SePh ⟶ 2 [ketone CH(SePh)R]

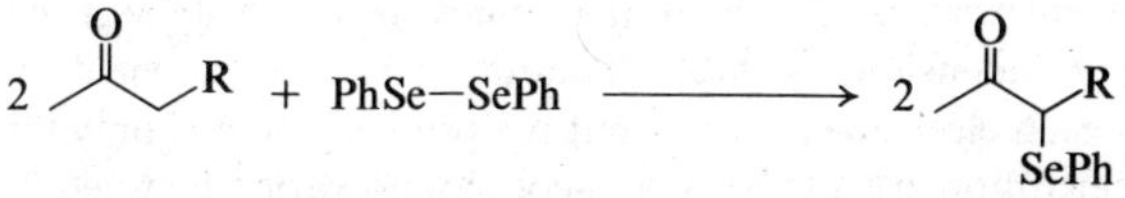

(60)

Some novel electrosynthetic reactions involving selenium compounds have been reported.[260–263] The oxyselenation of olefins[260] has been shown to be possible by electrolysis in a solution of diphenyl diselenide in methanol that contains catalytic quantities of bromide or chloride ions. The reaction mechanism is shown in Scheme 17 and the product is formed in 80—100% yield. A similar mechanism, *i.e.* attack of PhSeBr on the enol form of the ketone, is likely for the formation of α-phenylselenylcarbonyl compounds (60)[261] by another reaction which gives high yields (84—97%). The oxyselenation of olefins has been applied to the synthesis of the natural products *dl*-marmelactone (61), as shown in Scheme 18, and *dl*-rose oxide.[262]

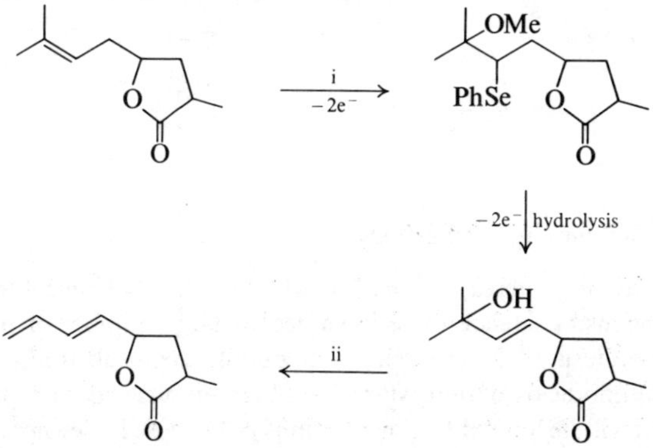

(61) [83%]

Reagents: i, (PhSe)₂, MeOH, Br⁻; ii, MeSO₂Cl, Et₃N, CH₂Cl₂

Scheme 18

[260] S. Torii, K. Uneyama, and K. Honda, *Tetrahedron Lett.*, 1980, **21**, 1863.
[261] S. Torii, K. Uneyama, and M. Ono, *Tetrahedron Lett.*, 1980, **21**, 2741.
[262] S. Torii, K. Uneyama, and M. Ono, *Tetrahedron Lett.*, 1980, **21**, 2653.

Seeber et al.[263] have discussed the oxidation of diphenyl selenide in acetonitrile, and they observed products formed by the routes shown in Scheme 19.

$$2\,Ph_2Se \rightarrow O \xleftarrow[H_2O]{-2e^-} Ph_2Se \xrightarrow[Ph_2Se]{-2e^-,\,-H^+} Ph_2\overset{+}{S}e-C_6H_4-SePh$$

Scheme 19

The attempts to carry out asymmetric synthesis electrochemically have been extended to the oxidation of phenyl methyl sulphide in wet acetonitrile at anode surfaces that were prepared by chemical modification of pyrolytic graphite[264] and at tin dioxide and dimensionally stable anodes.[265] The chemical yields of sulphoxide were good at each electrode (>90%) but the product showed only limited optical activity. The enantiomeric excess of (−)-compound varied between 0.3 and 2.5%, and depended on both the anode material and the potential.

The oxidation of a dithiocarboxylate leads to the formation of a S–S bond in high yields.[266] At more positive potentials, further oxidation of the product is observed, and it has been suggested that the product of this reaction is the S-heterocycle (62).

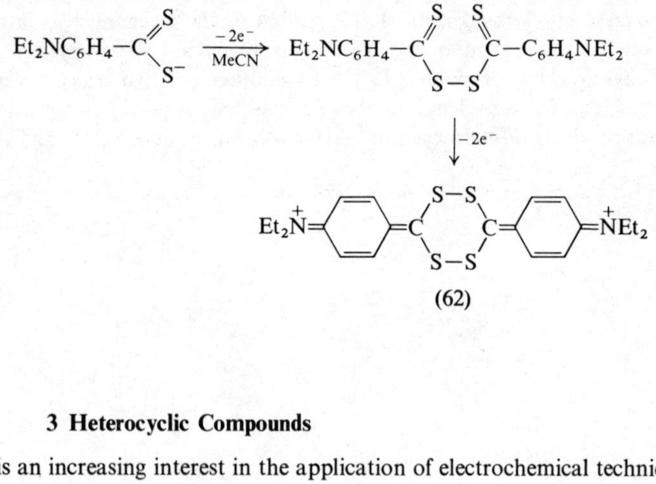

$$Et_2NC_6H_4-C\diagdown_{S^-}^{\diagup S} \xrightarrow[MeCN]{-2e^-} Et_2NC_6H_4-C\diagdown_{S-S}^{\diagup S\quad S\diagdown}C-C_6H_4NEt_2$$

$$\Big\downarrow -2e^-$$

$$Et_2\overset{+}{N}=\!\!\left\langle\!=\!\!=\right\rangle\!\!=\!\!C\diagdown_{S-S}^{\diagup S-S\diagdown}C=\!\!\left\langle\!=\!\!=\right\rangle\!\!=\overset{+}{N}Et_2$$

(62)

3 Heterocyclic Compounds

There is an increasing interest in the application of electrochemical techniques to heterocyclic chemistry. Hence there have been a large number of papers which describe the synthesis of heterocyclic compounds, their substitution, and their modification by anodic oxidation. Many have been discussed earlier in this Report, and this section will be limited to contributions not covered elsewhere. It might be noted that electroanalytical techniques have also been used for the investigation of the mechanism and kinetics of the decomposition of cationic intermediates and for the study of the electron-transfer properties of heterocycles; for example, of fused

---

[263] R. Seeber, A. Cinquantini, P. Zanello, and G. Mazzocchini, *J. Electroanal. Chem. Interfacial Electrochem.*, 1978, **88**, 137.
[264] B. E. Firth, L. L. Miller, M. Mitani, T. Rogers, J. Lennox, and R. E. Murray, *J. Am. Chem. Soc.*, 1976, **98**, 8271.
[265] B. E. Firth and L. L. Miller, *J. Am. Chem. Soc.*, 1976, **98**, 8272.
[266] G. Cauquis and A. Deronzier, *J. Chem. Soc., Chem. Commun.*, 1978, 809.

nitrogen heterocycles,[267] polyvinyl compounds with indolyl end-groups,[268] substituted dihydropyridines,[269] and 2,4,5-triaryl-imidazoles.[270]

Tabakovic and co-workers have reported the synthesis, in high yield, of several nitrogen heterocycles[271,272] and also of 1,3-thiazoles[273] by anodic oxidation in acetonitrile. Typical reactions are shown in Scheme 20. In contrast, they report that the oxidation of chalcone phenylhydrazones leads to a mixture of products.[274] Electroanalytical techniques have been used to investigate the mechanism of several of these reactions. Cauquis et al.[275] have described the oxidation of N-amino-1,2,3,4-tetrahydroquinoline (63) in acetonitrile. In the presence of the base 2,6-lutidine, the product is the dimeric azo-compound; with added acid, the product is the stable cation (64). In acid medium, in the presence of olefin, however, coupling occurs to give a new heterocycle (65), in a yield of 66%.

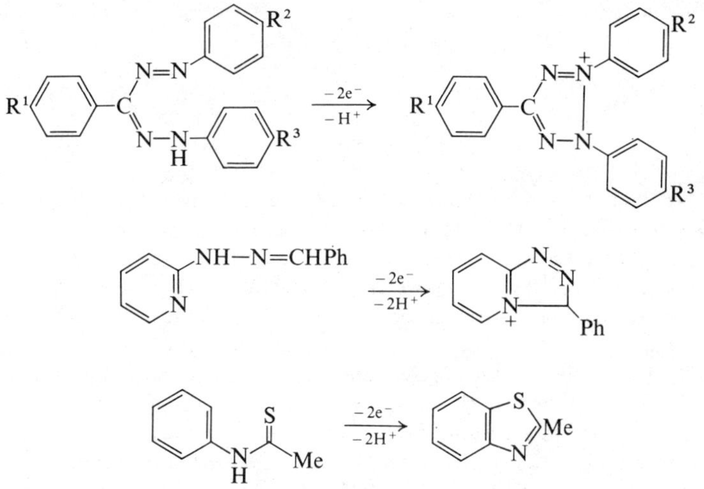

Scheme 20

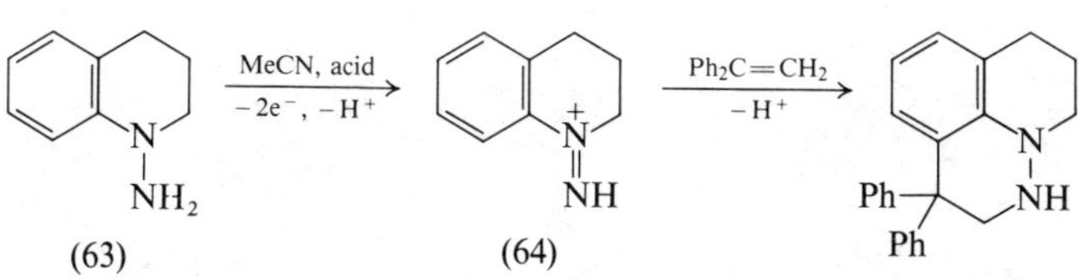

[267] S. Hunig and H. C. Steinmetzer, *Liebigs Ann. Chem.*, 1976, 1090.
[268] S. Hunig and H. C. Steinmetzer, *Liebigs Ann. Chem.*, 1976, 1060.
[269] J. Klima, A. Kürfurst, J. Kathan, and J. Volke, *Tetrahedron Lett.*, 1977, 2725.
[270] P. Hülnhagen and H. Baumgartel, *J. Electroanal. Chem. Interfacial Electrochem.*, 1979, **98**, 119.
[271] I. Tabakovic, M. Trkovnik, and Z. Grujic, *J. Chem. Soc., Perkin Trans. 2*, 1979, 166.
[272] I. Tabakovic, M. Trkovnik, and D. Galijas, *J. Electroanal. Chem. Interfacial Electrochem.*, 1978, **86**, 241.
[273] I. Tabakovic, M. Trkovnik, M. Batusic, and K. Tabakovic, *Synthesis*, 1979, 590.
[274] I. Tabakovic, M. Lacan, and Sh. Damoni, *Electrochim. Acta*, 1976, **21**, 621.
[275] G. Cauquis, B. Chabaud, and Y. Gohee, *Tetrahedron Lett.*, 1977, 2583.

The electrogeneration and the coupling *in situ* of orthoquinone with nucleophiles, in aqueous acetate, has also been shown to be a successful route to oxygen-heterocycles (yields were 90 and 95%); the two reactions that were reported[276] are shown in Scheme 21.

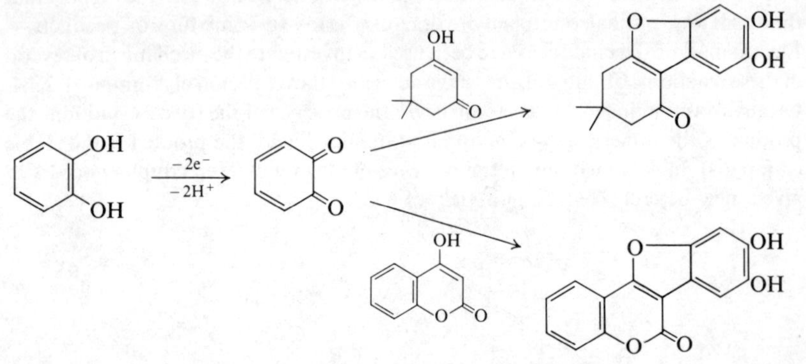

Scheme 21

Miller and co-workers have continued their studies of the oxidation of laudanosine (66; X=H, R=Me) to *o*-methylflavinantine (67; X=H, R=Me) in acetonitrile and, indeed, have shown that, by carrying out the electrolysis in acetonitrile that contains 1% of water and sodium bicarbonate (a weak acid, to protonate the amine), the yield is increased to 90% as compared with 55% in dry unbuffered acetonitrile.[277] Model compounds have been used to show that, in the absence of acid, initial transfer of electrons occurs at the nitrogen atom, and it is the oxidation of the amine which leads to electrode filming and to loss of yield. On protonation of the nitrogen, the mechanism of the reaction changes, and although the oxidation potential is shifted to a more positive value, the yield increases and the electrolysis proceeds at a higher current density.[278,279] A further paper extends the reaction to a series of compounds that are related to laudanosine, *i.e.* (66; X=H or

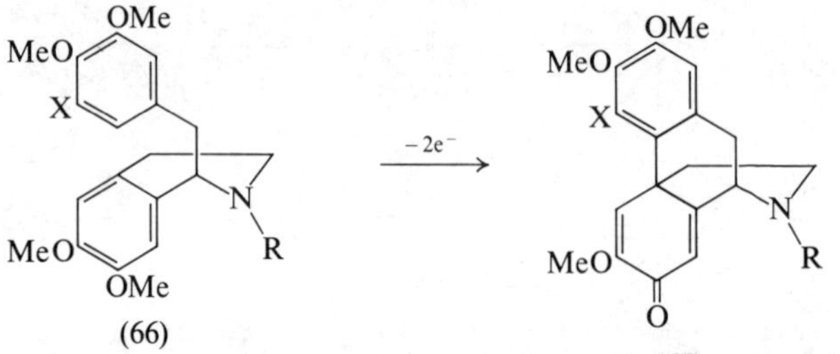

[276] Z. Grujic, I. Tabakovic, and M. Trkovnik, *Tetrahedron Lett.*, 1976, 4823.
[277] J. Y. Becker, L. L. Miller, and F. R. Stermitz, *J. Electroanal. Chem. Interfacial Electrochem.*, 1976, **68**, 181.
[278] J. B. Kerr, T. C. Jempty, and L. L. Miller, *J. Am. Chem. Soc.*, 1979, **101**, 7338.
[279] L. L. Miller, R. F. Stewart, J. P. Gillespie, V. Ramachandran, Y. H. So, and F. R. Stermitz, *J. Org. Chem.*, 1978, **43**, 1580.

of developers for colour photography, the 1-phenylpyrazolidin-3-ones, in both aqueous and non-aqueous solvents. The behaviour of some 4-substituted derivatives in water, at high pH, has been described.[309] All show reversible one-electron oxidation on cyclic voltammograms at high scan rates, but of interest is the coupled chemistry on a longer time-scale; when one of the substituents is the hydroxymethyl group, the final product arises by the sequence shown in Scheme 24.

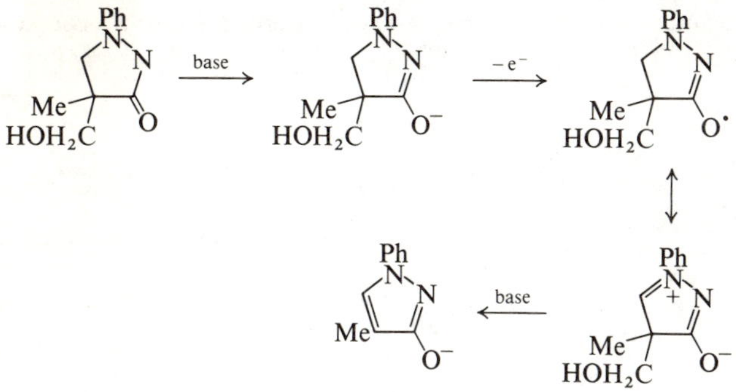

Scheme 24

In unbuffered acetonitrile,[310] the anode reactions of 1-phenylpyrazolidin-3-one are complicated by the tendency of the substrate to act as a base, and hence separate waves are observed for the unprotonated and (at more positive potentials) the protonated forms. The major process is, however, as shown in Scheme 25, the 1-phenylpyrazolin-3-one being formed in 75% yield. The other product is a complex dimeric species. When excess chloride ion is present (as is the case in colour photography), this ion can act as the base, and the reaction is very straightforward;[311] the yield increases to over 95%. The corresponding imine shows quite different coupled chemistry; the cation radical decays to form a dimer (78).[312]

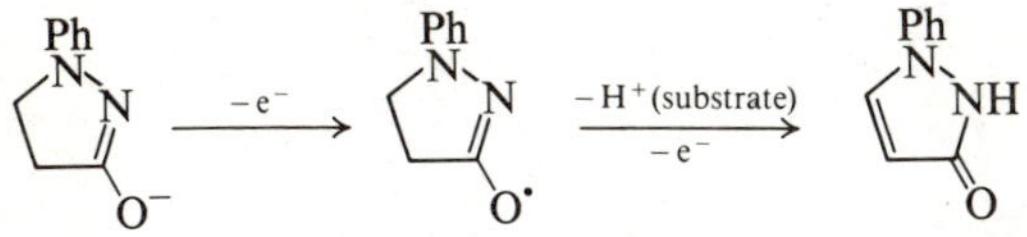

Scheme 25

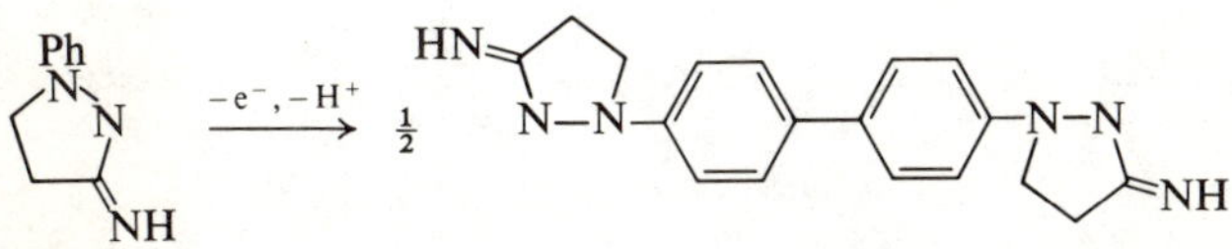

(78)

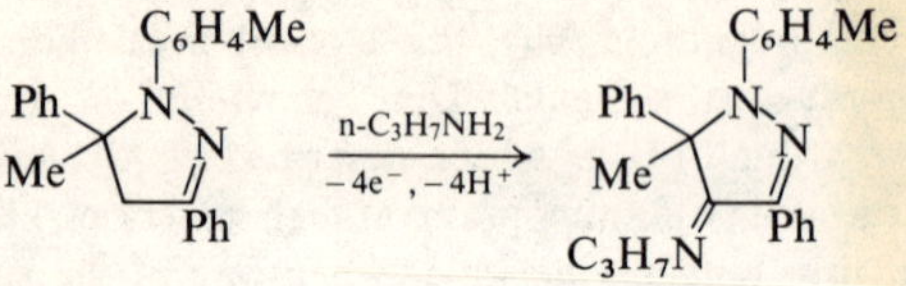

(79)

Pragst and Nastke[313] have reported the oxidation of a 5,5'-disubstituted 2-pyrazoline and showed that, in the presence of a primary amine, a coupled product (79) results.

---

[313] F. Pragst and R. Nastke, *Z. Chem.*, 1976, **16**, 487.

Br; R = H, Me, or $H_2C=CHCH_2$), and it was demonstrated that several morphinandiones can be obtained in yields above 80%. Other groups have studies the analogous aryl–aryl coupling of the derivatives of isoquinoline (68; R = Me or H, $n = 1$ or 2, Z = NH) and isochroman (68; R = Me or H, $n = 1$ or 2, Z = O)[280] and isochromanone,[281] but the yields of coupled products were lower (20—55%). Sainsbury and co-workers[282—285] have also attempted a number of other biphenyl-coupling reactions in order to give precursors to natural products. In general, the anode reactions that were investigated, e.g. the oxidation of diaryl amides and diaryl esters, did not lead to the desired couplings, and the authors have sought to define the structural features which would allow the aryl–aryl coupling to occur.

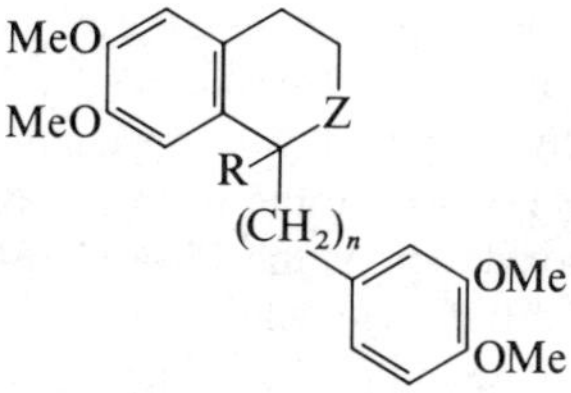

(68)

A number of papers have described simple addition and/or substitution processes, particularly cyanation, methoxylation, and acetoxylation. Yoshida[286—288] has reported the cyanation of furans, pyrroles, and indoles. The yields of cyanated products from the oxidation of N-alkylated or N-arylated pyrroles and indoles in MeOH–NaCN are good.[286] With the pyrroles, ring-substituted compounds are the major products provided that the 2- or 5-positions are free; e.g. as for (69), but chain substitution is observed with 2,5-dialkylated substrates. Eberson[289] has shown that this occurs by a base-catalysed rearrangement of the 2,5-dicyano-adduct, as shown in Scheme 22. With indoles, substitution occurs in the nitrogen ring. Yoshida[287,288] has suggested that the cis/trans ratio of the addition products from the oxidation of 2,5-dimethylfuran in MeOH–NaCN provides strong evidence for reaction in the adsorbed state.

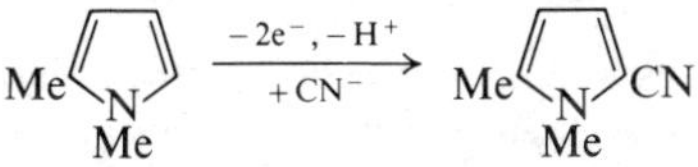

(69)   [60%]

[280] U. Palmquist, A. Nilsson, T. Petterson, A. Ronlan, and V. D. Parker, J. Org. Chem., 1979, **44**, 196.
[281] I. W. Elliot, J. Org. Chem., 1977, **42**, 1090.
[282] M. Sainsbury and J. Wyatt, J. Chem. Soc., Perkin Trans. 1, 1976, 661.
[283] M. Sainsbury and J. Wyatt, J. Chem. Soc., Perkin Trans. 1, 1977, 1750.
[284] M. Sainsbury and J. Wyatt, J. Chem. Soc., Perkin Trans. 1, 1979, 108.
[285] M. P. Carmody, M. Sainsbury, and M. D. Johnson, J. Chem. Soc., Perkin Trans. 1, 1980, 2013.
[286] K. Yoshida, J. Am. Chem. Soc., 1977, **99**, 6111.
[287] K. Yoshida, J. Chem. Soc., Chem. Commun., 1978, 20.
[288] K. Yoshida, J. Am. Chem. Soc., 1976, **98**, 254.
[289] L. Eberson, Acta Chem. Scand., Ser. B, 1980, **34**, 747.

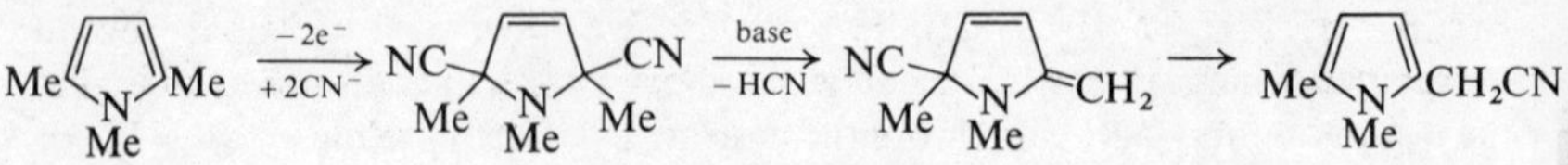

Scheme 22

The methoxylation of N-formyl derivatives of cyclic amines, e.g. (70), is another reaction which has been carried out on a large laboratory scale, and it gives extremely good yields.[290] Srogl et al.[291] have studied the methoxylation of 2,2'-difurylmethane, 2,2'-dithienylmethane, and the mixed compound; the 2,5-dimethoxylated addition product is obtained in good yield from the compounds with furan rings, but only methoxylation of the methylene bridge occurs with 2,2'-dithienylmethane. Other papers have reported the application of methoxylation in synthetic sequences, i.e. the 2,5-dimethoxylation of a furylacetate ester has been used in the synthesis of 4-hydroxy-2-alkyl-cyclopentanones,[292] the methoxylation of benzothiophens such as (71) as a route to the corresponding monoketal and quinones,[293] and the methoxylation at the 15- and 16-positions of bilindione.[294] Torii et al.[295] have reported the acetoxylation of indoles and indolines in acetic acid–triethylamine as a route to indigo. The reactions shown in Scheme 23 can be completed in a single pot, and indigo is formed by treatment of the product with base. The mechanism of the oxidation of a tetrasubstituted pyrrole in acidic, basic, and non-buffered acetonitrile has been studied and several stable intermediates have been identified by electrochemical and spectroscopic techniques.[296]

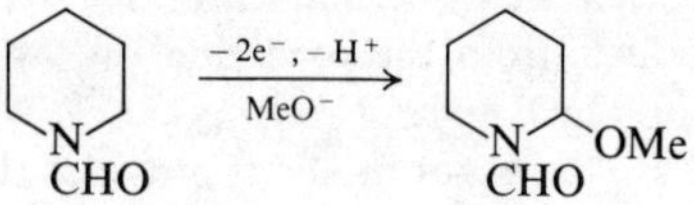

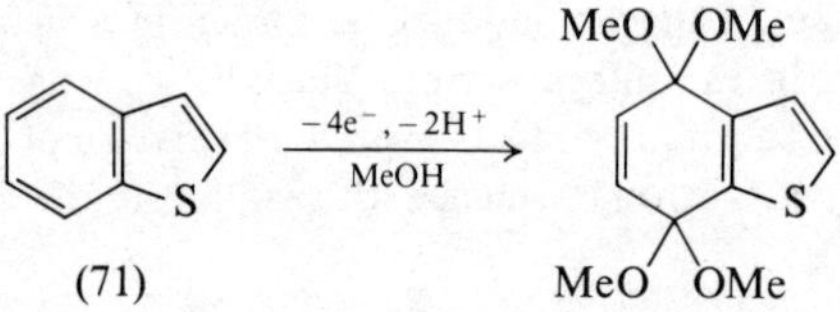

Several anode reactions with heterocyclic substrates lead to dimeric products. Bobbitt et al.[297] have discussed the stereochemistry of the reaction (in wet acetonitrile) to produce (72) and showed it to be much more specific than catalytic oxidation. Only one C–C dimer is formed, and since only molecules of substrate

---

[290] K. Nyberg and R. Servin, *Acta Chem. Scand., Ser. B*, 1976, **30**, 640.
[291] J. Srogl, M. Janda, I. Stibor, and Z. Salajka, *Collect. Czech. Chem. Commun.*, 1977, **42**, 1361.
[292] T. Shono, H. Hamaguchi, and K. Aoki, *Chem. Lett.*, 1977, 1053.
[293] B. L. Chenard and J. S. Swenton, *J. Chem. Soc., Chem. Commun.*, 1979, 1172.
[294] F. Eivazi, W. M. Lewis, and K. M. Smith, *Tetrahedron Lett.*, 1977, 3083.
[295] S. Torii, T. Yamanaka, and H. Tanaka, *J. Org. Chem.*, 1978, **43**, 2882.
[296] P-J. Grossi, L. Marchetti, R. Ramasseul, A. Rassat, and D. Serve, *J. Electroanal. Chem. Interfacial Electrochem.*, 1978, **87**, 353.
[297] J. M. Bobbitt, I. Noguchi, H. Yagi, and K. H. Weisgraber, *J. Org. Chem.*, 1976, **41**, 845.

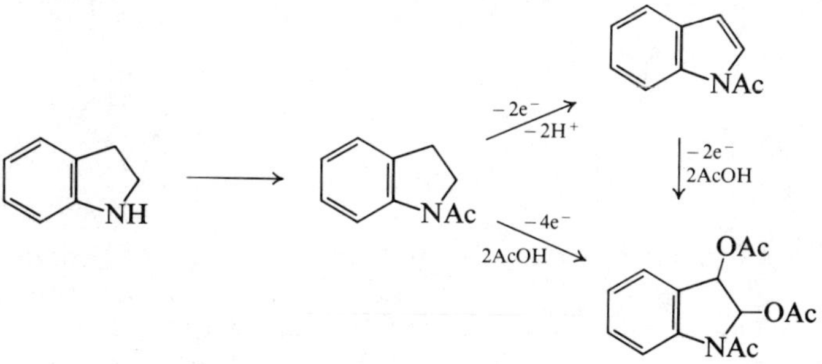

Organic Electrochemistry – Synthetic Aspects: Oxidations

Scheme 23

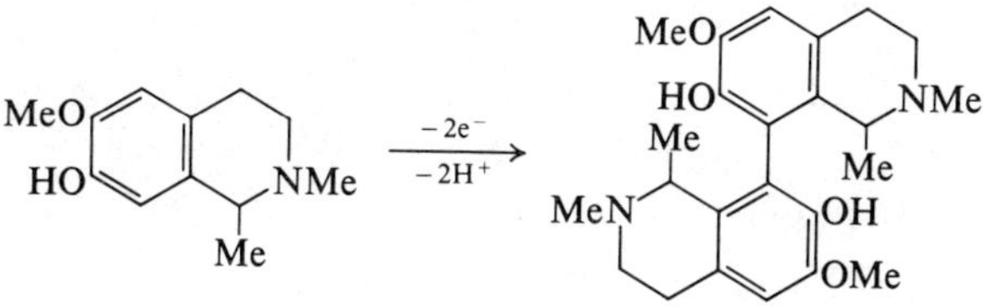

(72)

with the same configuration at C-1 couple (R with R; S with S), it is proposed that the dimerization takes place on the surface of the anode. Other dimerizations which, in aprotic media, have been reported to give good yields include the oxidation of 2,3-diphenylindole (73),[298] a substituted thiazole (74),[299] and a sulphur compound (75).[300] Diaz et al.[301] have also reported the polymerization of pyrrole and showed

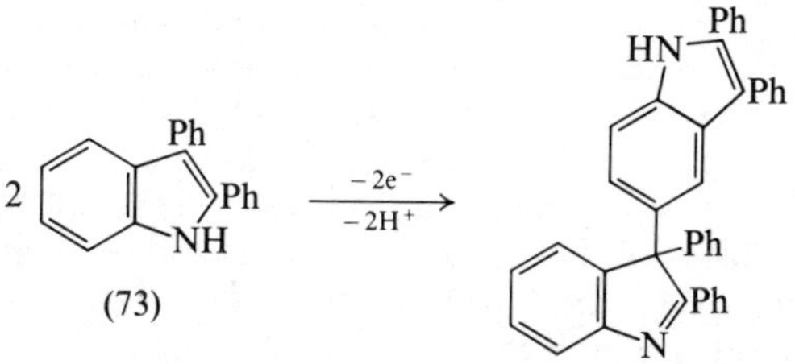

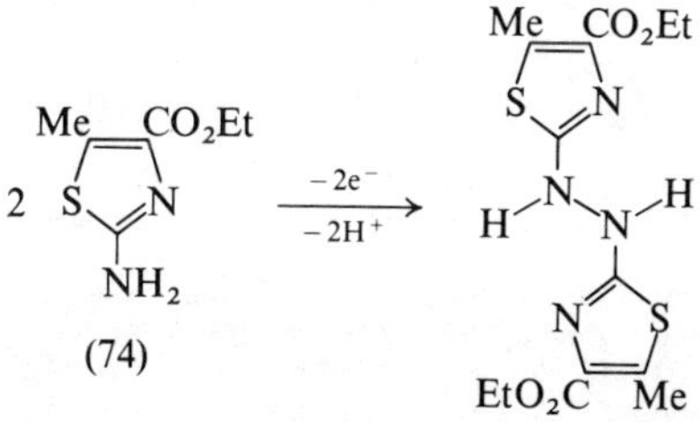

---

[298] G. T. Cheek and R. F. Nelson, *J. Org. Chem.*, 1978, **43**, 1230.
[299] G. Cauquis, H. M. Fahmy, G. Pierre, and M. H. Elnagdi, *Electrochim. Acta*, 1979, **24**, 391.
[300] C. T. Pedersen, V. D. Parker, and O. Hammerich, *Acta Chem. Scand., Ser. B*, 1976, **30**, 478.
[301] A. F. Diaz, K. K. Kanazawa, and G. P. Gardini, *J. Chem. Soc., Chem. Commun.*, 1979, 635.

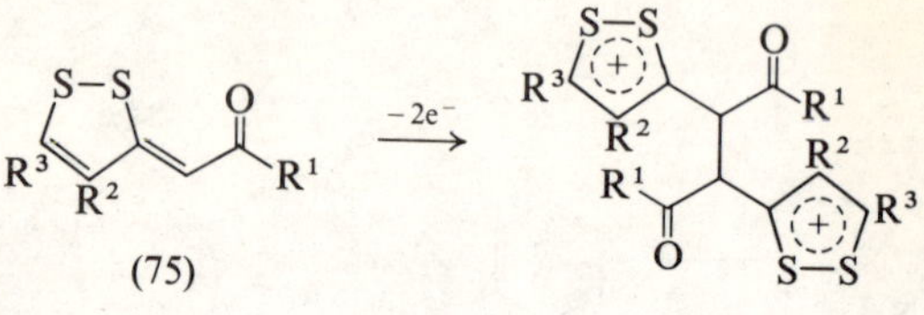

that it forms a conducting surface film which may be used as a modified electrode.

Moses and co-workers[302,303] have described the electrochemical oxidation of some cyclic sulphur compounds in acetonitrile and have used cyclic voltammetry and product identification to show that the cationic intermediates undergo rearrangement. The reactions investigated where those of (76) and (77).

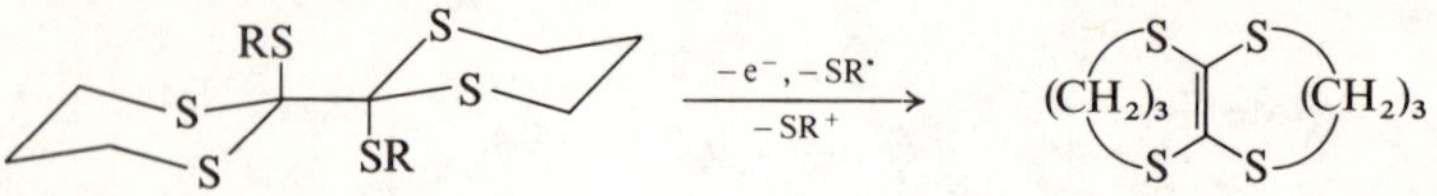

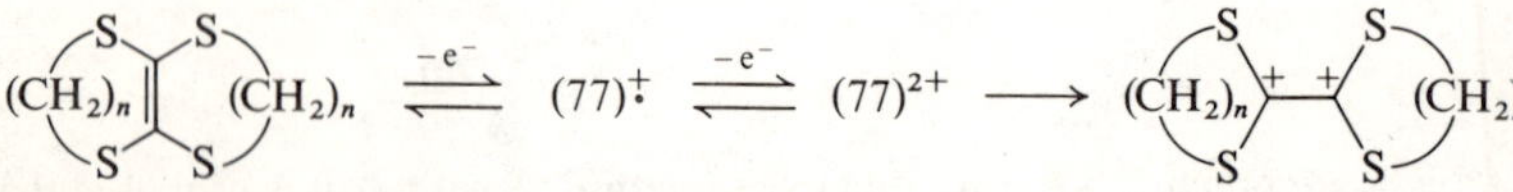

(77) $n = 2$ or 3

Electrode reactions have been used to increase the unsaturation of heterocyclic molecules, and the reactions which have been reported include the oxidation of 1,4-dihydropyridines that are substituted by CN, CO_2Et, or Ac in the 3- or 5-positions,[304] the cleavage of N-acetyl linkages to give diazines,[305] and the decarboxylation of a vicinal dicarboxylic acid to introduce a double-bond and form the metastable bicyclic compound 2-azabicyclo[2.2.2]octa-5,7-diene.[306] The electrochemistry of some bicyclic amines has also been investigated by Nelson and Kessel,[307] who have demonstrated that the oxidation of 9-t-butylazabicyclo[3.3.1]nonane in acetone leads to a stable cation radical because of the absence of a hydrogen atom that is α to the nitrogen and hence capable of being lost; this is the first report of a stable cation radical of an aliphatic amine. Furthermore, a study of the products from the oxidation of tropanes in acetonitrile has been reported.[308]

Joslin and co-workers[309—312] have continued to report on the electrochemistry

---

[302] R. M. Harnden, P. R. Moses, and J. Q. Chambers, *J. Chem. Soc., Chem. Commun.*, 1977, 11.
[303] P. R. Moses, R. M. Harnden, and J. Q. Chambers, *J. Electroanal. Chem. Interfacial Electrochem.*, 1977, **84**, 187.
[304] V. Skala, J. Volke, V. Ohanka, and J. Kathan, *Collect. Czech. Chem. Commun.*, 1977, **42**, 292.
[305] P. Martigny, H. Lund, and J. Simonet, *Electrochim. Acta*, 1976, **21**, 345.
[306] H. Sliwa and Y. Le Bot, *Tetrahedron Lett.*, 1977, 4129.
[307] S. F. Nelson and C. R. Kessel, *J. Chem. Soc., Chem. Commun.*, 1977, 490.
[308] B. L. Laube, M. R. Asirvathem, and C. K. Mann, *J. Org. Chem.*, 1977, **42**, 670.
[309] H. H. Adam and T. Joslin, *J. Electroanal. Chem. Interfacial Electrochem.*, 1976, **72**, 197.
[310] H. H. Adam, T. Joslin, and B. D. Baigrie, *J. Chem. Soc., Perkin Trans. 2*, 1977, 1287.
[311] B. D. Baigrie and T. Joslin, *J. Electroanal. Chem. Interfacial Electrochem.*, 1978, **87**, 405.
[312] B. D. Baigrie, T. Joslin, and D. W. Sopher, *J. Chem. Soc., Perkin Trans. 2*, 1979, 77.

JUL 05 1983